Dinosaur Paleobiology

Books in the **Topics in Paleobiology** series will feature key fossil groups, key events, and analytical methods, with emphasis on paleobiology, large-scale macroevolutionary studies, and the latest phylogenetic debates.

The books will provide a summary of the current state of knowledge and a trusted route into the primary literature, and will act as pointers for future directions for research. As well as volumes on individual groups, the Series will also deal with topics that have a cross-cutting relevance, such as the evolution of significant ecosystems, particular key times and events in the history of life, climate change, and the application of new techniques such as molecular paleontology.

The books are written by leading international experts and will be pitched at a level suitable for advanced undergraduates, postgraduates, and researchers in both the paleontological and biological sciences.

The Series Editor is *Mike Benton*, Professor of Vertebrate Palaeontology in the School of Earth Sciences, University of Bristol.

The Series is a joint venture with the *Palaeontological Association*.

COMPANION WEBSITE

This book has a companion website:

www.wiley.com/go/brusatte/dinosaurpaleobiology

with Figures and Tables from the book for downloading

Dinosaur Paleobiology

Stephen L. Brusatte

WILEY-BLACKWELL

A John Wiley & Sons, Ltd., Publication

This edition first published 2012 © 2012 by John Wiley & Sons, Ltd.

Wiley-Blackwell is an imprint of John Wiley & Sons, formed by the merger of Wiley's global Scientific, Technical and Medical business with Blackwell Publishing.

Registered office: John Wiley & Sons, Ltd, The Atrium, Southern Gate, Chichester, West Sussex, PO19 8SQ, UK

Editorial offices: 9600 Garsington Road, Oxford, OX4 2DQ, UK
 The Atrium, Southern Gate, Chichester, West Sussex, PO19 8SQ, UK
 111 River Street, Hoboken, NJ 07030-5774, USA

For details of our global editorial offices, for customer services and for information about how to apply for permission to reuse the copyright material in this book please see our website at www.wiley.com/wiley-blackwell.

The right of the author to be identified as the author of this work has been asserted in accordance with the UK Copyright, Designs and Patents Act 1988.

Library of Congress Cataloging-in-Publication Data

Brusatte, Stephen.
  Dinosaur paleobiology / Stephen L. Brusatte.
    p. cm. – (Topics in paleobiology)
  Includes bibliographical references and index.
  ISBN 978-0-470-65657-0 (hardcover : alk. paper)—ISBN 978-0-470-65658-7 (pbk. : alk. paper)
1. Dinosaurs. 2. Paleobiology. I. Title.
  QE861.4.B786 2012
  567.9–dc23   2011050466

A catalogue record for this book is available from the British Library.

Wiley also publishes its books in a variety of electronic formats. Some content that appears in print may not be available in electronic books.

Set in 9/11.5pt Trump Mediaeval-Roman by Thomson Digital, Noida, India

1   2012

Dedication

To my wife, Anne

Contents

Contents

Foreword

Paleobiology is a vibrant discipline that addresses current concerns about biodiversity and about global change. Further, paleobiology opens unimagined universes of past life, allowing us to explore times when the world was entirely different and when some organisms could do things that are not achieved by anything now living.

Much current work on **biodiversity** addresses questions of origins, distributions, and future conservation. Phylogenetic trees based on extant organisms can give hints about the origins of clades and help answer questions about why one clade might be more species-rich ("successful") than another. The addition of fossils to such phylogenies can enrich them immeasurably, thereby giving a fuller impression of early clade histories, and so expanding our understanding of the deep origins of biodiversity.

In the field of **global change**, paleobiologists have access to the fossil record and this gives accurate information on the coming and going of major groups of organisms through time. Such detailed paleobiological histories can be matched to evidence of changes in the physical environment, such as varying temperatures, sea levels, episodes of mid-ocean ridge activity, mountain building, volcanism, continental positions, and impacts of extraterrestrial bodies. Studies of the influence of such events and processes on the evolution of life address core questions about the nature of evolutionary processes on the large scale.

As examples of **unimagined universes**, one need only think of the life of the Burgess Shale or the times of the dinosaurs. The extraordinary arthro-pods and other animals of the Cambrian sites of exceptional preservation sometimes seem more bizarre than the wildest imaginings of a science fiction author. During the Mesozoic, the sauropod dinosaurs solved basic physiological problems that allowed them to reach body masses ten times those of the largest elephants today. Further, the giant pterosaur *Quetzalcoatlus* was larger than any flying bird, and so challenges fundamental assumptions in biomechanics.

Books in the **Topics in Paleobiology** series will feature key fossil groups, key events, and analytical methods, with emphasis on paleobiology, large-scale macroevolutionary studies, and the latest phylogenetic debates.

The books will provide a summary of the current state of knowledge, a trusted route into the primary literature, and will act as pointers for future directions for research. As well as volumes on individual groups, the Series will also deal with topics that have a cross-cutting relevance, such as the evolution of significant ecosystems, particular key times and events in the history of life, climate change, and the application of new techniques such as molecular paleontology.

The books are written by leading international experts and will be pitched at a level suitable for advanced undergraduates, postgraduates, and researchers in both the paleontological and biological sciences.

Michael Benton,
Bristol,
November 2011

Preface

Dinosaurs are everywhere these days. They are the most popular exhibits in many museums, the stars of movies and the focus of television documentaries, the pitchmen in advertising campaigns and the subject of gushing articles in magazines and newspapers. Looking at how dinosaurs are portrayed in the popular press, it is easy to lump them together with leprechauns, unicorns, and dragons – creatures of myth and iconic lore that only exist in the imaginations of children and the whimsy of pop culture. But dinosaurs are not creatures of fantasy – they were real animals, of many fantastic shapes and sizes, that dominated terrestrial ecosystems for an astounding span of over 160 million years. They were living, breathing, feeding, moving, reproducing, evolving organisms that originated in the aftermath of the worst mass extinction in earth history, rose to dominance as a supercontinent was splitting and climates were fluctuating, evolved into some of the largest and most fearsome animals the planet has ever seen, and then suddenly went extinct right at the same time that a giant comet or asteroid slammed into the earth and supervolcanoes were belching rivers of lava. And perhaps most astonishing of all, these ancient creatures, so often symbols of lethargy and failure, were the ancestors of one of the most successful groups of living animals: the birds.

The scientific study of dinosaurs has been experiencing a remarkable renaissance over the past couple of decades. Scientific understanding of dinosaur anatomy, biology, and evolution has advanced to such a degree that paleontologists often know more about 100 million-year-old dinosaurs than many species of living organisms. Research is proceeding at a frenetic pace, as illustrated by a simple statistic: during 2010, the year I proposed and began writing this book, some 63 new species of dinosaurs were discovered. That's a stupendous rate of over one new species per week, which has largely been fueled by a great increase in fieldwork exploration (especially in China and South America) and an ever-expanding roster of graduate students and other young researchers choosing to study dinosaurs. And not only is our stockpile of dinosaur fossils growing at an exponential pace, but so is the development of new research techniques. It used to be that paleontologists could pontificate on the biology, evolution, and extinction of dinosaurs based only on the flimsiest scraps of evidence, interpreted with a healthy dose of imagination and a snickering dismissal of the explicit, quantitative, repeatable methodologies that have long been the norm in most other sciences. Those days are long gone. Today, dinosaur paleontology is a dynamic science that demands evidence-based rigor and is firmly integrated with many other scientific disciplines. Indeed, researchers often draw from a diverse repertoire of anatomy, geology, chemistry, physics, mathematics, and statistics when studying dinosaurs. It is not uncommon to see advanced calculus used to estimate dinosaur body masses, computerized engineering analyses marshaled to test whether certain dinosaurs were capable of feeding or moving a certain way, or statistics utilized to explicitly assess whether some dinosaurs were evolving faster or slower than others.

The breadth of current dinosaur research is vast. Some scientists spend their careers discovering and describing new species, others may focus solely on anatomy or genealogy, and others concentrate on studying dinosaur locomotion or feeding. In general, though, all contemporary work on dinosaurs provides evidence for addressing two main questions. First, how did dinosaurs function as living animals? Second, what is the grand narrative of dinosaur evolution across the Mesozoic? The only way to attack these questions in a defensible competent manner is to interpret the primary evidence – the actual dinosaur fossils that provide a bedrock for the entire enterprise of dinosaur research – using explicit quantitative methodologies. The emerging answers to these questions, and the evidence and methods that are revealing them, are the focus of this book. Like any science, dinosaur paleontology is constantly changing as new fossils are found and new research techniques are developed and refined. Our knowledge of dinosaur biology and evolution is shifting fast, and this book is an attempt to capture what is currently known about this remarkable group of ancient creatures that dominated our planet for so long.

From a more personal standpoint, this book is also a young, perhaps brazen, researcher's examination of his field of study. I am in the somewhat unusual position of writing this book as a PhD student – a scientist without an advanced degree, with less than a decade of research experience, who has not had the time and wisdom to (at least yet) make a substantial mark on the field. But although I may not be the most traditional author of a technical dinosaur book, and although perhaps I should be focusing more on my thesis than on writing books, I feel that I am able to present a

perspective that has yet to be tapped by the oversaturated dinosaur book market. I have been brought up and trained within the dynamism of contemporary dinosaur research, and have been experiencing the explosive growth of this field as a dizzying cocktail of new discoveries and techniques have enabled modern scientists to understand dinosaurs in unprecedented detail. In many ways this book is a personal journey. I do not pretend that this book is an exhaustive encyclopedia of everything that is currently known about dinosaurs, or a technical critique of the minutiae of every method and each piece of evidence. Instead, what I present is my understanding of dinosaur biology and evolution – the understanding of a student actively learning about dinosaurs and in the midst of planning his own research program and career. I present what I find interesting and empowering, what I think is most important and exciting about contemporary research, and where I think the field is heading.

So, then, what do scientists actually know about dinosaurs? As it turns out, the truth about dinosaur biology and evolution is surely more fascinating than even the most sensational dinosaur documentary or movie, and more than fascinating enough to fuel the passion of this member of the MTV generation. Indeed, without even a hint of hyperbole, the story of dinosaur evolution is one of the greatest stories ever told.

Steve Brusatte
New York, USA
September 1, 2011

Acknowledgments

Writing a book is a lot of fun, but also a lot of work. Although my name may appear on the cover, this book could never have been written without the help of so many friends and colleagues across the globe. One of my favorite aspects of paleontological research is the friendships and collaborations that are fostered, and I'm pleased to be able (even if in a small way) to showcase the work of many of my colleagues and include their contributions in the form of photos and illustrations. This book relects my personal journey studying paleontology, and building a career in research, over the past decade. I have tried, wherever possible, to include photographs and figures that I have compiled during my research work, fieldwork, and museum visits, or those provided by trusted colleagues and friends. These colleagues are too numerous to thank here, but individuals who helped with images are credited in the figure captions. To all of them, let me apologize one final time for all my nagging questions.

I have been very fortunate in my young career to have been mentored by three very excellent advisors: my undergraduate advisor Paul Sereno at the University of Chicago, my Master's advisor Mike Benton at the University of Bristol, and my PhD advisor Mark Norell at Columbia University and the American Museum of Natural History. I recognize how lucky I have been to study under three of the most prominent luminaries in the field, and thank them for all of heir guidance, advice, and support over the years. Specific to this book, I would like to offer my sincere thanks to Mike Benton, the editor of the Topics in Paleobiology series, who invited me to write a book on dinosaurs for Wiley-Blackwell. I am humbled that he would place such trust in me at such a young stage in my career, and I hope that I have seized this remarkable opportunity and written a book that justifies his confidence in me. And to Mark Norell, my current advisor, please know that I will always appreciate the freedom that you have provided me as a student to pursue whatever interests me. Not every PhD advisor would allow his student to put a doctoral thesis on the backburner to write a book.

Although many colleagues helped with images and advice, a few people deserve special mention here. I am very pleased to feature the skeletal reconstructions of Scott Hartman, the photographs of Mick Ellison, and the artistic life reconstructions of Jason Brougham. Scott, Mick, and Jason are three of the best artists in the business, and are consistently producing beautiful and scientifically accurate work that, at least in my opinion, sets a benchmark for the field. Without their contributions this book would be little more than a jumble of 130,000 words; if this book does succeed in bringing dinosaurs to life, it is largely due to their reconstructions, photos, and illustrations. A lot of the work they have provided here has not been reproduced before, and all three worked tirelessly to help make this book something more than just a run-of-the-mill dinosaur tome. Several trusted colleagues also read large portions of this book, including Roger Benson, Mike Benton, Richard Butler, Matt Carrano, Greg Erickson, Paul Gignac, John Hutchinson, and Pat O'Connor, as well as the formal reviewers (Paul Barrett and Larry Witmer). Their advice was instrumental, and I thank them for their encouragement, suggestions, and frank criticism that helped tighten my writing and improve the text. All mistakes, however, are of course mine.

My continuing development as a scientist has been facilitated by the friendship, collaboration, advice, and assistance of many trusted colleagues. I would like to especially acknowledge my two closest colleagues, Roger Benson and Richard Butler, whom I consider something of scientific blood brothers. I've been fortunate to work on many projects with Richard and Roger, and have shared many long car journeys, evenings over beers, and hours in museum collections learning from them. They are two of the most dynamic, thoughtful young researchers in the field, and I have no doubt that they will emerge as among the most respected voices in dinosaur research as their careers unfold. I've also enjoyed fruitful collaborations and friendships with many other close colleagues, including Thomas Carr, Zoltán Csiki, Phil Currie, Graeme Lloyd, Octavio Mateus, Josh Mathews, Grzegorz Niedźwiedzki, Marcello Ruta, Steve Wang, Scott Williams, and Tom Williamson. My fellow graduate students and advisors have provided constant inspiration, including Carol Abraczinskas, Marco Andrade, Amy Balanoff, Robin Beck, Mark Bell, Gabe Bever, Jianye Chen, Jonah Choiniere, John Flynn, Andres Giallombardo, Christian Kammerer, Mike LaBarbera, Shaena Montanari, Sterling Nesbitt, Paul Olsen, Rui Pei, Albert Prieto-Márquez, Manabu Sakamoto, Michelle Spaulding, Mark Webster, Hongyu Yi, and Mark Young. Three colleagues that I have only worked with briefly, but have long admired for their adherence to quantitative rigor and ability to ask and answer interesting questions, are Matt Carrano, Greg Erickson, and John Hutchinson. Many other colleagues also have helped me along, including Robert Bronowicz, Dan Chure, Julia Desojo, Phil Donoghue, Gareth Dyke, Jerzy Dzik, Martin Ezcurra, Dave Gower, Mike Henderson, Dave Hone, Steve Hutt, Randy Irmis, Max Langer, Pete Makovicky, Darren Naish, Chris Organ, Emily Rayfield, Nate Smith, Tomasz Sulej, Corwin Sullivan, Alan Turner, Mátyas Vremir, Anne Weil, Jessica Whiteside, Zhao Xijin, and Xu Xing.

And finally, but most importantly, I must thank my family and close personal friends. My parents (Jim and Roxanne Brusatte) have long fostered my passion for paleontology and writing, even going as far as letting me plan whole days of family vacations dedicated to museum hopping (which my brothers, Mike and Chris, must have really enjoyed). My parents and in-laws (Peter and Mary Curthoys) have generously provided space in their homes for me to write this book. A handful of good personal friends have helped fuel my passion for writing and have molded me (gradually, and surely with much pain) into a competent author: Fred Bervoets, Lonny Cain, Lynne Clos, Allen Debus, Mike Fredericks, Richard Green, Joe Jakupcak, Mike Murphy, and Dave Wischnowsky. I sincerely thank all of the help that my editor, Ian Francis, has provided with this book. And last, but certainly not least, I dedicate this book to my patient and beautiful wife Anne, who someday (I hope) will understand that it isn't too strange for a grown man to spend his days thinking about 65 million-year-old, 6-tonne, killer-toothed megapredators.

1 An Introduction to Dinosaurs

It is necessary to begin with a straightforward, if not pedantic, question: what is a dinosaur? In popular parlance a dinosaur is often anything that is old, big, or frightening. Any kindergartner could identity *Tyrannosaurus* or *Triceratops* as dinosaurs, and they would be correct, but newspapers will often sloppily use the term "dinosaur" to refer to flying reptiles (pterosaurs), marine reptiles (plesiosaurs, ichthyosaurs, etc.), or even large mammals (such as the woolly mammoth). "Dinosaur" has become a cultural and political idiom as well: out-of-touch politicians or washed-up celebrities are often mockingly ridiculed as "dinosaurs," a synonym for lethargy, obsolescence, and inevitable extinction.

Although the term "dinosaur" is firmly established in the popular lexicon, it is also a scientific term that refers to a specific group of organisms that shared particular anatomical features and lived during a certain period of time. While the popular definition of "dinosaur" is amorphous, the scientific definition is precise. We will get to that definition in a moment, but first it is necessary to review exactly where dinosaurs fit in the tree of life – when they evolved, what they evolved from, and who their closest relatives are – so it is easier to comprehend the explicit distinction between dinosaur and non-dinosaur. Some of the following discussion may seem elementary to more advanced readers, and I intentionally use a more conversational tone in this introduction to appeal to non-specialists and younger students. It is important, however, to set the stage for this book by first painting in broad strokes, before progressing to a more nuanced discussion of dinosaur anatomy, ecology, behavior, and function.

Dinosaur Paleobiology, First Edition. Stephen L. Brusatte.
© 2012 John Wiley & Sons, Ltd. Published 2012 by John Wiley & Sons, Ltd.

Dinosaurs: A Brief Background

Dinosaurs are one of the best-known, most intensively studied, and most successful groups of tetrapods: animals with a backbone that have limbs with digits (fingers and toes) (Fig. 1.1). Within the tetrapod group, dinosaurs are members of a speciose subgroup of reptiles called the Archosauria, which literally means "ruling reptiles" in Greek (Cope, 1869; Romer, 1956; Carroll, 1988; Benton, 2005) (Figs 1.1–1.6). This is a fitting moniker, as archosaurs have been a major component of terrestrial ecosystems since the early Mesozoic, and for large swaths of time have been ecologically dominant and incredibly diverse (Benton, 1983; Fraser, 2006). Living archosaur subgroups include two major clades, birds and crocodylomorphs, which are among the most familiar and successful groups of extant vertebrates (note that a "clade" refers to a group of animals that includes an ancestor and all of its descendants; Fig. 1.5) (Gauthier, 1986; Sereno, 1991a; Nesbitt, 2011). However, the great majority of archosaur diversity is extinct, and the two main

living groups merely represent two highly aberrant body types (fliers and semiaquatic sprawlers) that were able to endure several mass extinction events that pruned most other lineages on the archosaur family tree. Dinosaurs, without a doubt, are the most familiar of these extinct archosaurs.

The archosaur clade is an ancient group that originated approximately 250 million years ago (Nesbitt, 2003, 2011; Brusatte et al., 2010a, 2011a; Nesbitt et al., 2011). Some of the closest archosaur relatives are known from the Late Permian (e.g. Dilkes, 1998; Nesbitt et al., 2009a), and archosaurs themselves arose within the first few million years after the devastating Permo-Triassic mass extinction, the largest instance of mass death in earth history, estimated to have eradicated 75–95% of all species (Raup, 1979; Stanley and Yang, 1994; Benton, 2003; Erwin, 2006; Clapham et al., 2009). The Permo-Triassic extinction interval was a time of death and destruction on a massive scale, but its aftermath was a time of equally large-scale rebirth: ecosystems were reshuffled, organisms that were once overshadowed had the freedom to flower, and entirely new groups originated and

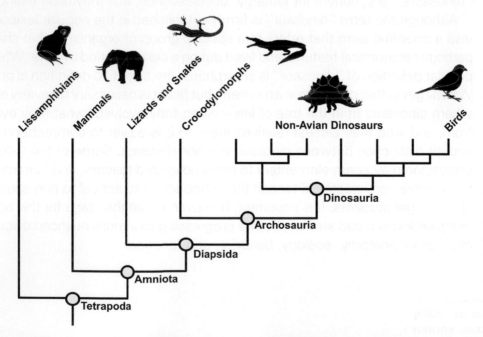

Figure 1.1 A simplified genealogical tree (cladogram) of tetrapods (limbed vertebrates) showing the position of dinosaurs and their closest relatives. Artwork by Simon Powell, University of Bristol.

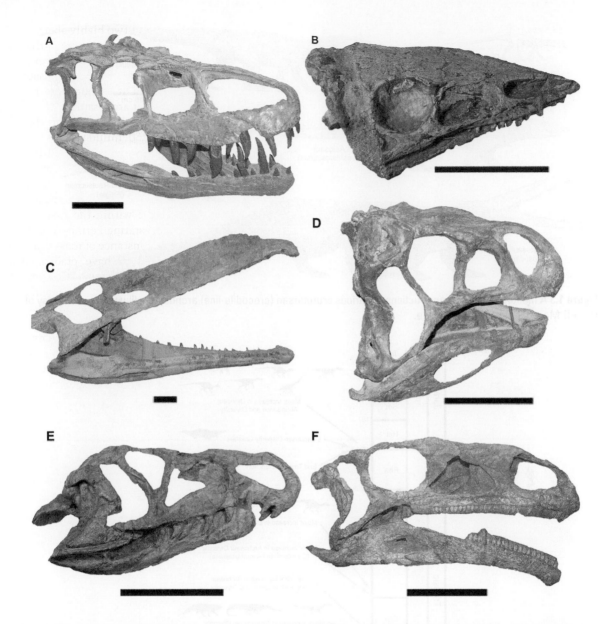

Figure 1.2 A montage of the skulls of various archosaurs, including the rauisuchian crurotarsan *Batrachotomus* (A), the aetosaurian crurotarsan *Aetosaurus* (B), the phytosaurian crurotarsan *Nicrosaurus* (C), the poposauroid crurotarsan *Lotosaurus* (D), the ornithosuchid crurotarsan *Riojasuchus* (E), and the sauropodomorph dinosaur *Plateosaurus* (F).

diversified in the barren, post-extinction landscape (Benton et al., 2004; Sahney and Benton, 2008). Among these entirely new groups were "modern" lineages such as turtles, mammals, lepidosaurs (lizards and their relatives), lissamphibians (frogs and salamanders), and archosaurs. It is no wonder that the Triassic Period is often called the "birth of modern ecosystems," as so many of today's most distinctive and successful clades originated during this time.

Figure 1.3 A montage of life reconstructions of various crurotarsan (crocodile-line) archosaurs. Illustrations courtesy of Dr Jeff Martz, National Park Service.

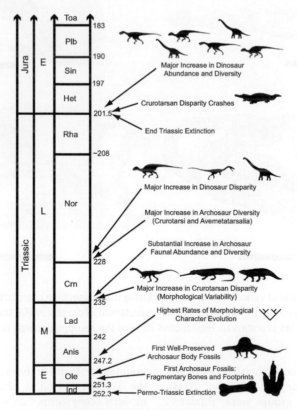

Figure 1.4 A general timeline of important events during the first 70 million years of archosaur evolution during the Triassic and early Jurassic. Image based on illustration in Brusatte et al. (2011a).

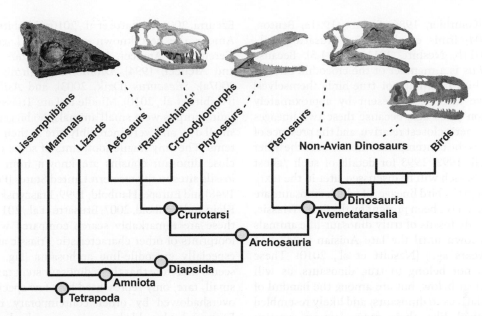

Figure 1.5 A simplified genealogical tree (cladogram) of archosaurs, showing the position of dinosaurs and their closest relatives. Artwork by Simon Powell, University of Bristol.

The archosaur clade diversified rapidly after its origination, as most of the major archosaur sub-clades and body plans were established by the end of the Early Triassic, a mere 5 million years after the mass extinction (Brusatte et al., 2011b) (Fig. 1.4). The oldest unequivocal archosaur body fossil with a well-constrained age and phylogenetic position is *Xilousuchus*, from the late Olenekian/early Anisian (c.247–248 million years ago) of China

(Nesbitt et al., 2011). This species is a derived member of the "crocodile line" of archosaur phylogeny, which is properly referred to as Crurotarsi (also sometimes called Pseudosuchia). Crurotarsi includes crocodylomorphs and their closest extinct relatives, whereas the other half of the archosaur clade, the "bird-line" group Avemetatarsalia (sometimes also called Ornithodira), includes birds, dinosaurs, and pterosaurs (the familiar flying

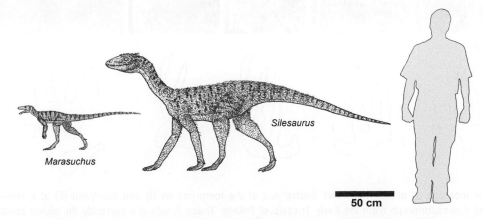

Figure 1.6 Life reconstructions of the basal non-dinosaurian dinosauromorphs *Marasuchus* and *Silesaurus*, two of the closest relatives to dinosaurs. Illustrations courtesy of Dr Jeff Martz, National Park Service.

reptiles) (Gauthier, 1986; Sereno, 1991a; Benton, 1999, 2004; Irmis et al., 2007a; Brusatte et al., 2010a, 2011b; Nesbitt, 2011) (Fig. 1.5). Because *Xilousuchus* is a member of the crocodile lineage, then the bird line (but not true birds themselves) must have also been present by approximately 248 million years ago, because these two lineages are each other's closest relative, and the presence of one implies the contemporary existence of the other (see Norell, 1992, 1993 for details of such "ghost lineages," which will be discussed later in the text).

Although the bird lineage, of which dinosaurs are a part, must have been present by the Early Triassic, the first body fossils of truly dinosaur-like animals are not known until the late Anisian (c.243–244 million years ago) (Nesbitt et al., 2010). These fossils do not belong to true dinosaurs, as will become clear below, but are among the handful of closest relatives to dinosaurs, and likely resembled and behaved like their more famous cousins (Fig. 1.6). More properly, they are members of the "dinosaur stem clade," technically known as Dinosauromorpha (Sereno, 1991a; Benton, 1999, 2004;

Ezcurra, 2006; Brusatte et al., 2010a; Nesbitt, 2011). Among the best known species are *Lagerpeton* (Sereno and Arcucci, 1993), *Marasuchus* (Sereno and Arcucci, 1994), *Dromomeron* (Irmis et al., 2007a), *Silesaurus* (Dzik, 2003), and *Asilisaurus* (Nesbitt et al., 2010). Middle to Late Triassic dinosauromorphs were small animals, no bigger than a small dog, and were incredibly rare in their ecosystems. The tiny fragile footprints of some of these close dinosaur cousins are known from several fossils sites in the western United States (Peabody, 1948) and Europe (Haubold, 1999; Ptaszynski, 2000; Klein and Haubold, 2007; Brusatte et al., 2011a), and these are remarkably scarce compared with the footprints of other characteristic Triassic animals, especially crocodile-line archosaurs (Fig. 1.7). It seems therefore that these dinosaur stem taxa were small, rare, only represented by a few species, and overshadowed by other contemporary reptiles. From such a humble beginning came the dinosaurs.

True dinosaurs likely originated some time in the Middle Triassic, although it is difficult to pinpoint the exact time. The first dinosaur body fossils

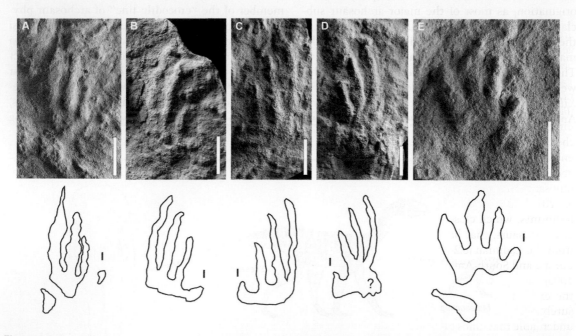

Figure 1.7 A montage of photographs and illustrations of the footprints (A–D) and handprint (E) of a small-bodied quadrupedal dinosauromorph from the Early Triassic of Poland. These fossils are currently the oldest known fossil evidence of the dinosauromorph lineage. Scale bars equal 1 cm. Images by Grzegorz Niedźwiedzki and modified from Brusatte et al. (2011a).

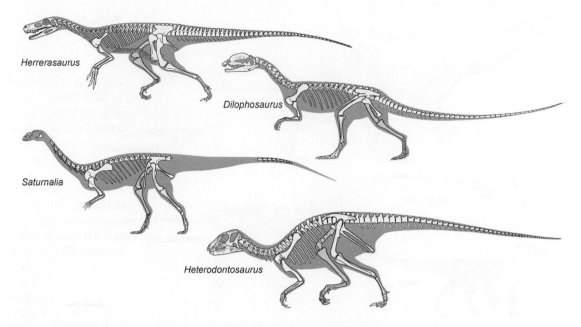

Figure 1.8 Skeletal reconstructions of four early dinosaurs from the Late Triassic to Early Jurassic: the theropod *Herrerasaurus*, the theropod *Dilophosaurus*, the sauropodomorph *Saturnalia*, and the ornithischian *Heterodontosaurus*. Illustrations by Frank Ippolito (American Museum of Natural History) and modified from Brusatte et al. (2010b).

are known from rocks that were deposited in Argentina at approximately the Carnian–Norian boundary (c.228 million years ago) (Rogers et al., 1993; Shipman, 2004; Brusatte et al., 2010b; Ezcurra, 2010a; Langer et al., 2010; Martinez et al., 2011) (Figs 1.8 and 1.9). However, it is almost certain that dinosaurs arose several million years earlier. First, the closest relatives of dinosaurs were clearly present by at least 243 million years ago, as outlined above, and it is reasonable to hypothesize that dinosaurs originated around this time (Nesbitt et al., 2010). Second, there are a number of provocative footprints, which closely match the feet of primitive dinosaurs, that have recently been described from the Ladinian (c.242–235 million years ago) of Europe and South America (Gand and Demathieu, 2005; Melchor and de Valais, 2006). Regardless of the exact timing of dinosaur origins, which will surely become clearer as new fossils are found, it is undeniable that dinosaurs began to diversify quickly once they originated. By the time the first dinosaur body fossils appear in the fossil record, representatives of the three major subgroups of dinosaurs – the carnivorous theropods, long-necked

sauropodomorphs, and herbivorous and often armored or crested ornithischians – are already present (Sereno and Novas, 1992; Sereno et al., 1993; Langer et al., 1999, 2010; Butler et al., 2007; Martinez and Alcober, 2009; Brusatte et al., 2010b; Ezcurra and Brusatte, 2011; Martinez et al., 2011).

Therefore, by the Late Triassic, the Age of Dinosaurs was in full swing, and over the course of the next 50 million years dinosaurs would continue to diversify into new species and body types, before ultimately becoming the dominant mid-to-large size vertebrates in terrestrial ecosystems globally in the Early Jurassic, about 176 million years ago (Benton, 1983; Brusatte et al., 2008a, 2008b, 2010b) (Fig. 1.4). From this point on, throughout the remainder of the Jurassic and the Cretaceous, from approximately 175 to 65 million years ago, dinosaurs truly were "ruling reptiles" in every sense of the phrase. They lived in all corners of the globe, including the Arctic highlands, and reached some of the most stupendous sizes ever seen in land-living animals. Some species developed absurdly long necks, others extravagant horns and armor that would make a medieval

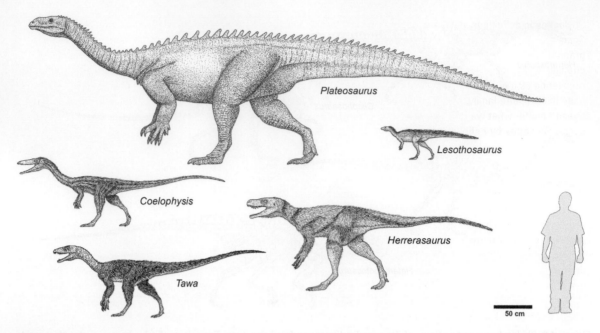

Figure 1.9 Life reconstructions of early dinosaurs from the Late Triassic. Illustrations courtesy of Dr Jeff Martz, National Park Service.

knight blush, and yet others grotesque skulls, longer than an average man is tall and packed with dagger-like teeth, perfect for delivering bone-crunching bites. This fantastic array of dinosaurs – predators and herbivores, dwarves and 50-m long behemoths and all sizes in between – continued to evolve in concert with the slow drift of the continents and the roller-coaster wiggles of climate change, until an unexpected visitor from outer space smashed into the planet 65 million years ago, snuffing out the Age of Dinosaurs and permitting the survival of only one marginal, aberrant dinosaur subgroup: the birds.

The Scientific Definition of Dinosaurs

The above review liberally used terms like "true dinosaur" and "close dinosaur cousin." Vague terminology like this can often be maddening, and can sadly obstruct communication between scientists. Thankfully, however, there is an explicit definition of what constitutes a dinosaur (the "true dinosaurs"). Dinosaurs are defined by

scientists as "members of the least inclusive clade containing *Triceratops horridus* and *Passer domesticus* (the living house sparrow)" (Padian and May, 1993; Sereno, 1998; Sereno et al., 2005). At first this definition may seem confusing, and perhaps even counterintuitive, but in fact it is quite straightforward.

Most modern biologists define groups of organisms, such as dinosaurs or mammals or birds, based on ancestry, not on the possession of certain characteristics (e.g. de Queiroz and Gauthier, 1990, 1992; Sereno, 2005). An animal is a dinosaur if it falls in a certain place on the family tree of life, in this case that group of organisms that includes *Triceratops*, the living sparrow (*Passer*), and all descendants of their common ancestor. This hypothetical common ancestor can be visually traced on a family tree (properly called a cladogram, or a phylogeny) of reptiles: simply find *Triceratops*, then *Passer*, and then trace the branches leading to both species down to their common meeting point (Fig. 1.10). Any species that can also be traced down to this common ancestor – in other words, any species that descended from this ancestor – is by definition a dinosaur.

Phylogenetic definitions may seem confusing, but they can be understood with analogies to our own family histories. Some of my ancestors, for instance, immigrated to the United States from northern Italy. As the story goes, my great grandfather, upon hearing distressing rumors of anti-Italian sentiment in his soon-to-be new homeland, decided to change his surname from the obviously Italian "Brusatti" to the somewhat more ambiguous "Brusatte" when registering as a new citizen. This name change can be thought of as the origin of a new group of organisms, in this case the Brusatte family, and anybody who has descended from my great grandfather is by definition a Brusatte. It doesn't matter what we look like – whether we are tall, short, fat, thin, or bald – or when or where we live. We are simply Brusattes by definition.

The definition of Dinosauria given above is called a phylogenetic definition, and it is a general definition that can be applied to any cladogram. Clearly, however, this definition needs a phylogeny for context, and it is unintelligible without a cladogram to refer to. The first scientists to study dinosaurs did not define them this way, which is unsurprising given that these pioneering paleontologists were working in a pre-Darwinian world in which evolution (and hence common ancestry) was regarded as heresy. The man who named Dinosauria, Richard Owen (1842), followed the custom of the time and defined dinosaurs as those

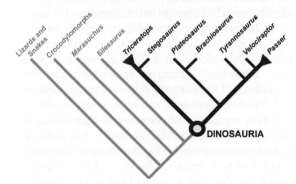

Figure 1.10 A schematic illustration showing how a group (such as Dinosauria) is defined in a phylogenetic sense. Dinosauria is formally defined as "members of the least inclusive clade containing *Triceratops horridus* and *Passer domesticus*." This definition requires a genealogical tree, or phylogeny, to make sense. In this case, locate *Triceratops* and *Passer* on the tree and then trace both branches back to their common ancestral meeting point (denoted by a circle). All species that also descended from this common ancestor are dinosaurs by definition (those species shown in black), whereas other species that fall outside this group are not dinosaurs by definition (those species shown in gray).

animals possessing a certain set of anatomical features, which included various traits relating to body size, posture, and locomotion (see below). Owen saw these features as essential characteristics – an unchangeable blueprint that set dinosaurs apart from other reptiles – but today we simply recognize them as products of common ancestry, as traits that all dinosaurs inherited from that distant ancestor that unites *Triceratops* and *Passer*. These are so-called synapomorphies: shared derived characters – evolutionary novelties – that unite a group on the tree of life.

This clarifies an important point: animals such as dinosaurs are not strictly defined by their anatomical features, but every group on the tree of life possesses a characteristic set of traits inherited from their common ancestor and thus absent in other organisms. These features are said to diagnose dinosaurs, rather than define them. An analogy can be seen in medicine: cancer is defined as a disease in which cells grow uncontrollably (a process), but is diagnosed by symptoms such as headaches, swelling, or abnormal breathing. Doctors never rigidly define a disease based on symptoms, but a certain disease usually has a characteristic set of symptoms, and by noticing and studying these symptoms a doctor can pinpoint the disease that is causing them. Dinosaurs, therefore, are defined based on ancestry, but share a common set of features, and by identifying and studying these features scientists can be sure that a certain specimen or organism is truly a dinosaur.

Characteristic Features of Dinosaurs

With the above semantics out of the way, we can now focus on those features that distinguish dinosaurs. After all, these anatomical features, and their biological and ecological significance, are much more interesting than the subtleties of cladograms, and the mundane quibbles about whether a certain species did or did not descend from a certain common ancestor. This criticism is not to trivialize phylogenetic definitions – their strength is in their explicitness and stability – but unfortunately tedious debates have raged over whether a certain species is a proper dinosaur or falls just outside of the group defined by *Triceratops* and *Passer*. These academic quarrels can be maddening, because the focus is on a technicality of nomenclature rather than much more illuminating discussions of biology, function, and evolution. And in one sense these debates miss the point, because even if an animal is not quite a dinosaur by definition, it may still have many features common to other dinosaurs, and may have resembled and behaved like true dinosaurs.

A prime example concerns a recently described group of peculiar Middle to Late Triassic archosaurs called the silesaurids. There is no question that these animals were very similar to dinosaurs, as they share several derived features with species that are unequivocally part of the *Triceratops–Passer* group. But there is debate over whether they are true dinosaurs: whether they descended from the common ancestor of *Passer* and *Triceratops*, or whether they are the closest relatives of true dinosaurs (i.e., are immediately outside the *Triceratops–Passer* group) (Dzik, 2003; Ferigolo and Langer, 2007; Irmis et al., 2007a; Brusatte et al., 2010a; Nesbitt et al., 2010; Nesbitt, 2011). This debate is indeed important for that narrow group of specialists which focuses on reptile phylogeny, and does have important ramifications for understanding patterns of character evolution, but is of little concern even for most dinosaur paleontologists. Therefore, in this section, I take a more catholic view of dinosaurs and focus not only on those features that precisely diagnose Dinosauria, but also features that are seen in a handful of the closest dinosaur relatives, which are not dinosaurs by definition but likely were very similar to dinosaurs in a biological sense. Throughout the remainder of this book the focus will be on true dinosaurs, but close dinosaur cousins ("stem dinosaurs") will sometimes be discussed for context or to flesh out exploration of biology, function, or large-scale evolutionary patterns.

When outlining features common to all dinosaurs, it is wise to begin with some historical background. Dinosauria was first established as a distinctive group by Owen (1842), who recognized that three extinct genera of large reptiles – *Megalosaurus*, *Iguanodon*, and *Hylaeosaurus* – shared several unusual features that were unknown in other reptiles, both living and extinct. These included features of the hips, limbs, and body posture, which generally indicated that dinosaurs had a more upright stance than other reptiles (see review in Cadbury, 2002). Discoveries of new fossils continued at a frenzied pace during the remainder of the 19th century, and by the dawn of the 20th century paleontologists had recognized that not only did all known dinosaurs share many features – including several additional hallmarks revealed by the new finds – but that they could be divided into two major subgroups: the "lizard-hipped" saurischians, which include theropods and sauropodomorphs, and the "bird-hipped" ornithischians (Seeley, 1887). These groups are recognized to this day as the two major subdivisions of dinosaurs. Over the next several decades, however, scientists gradually changed their conception of dinosaurs. For much of the 20th century, paleontologists considered saurischians and ornithischians to be separate lineages, which independently diverged long ago from separate "thecodont" (primitive archosaur) ancestors. Therefore, all the features common to saurischians and ornithischians were not seen as the product of common ancestry – characteristics that united all dinosaurs relative to other animals – but rather as insignificant nuances of the anatomy that evolved in parallel in both groups. The very idea of a single, distinctive dinosaur group had fallen out of favor.

This view began to change in the mid 1970s and within a few years was widely dismissed as outdated and incorrect. A new generation of paleontologists, motivated by new discoveries and conceptual advances, resurrected Owen's (1842) original notion of a single, unique group of Mesozoic reptiles – Dinosauria – that could be distinguished from all other organisms based on their possession of shared derived characters. This revolution in thinking was driven by two major factors. First, if saurischians and ornithischians were descended from separate ancestors, then the most primitive members of both groups should look very different from each other.

However, as new fossil finds of early saurischians and ornithischians were discovered in Triassic rocks across the world, this prediction was utterly rejected (Welles, 1954; Crompton and Charig, 1962; Reig, 1963; Colbert, 1970). Instead, primitive theropods, sauropodomorphs, and ornithischians were remarkably similar to each other, exactly as would be predicted if they diverged from a single common ancestor. Second, the advent of an explicit, numerical methodology for inferring genealogical relationships – cladistics – swept through the field of biology in the 1970s and 1980s (Hennig, 1965, 1966). Cladistic principles hold that a lengthy roster of shared anatomical features between two groups is much more likely to indicate close relationship than parallel evolution, and it would take quite a bit of special pleading to retain saurischians and ornithischians as separate entities that evolved so many eerily similar features independent of each other.

It was more plausible, therefore, that the myriad similarities between saurischians and ornithischians meant that these two groups descended from a common ancestor, and could be united as a single, larger group: Dinosauria. This view was persuasively articulated in a seminal 1974 paper by Robert Bakker and Peter Galton. In doing so, Bakker and Galton (1974: 168–169) highlighted a surprisingly long list of characteristic dinosaur features, many of which had been revealed by new discoveries during that long dark period when saurischians and ornithischians were assumed to be nothing but distant, convergent relatives. These features included an upright and fully erect posture, an enlarged deltopectoral crest on the humerus (which anchors large shoulder and chest muscles), a perforated hip socket for articulation with the head of the femur, a well-developed fourth trochanter and lesser trochanter on the femur (which anchor hindlimb muscles), and an ankle joint in which the proximal tarsals (astragalus and calcaneum) were "fixed immovably on the ends of the tibia and fibula [resulting in a] simple unidirectional hinge between the astragalus–calcaneum and distal tarsals." As is evident, many of these features have to do with the posture, strength, and range of motion of the forelimbs and hindlimbs: compared with their closest relatives, dinosaurs had a more upright stance and stronger, more muscular legs, which moved in a more restricted fore-aft direction, ideal for fast running and keen balance. Importantly, Bakker and Galton (1974) acutely recognized that many of these hallmark dinosaur features are also present in living birds, and thus support a close relationship between dinosaurs and birds. This was not a new idea, but one that was rapidly gaining traction in the field at the time. It had been proposed as early as the 1860s (Huxley, 1868, 1870a, 1870b), but had largely been ignored until the pioneering studies of John Ostrom in the 1960s and 1970s (Ostrom, 1969, 1973).

It is a great testament to the work of Bakker and Galton (1974) that many of the features they described as dinosaur trademarks are still considered valid today. This is no small feat, as the exact characteristics that diagnose a clade on the tree of life, such as Dinosauria, are constantly changing as new fossils are discovered and ideas are reinterpreted. At one point in time a certain character, such as a large deltopectoral crest, may only be known in one group, such as dinosaurs. It is easy to envision, however, how a single new fossil discovery, such as a new close dinosaur cousin with a large crest, could reveal that this feature is more widely distributed. This has, in fact, happened to several of Bakker and Galton's diagnostic characters but, importantly, most of the features they described are still only known in dinosaurs and a handful of their closest cousins, and their general argument that dinosaurs are distinguished from other reptiles by their posture and hindlimb anatomy still stands. But perhaps most important of all, Bakker and Galton's (1974) paper was a catalyst for future studies, and authors continue to actively debate exactly what characters unite dinosaurs.

Over the past four decades, beginning with Bakker and Galton's (1974) paper, approximately 50 characters have been identified as potential dinosaur synapomorphies. Many of these have emerged from detailed, higher-level cladistic analyses of archosaur phylogeny (Benton, 1984, 1999, 2004; Gauthier, 1986; Benton and Clark, 1988; Novas, 1989, 1992, 1996; Sereno, 1991a, 1999; Sereno and Novas, 1994; Fraser et al., 2002; Ezcurra, 2006, 2010a; Langer and Benton, 2006; Irmis et al., 2007a; Nesbitt et al., 2009b, 2010; Brusatte et al., 2010a; Martinez et al., 2011; Nesbitt, 2011). Of course, different phylogenies may imply different patterns of character evolution, and the exact

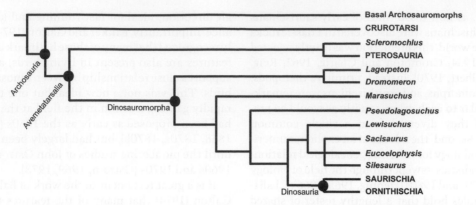

Figure 1.11 The genealogical relationships of "bird-line" archosaurs (Avemetatarsalia) based on the phylogenetic analysis of Brusatte et al. (2010a).

characters that diagnose Dinosauria often differ depending on the phylogeny being considered. To avoid the risk of getting mired in a tedious catalogue of different phylogenies, the discussion here uses the recent phylogeny of Brusatte et al. (2010a) and the review of dinosaur origins by Brusatte et al. (2010b) as guides. This phylogenetic context is graphically shown in Fig. 1.11.

Taking at first a reductionist view, seven features are currently recognized as unequivocal synapomorphies of Dinosauria. In other words, these characters are only known in true dinosaurs, and are absent even in the closest dinosaur cousins. These bona fide dinosaur hallmarks are known from across the skeleton, and include the following.

1 Temporal musculature that extends anteriorly onto the skull roof. The mandibular adductors (temporal muscles) are among the fundamental muscles of mastication in vertebrates: when they contract they elevate the lower jaw, allowing the mouth to close. Dinosaurs have an unusually large and extensive set of mandibular adductor muscles, which expand anteriorly onto the top of the skull (see Holliday, 2009 for review). Although muscle tissue is rarely preserved in dinosaur fossils, the location and size of the mandibular adductors can be deduced based on the position and size of a smooth fossa on the skull roof, to which these muscles attached. In most reptiles, including most archosaurs and even close dinosaur kin such as *Silesaurus*, the

fossa is restricted to the parietal bone, and is only expressed as a narrow depression in front of the supratemporal fenestra (one of the main diapsid skull openings, which will be described in more detail below) (Dzik, 2003). In dinosaurs, however, the fossa extends further anteriorly onto the frontal bone, and is a much deeper and more discrete depression (Fig. 1.12A,B). This indicates that the mandibular adductor muscles were larger and more powerful in dinosaurs than in close relatives, and probably implies that dinosaurs had a stronger bite than most other archosaurs.

2 Posterior process of the jugal bifurcates to articulate with the quadratojugal. The jugal bone forms the lateral "cheek" region of the skull underneath the eye and articulates posteriorly with the quadratojugal bone. Together these two bones define the ventral margin of the lateral temporal fenestra, the second of the two main diapsid skull openings. In all archosaurs other than dinosaurs, including *Silesaurus*, the posterior process of the jugal tapers and meets the quadratojugal at a simple overlapping joint (Dzik and Sulej, 2007). In dinosaurs, by contrast, the posterior process bifurcates into two prongs, which clasp the anterior process of the quadratojugal (Fig. 1.12C,D). The biological significance of these two conditions is uncertain, but it is likely that dinosaurs had a stronger jugal–quadratojugal articulation, and this may be functionally associated with their larger mandibular adductor musculature and inferred stronger bite force.

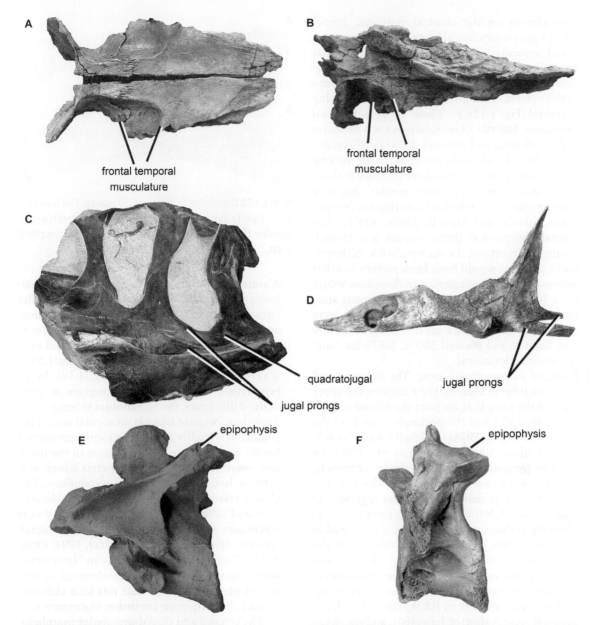

Figure 1.12 Distinctive features of dinosaurs. (A, B) Portions of the skulls of two theropod dinosaurs in dorsal view (*Dubreuillosaurus* and *Guanlong*) showing the anterior extension of the fossa for the temporal jaw muscles onto the frontal. (C, D) The bifurcated posterior process of the jugal, for articulation with the quadratojugal (jugal and quadratojugal of the theropod *Allosaurus* shown in articulation in C, only the jugal of the tyrannosaurid theropod *Alioramus* shown in D). (E, F) The epipophysis, a bump-like projection of bone on the dorsal surface of the postzygapophysis of the cervical vertebrae of the large theropod *Aerosteon* (E) and the tyrannosaurid *Alioramus* (F). Photographs (D) and (F) by Mick Ellison; image (E) courtesy of Dr Roger Benson.

3 Epipophyses on the cervical vertebrae. Epipophyses are projections of bone, which range from small mounds to more elaborate flanges, that protrude from the dorsal surfaces of the postzygapophyses of the cervical vertebrae (those parts of the vertebra that articulate with the following vertebra) (Fig. 1.12E,F). These are present in all dinosaurs, but not close relatives such as *Marasuchus* (Sereno and Arcucci, 1994) or *Silesaurus* (Dzik, 2003; Piechowski and Dzik, 2010). Various muscles of the neck would have attached to these structures, as well as some muscles that may have extended onto the back and thorax (Tsuihiji, 2005; Snively and Russell, 2007a, 2007b). The primary function of these muscles is to extend, rotate, and reinforce the neck and back. Although these muscles would have been present in other archosaurs, the epipophyses in dinosaurs would have increased their available attachment area, perhaps indicating that these muscles were stronger or capable of a greater range of motion (see Snively and Russell 2007a, 2007b for functional considerations).

4 Elongate deltopectoral crest. The deltopectoral crest is a ridge of bone on the humerus, the upper bone of the arm, that anchors the deltoid muscle of the shoulder and the pectoralis muscle of the chest (Coombs, 1978a; Nicholls and Russell, 1985; Dilkes, 2000; Jasinoski et al., 2006). Its primary purpose is to support the latter muscle, whose contraction brings the arm closer to the body. A discrete deltopectoral crest is present in many animals, but it is especially prominent and elongate in dinosaurs, in which it is expressed as an offset flange that extends for 30–40% of the length of the entire humerus (Fig. 1.13). In most other archosaurs, including close dinosaurian relatives such as *Marasuchus* (Sereno and Arcucci, 1994) and *Silesaurus* (Dzik, 2003), the deltopectoral crest is shorter, less offset, and restricted to the proximal portion of the humerus. The large deltopectoral crest of dinosaurs indicates that forelimb motion, particularly adduction towards the body, was especially powerful.

5 Open acetabulum in the pelvis. The acetabulum is the joint surface on the pelvis that articulates with the femur (thigh bone). In humans this is a ball-and-socket joint: the globular head of the femur fits into a deep depression on the pelvis.

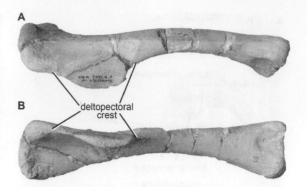

Figure 1.13 Distinctive features of dinosaurs. The humerus of the Late Triassic theropod *Liliensternus* in lateral (A) and anterior (B) views showing the expanded deltopectoral crest.

A similar condition, although with a much shallower socket and a less spherical head of the femur, is present in most reptiles, including most archosaurs. In these animals, the acetabulum is always a discrete socket, which is backed by a medial wall of bone. Dinosaurs, by contrast, have a very different morphology (Fig. 1.14). In all primitive dinosaurs, and most species of more derived dinosaurs, the acetabulum is "open" like a window, because there is no medial wall. This condition is readily apparent in even fragmentary fossils, as a concave ventral margin of the ilium (the most dorsal of the three pelvis bones) is a surefire hallmark of an open acetabulum. The closest relatives of dinosaurs, including *Marasuchus* and *Silesaurus*, have a ventral ilium that is essentially straight, but punctuated by a small concave divot (Sereno and Arcucci, 1994; Dzik, 2003). This is often referred to as an "incipiently open" acetabulum, and is hypothesized to be a transitional morphology that was later elaborated into the fully open condition of dinosaurs.

The opened and closed acetabular morphologies have clear functional significance (Fig. 1.15). Many reptiles, including primitive archosaurs, have a sprawling posture. In these sprawling forms, of which crocodiles are a prime example, the femur is angled outwards to a near horizontal inclination, and during locomotion the full weight of the body is transmitted medially, directly between the femur and the medial wall of the acetabulum. Therefore, it is no surprise that

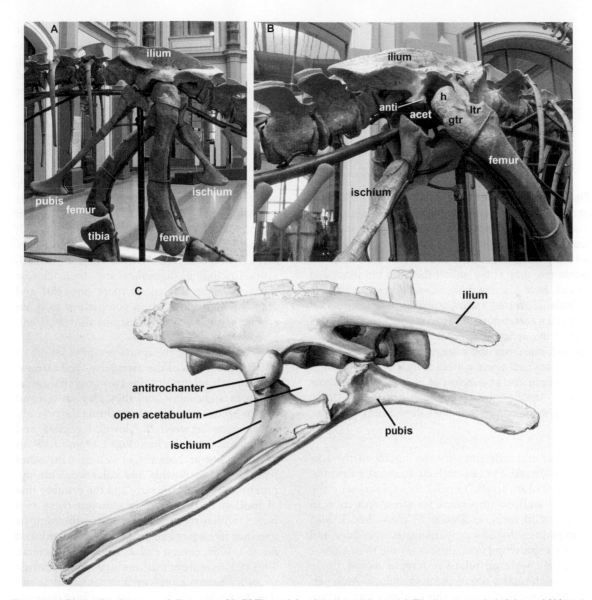

Figure 1.14 Distinctive features of dinosaurs. (A, B) The pelvis of the basal theropod *Elaphrosaurus* in left lateral (A) and oblique right lateral/posterior (B) views showing the articulation of the various bones of the pelvis and the distinctive open acetabulum of dinosaurs. (C) The articulated pelvis of the ornithischian dinosaur *Thescelosaurus* showing the open acetabulum and antitrochanter. Image (C) courtesy of the American Museum of Natural History Library (image #338613). acet, acetabulum; anti, antitrochanter; gtr, greater trochanter; h, head; ltr, lesser trochanter.

the acetabulum has a bony medial wall to provide reinforcement and dissipate stress. Dinosaurs and close relatives, however, have a more upright posture in which the hindlimbs are positioned directly underneath the body. This is facilitated by a modified femur, which has a head offset approximately 90° from the shaft, thus allowing the shaft to reposition itself in a vertical orientation (Fig. 1.15). As a result, the brunt of the body weight is transmitted between the top of the

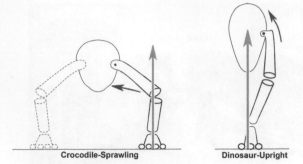

Crocodile-Sprawling   Dinosaur-Upright

Figure 1.15 Schematic of force vectors in a sprawling animal (a crocodile) and an upright-walking dinosaur. The gray arrows indicate the major forces created when the foot impacts the ground during locomotion and the black arrows indicate the direction that this force is transmitted within the body of the animal (between the femur and pelvis). Note that the main internal force in sprawling animals is directed inward, explaining the bony medial wall of the acetabulum (used to dissipate stress), whereas that of the upright-walking dinosaur is directed mostly upward, explaining why a bony medial wall is not necessary to dissipate stress in these animals (but a robust lip of bone above the acetabulum is necessary to dissipate stress). Modified from Hutchinson and Gatesy (2000). Used with permission from the Paleontological Society.

femur and only the very top surface of the acetabulum, and is not deflected medially into the acetabulum itself. This likely explains why the bony wall, so important for stress reduction in sprawling taxa, is absent in dinosaurs. It also explains why, in comparison to crocodiles and other sprawling taxa, dinosaurs and their closest upright-walking relatives have a robust lip of bone along the top of the acetabulum. And, perhaps most reaffirming, this explains why a handful of aberrant crocodile-line archosaurs that stood fully upright in a dinosaur-like fashion, such as *Effigia* (Nesbitt and Norell, 2006) and *Poposaurus* (Weinbaum and Hungerbühler, 2007; Schachner et al., 2011a), have an almost identical open acetabulum in their pelvis.

6 Fourth trochanter of the femur asymmetrical. The fourth trochanter is a ridge on the posterior surface of the femur and is present in all archosaurs. It anchors the caudofemoralis musculature, a large set of muscles that extends from the

tail to the femur and primarily acts to retract, or pull back, the leg during locomotion. Many archosaurs have either a subtle trochanter, indicating weak caudofemoral muscles, or a symmetrical and rounded trochanter. Dinosaurs, on the other hand, possess an asymmetrical crest-like trochanter in which the more distal portion of the crest is expanded relative to the proximal portion (Fig. 1.16D). This asymmetry is best seen in lateral or medial views, where it is apparent that the distal part of the trochanter forms a steeper angle to the femoral shaft than the more proximal trochanter. The functional significance of an asymmetrical, as opposed to symmetrical, fourth trochanter is unclear. In general, it is hypothesized that the large trochanters of many archosaurs, including dinosaurs and their closest relatives, were related to more powerful and efficient limb motions, in comparison to other species with delicate trochanters (Gatesy, 1990; Farlow et al., 2000).

7 Articular facet for the fibula occupies less than 30% of the width of the astragalus. The astragalus and calcaneum are the two proximal tarsal bones in archosaurs, and they play an integral role in hindlimb motion by forming the primary articulation between the lower leg (tibia and fibula bones) and the foot (Figs 1.17 and 1.18). In crocodile-line archosaurs, as well as many other reptiles, the astragalus and calcaneum are approximately the same size, and the primary line of motion in the ankle is between these two bones, which rotate against each other and fit together like a peg and socket (Cruickshank and Benton, 1985; Sereno and Arcucci, 1990; Sereno, 1991a). This suite of features is generally referred to as a "rotary joint" or a "crurotarsal joint" (Fig. 1.18A,B).

However, dinosaurs and their closest relatives are immediately recognized by a modified condition, in which the astragalus is much larger than the calcaneum (Fig. 1.18C). In these taxa, the astragalus is firmly braced against both the calcaneum and the tibia and fibula, and these four bones essentially form a single functional complex with no rotary motion between any of the individual elements (Fig. 1.17). Most strikingly, the astragalus has a long, thin, tongue-like flange called the ascending process that sits

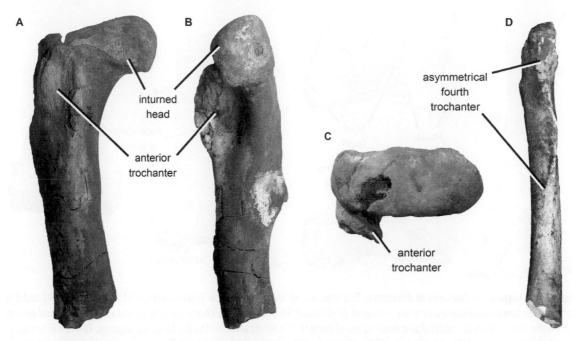

Figure 1.16 Distinctive features of dinosaurs. The left femur of the tyrannosaurid theropod *Tarbosaurus* in anterior (A), medial (B), and proximal (C) views illustrating the diagnostic inturned femoral head and anterior trochanter of dinosaurs. (D) The left femur of the small theropod *Miragaia* in posterior view illustrating the asymmetrical fourth trochanter of dinosaurs.

against the anterior surface of the distal tibia, firmly uniting the two bones. As a result, the primary line of motion is between the leg plus proximal tarsals complex and the foot itself. This is the so-called "hinge joint," or "mesotarsal" condition of bird-line archosaurs, which clearly differs from the crurotarsal condition of crocodile-line archosaurs.

Although these general ankle types – crurotarsal versus mesotarsal – distinguish crocodile-line archosaurs from bird-line archosaurs, true dinosaurs do have a further unique condition of the ankle that even mesotarsal "stem dinosaurs" such as *Silesaurus* do not possess. In dinosaurs, the fibula makes only a restricted contact with the astragalus, such that the smooth articular facet for the fibula on the astragalus is less than 30% of the width of the astragalus (Fig. 1.17B). In functional terms, this means that the fibula of dinosaurs is reduced and the tibia is the dominant bone of the lower leg. This probably relates to the general dinosaurian condition of upright

posture and fast locomotion, as a limb can move faster and more efficiently as a simpler structure, with less range of motion between individual bones and one dominant bone to the expense of others.

In summary, true dinosaurs are distinguished from all other reptiles by the seven features discussed above. There are also a number of additional features that, while not strictly diagnostic of Dinosauria itself, are only seen in dinosaurs and a few of their closest relatives: "stem dinosaurs" such as *Lagerpeton*, *Marasuchus*, *Silesaurus*, and *Asilisaurus*, which lived during the Early to Late Triassic (Sereno and Arcucci, 1993, 1994; Dzik, 2003; Nesbitt et al., 2010). Unsurprisingly, most of these features are also indicative of upright posture, fast locomotion, and a muscular reinforced skeleton. The most distinctive and important of these features include the following.

1 Three or more sacral vertebrae. The sacral vertebrae articulate with the pelvis, fitting tightly

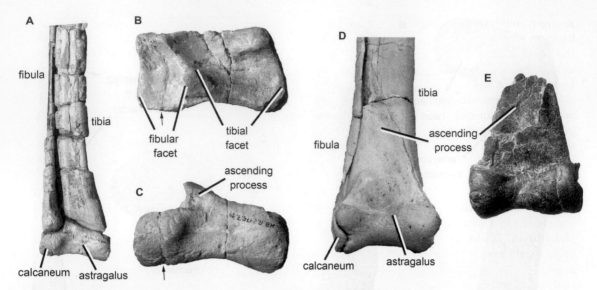

Figure 1.17 Distinctive features of dinosaurs. The articulated ankle of the Late Triassic theropod *Liliensternus* (A) and the Late Cretaceous ornithomimosaurian theropod *Gallimimus* (D) showing the characteristic mesotarsal ankle condition in which the calcaneum is reduced to a small spool of bone that is firmly attached to the large astragalus. The ankle bones of *Liliensternus* are also shown in proximal (B) and anterior (C) views, and the ankle of *Deinonychus* is shown in anterior (E) view. Note that the arrow in (B) points to the calcaneum–astragalus contact: only a small portion of the astragalus contributes to the fibular facet (another distinctive feature of dinosaurs). Image (E) taken by the author but copyright of the Peabody Museum of Natural History.

between the left and right ilium bones, and therefore are important in rigidly connecting the vertebral column and the hindlimbs. Whereas many archosaurs have only two sacral vertebrae, dinosaurs and some of their closest relatives have three or more sacral vertebrae. In fact, some derived dinosaurs have more than six sacral vertebrae. The increased sacral count is reflective of a

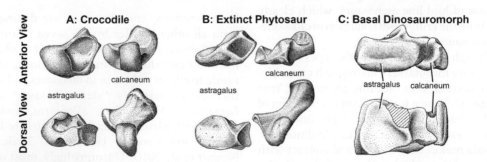

Figure 1.18 Mesotarsal vs. crurotarsal ankles. Crurotarsal ankles in a crocodile (A) and a phytosaur (an extinct member of the crocodile-line of archosaur phylogeny) (B), and a mesotarsal ankle in the basal dinosauromorph *Marasuchus* (C). In crurotarsal ankles the astragalus and calcaneum are approximately equal in size and articulate together at a mobile ball-in-socket joint. Mesotarsal ankles, however, are characterized by a proportionally enlarged astragalus and a tiny calcaneum, which articulate at a firm contact that permits no motion between them. Reproduced with permission from Sereno (1991a).

stronger, reinforced articulation between the backbone and pelvis, and would have helped brace the more muscular and swifter hindlimbs of dinosaurs.

2 Elongate pubis and enlarged antitrochanter. The pubis is the most anterior of the three bones of the pelvis, and in dinosaurs and close relatives this bone is extremely elongate relative to the shorter squatter morphology in other archosaurs. A major pelvic muscle, the puboischiofemoralis, originated along the pubis, and stretched onto the ischium (the most posterior pelvic bone) and the proximal part of the femur (Hutchinson and Gatesy, 2000; Carrano and Hutchinson, 2002; Hutchinson, 2001a, 2002). Therefore, a longer pubis would have increased the attachment area of this muscle, a consequence whose functional significance is not immediately clear. It is likely, however, that this muscle rearrangement was related to the upright posture of dinosaurs and the general strengthening of their legs (see Fig. 1.14). Another plausibly related feature is seen in the posterior part of the pelvis. The antitrochanter is a rough articular surface where the greater trochanter of the femur, the spur that the puboischiofemoralis muscle attaches to, makes contact with the pelvis (Hutchinson and Gatesy, 2000; Hutchinson, 2001b, 2002). This contact zone is limited to a narrow corner of the ischium in most archosaurs, but in dinosaurs and close relatives it is greatly expanded, and extends across parts of both the ischium and ilium (see Fig. 1.14). Once again, the precise functional significance of the enlarged antitrochanter is unclear, but at the very least it is indicative of a broader, and perhaps tighter, articulation between the femur and pelvis. This may have helped strengthen and rigidify the leg, and would have restricted the range of motion of the femur such that it primarily moved in a single plane, which is important for fast-running animals.

3 Anterior trochanter on the femur. The puboischiofemoralis muscles, which bring the hindlimb forward and towards the body, attach to the anterior surface of the femur, immediately below the head that articulates with the pelvis (Hutchinson and Gatesy, 2000; Hutchinson, 2001b, 2002). This attachment site is generally smooth in most archosaurs, but in dinosaurs and close

relatives is expanded into a rugose flange called the anterior trochanter (see Fig. 1.16). Therefore, dinosaurs and their kin would have had larger and more powerful hindlimb muscles, consistent with their upright posture and rapid locomotion.

4 Elongate, compact metatarsus with reduced lateral and medial digits. All bird-line archosaurs, including dinosaurs, stem dinosaurs, and pterosaurs, share a unique condition in which the metatarsal bones of the foot are bunched together and elongated (Gauthier, 1986) (Fig. 1.19). Furthermore, dinosaurs and their closest relatives have greatly reduced the size of the first and fifth metatarsals and their corresponding digits – the toes on the inside and outside of the feet – such that the central three metatarsals dominate the foot and form a simplified, paddle-like structure (Fig. 1.19). In sum, these modifications allow the metatarsus to act as a single unified structure, which is essentially a third major long bone of the hindleg (along with the femur and tibia). Unlike the metatarsals of humans and many other animals, including most archosaurs, these bones did not contact the ground during locomotion in bird-line archosaurs. Instead, only the toes themselves would have touched the substrate. Similar lengthening, strengthening, and simplifying of the metatarsus is seen in living animals that run rapidly, such as horses and gazelles. The bunched metatarsus of bird-like archosaurs is so unusual, and functionally significant, that it clearly registers in footprints of these species, including the oldest known members of the clade from the early Olenekian of Poland (Brusatte et al., 2011a) (see Fig. 1.7). In these footprints, as well as hundreds of other footprints from later in the Triassic and throughout the Mesozoic, only the toes are impressed in the sediment, and the digits themselves are nearly parallel due to the bunched construction of the foot (see also Ptaszynski, 2000; Carrano and Wilson, 2001).

With the above exhaustive list in mind, there should be no confusion between dinosaurs and other archosaurs. The dinosaur clade, as well as slightly more inclusive clades that also include "stem dinosaurs" and pterosaurs, are strikingly modified relative to crocodiles and other reptiles. More than 10 distinctive features are known only in

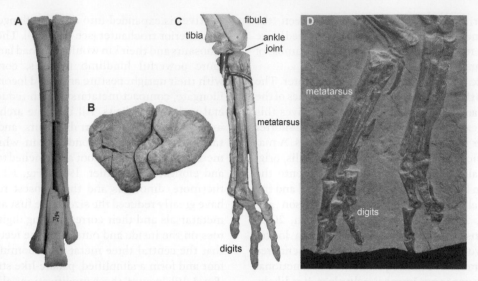

Figure 1.19 Distinctive features of dinosaurs. The metatarsus of the ornithomimosaurian theropod *Gallimimus* in anterior view (A), the metatarsus of the tyrannosaurid *Tarbosaurus* in proximal view (B), the metatarsus of early theropod *Elaphrosaurus* in posterior view (C), and the feet of the small theropod *Compsognathus* (D). All illustrate the "bunched" morphology characteristic of dinosaurs, in which the individual metatarsals are clustered close together and not splayed apart (as in crocodiles and other close dinosaurian relatives).

dinosaurs and their closest relatives, and possession of these characters is a surefire indication that a fossil specimen is a dinosaur. But these characters are not merely typological badges of honor that label specimens as dinosaurs, but also dynamic features of the skeleton that had clear functional and biological significance. Most of these characters are related to posture and muscle function: dinosaurs walked more upright and were faster than their closest relatives, and they had particularly strong jaw, neck, and forearm muscles and a rigid skeleton ideal for withstanding the rigors of a fast, active lifestyle.

The Major Dinosaur Subgroups

All dinosaurs evolved from a common ancestor and share a common set of skeletal features. With this said, a mere glance at any dinosaur museum exhibit – say, a *Tyrannosaurus* locked in aggressive battle with the deadly horns of *Triceratops* – is a dazzling reminder of just how different dinosaurs can be from one another. This remarkable diversity of size,

shape, diet, ornamentation, and lifestyle is one reason that the public is so fascinated with dinosaurs, and is surely a primary reason why dinosaurs were able to dominate terrestrial ecosystems for over 100 million years.

As briefly touched on above, dinosaurs can principally be divided into two major subgroups, the "lizard-hipped" saurischians and "bird-hipped" ornithischians, each of which can be further divided, finer and more finely, into less inclusive subgroups (see Fig. 1.21). This is, of course, because the family tree of life is hierarchical: a human is a primate, primates are one of many groups of mammals, mammals are tetrapods, tetrapods are vertebrates, vertebrates are one of numerous groups of animals, and so on. Chapter 4 will provide a detailed overview of the dinosaur family tree, and a discussion of the most comprehensive and up-to-date phylogenetic analyses used to construct it. First, however, it is necessary to provide a summary outline of the major dinosaur subgroups, to introduce the fundamental splits in dinosaur evolution and build a framework for the remainder of this narrative.

One of the great ironies of dinosaur paleontology is that the name Ornithischia is derived from the

Greek for "bird hip," even though birds are actually direct descendants of saurischian dinosaurs, and thus technically members of the "lizard-hipped" clade itself! This confusion, however annoying, is justified with a bit of historical context. When Seeley (1887) first recognized and named the ornithischian clade, the idea that birds evolved from dinosaurs, and particularly the more specific hypothesis that they derived from small carnivorous saurischians, was little more than a fringe speculation. Seeley was one of the first scientists to present a detailed classification of dinosaurs: an attempt to make order out of the exasperating diversity of shape, size, and diet among the ever-expanding pantheon of dinosaur species. As a keen anatomist, Seeley recognized a fundamental difference between two clusters of dinosaurs. Many species, such as the large carnivore *Allosaurus* and the long-necked herbivore *Diplodocus*, had a pelvis in which the pubis bone projected forward, just as in most living reptiles (Fig. 1.20A). He referred to these dinosaurs as saurischians: the "lizard-hipped" group. A few other species, though, had a bizarre condition in which the pubis was rotated backwards, so that it paralleled the ischium (Fig. 1.20B). These dinosaurs included the plated *Stegosaurus* and herbivorous *Iguanodon*, and Seeley collectively referred to them as ornithischians,

or "bird-hipped" dinosaurs, because living birds have a similar pelvic configuration. Seeley, therefore, was not proposing that ornithischian dinosaurs evolved from birds, but merely that their most distinguishing feature – their retroverted pubis bone – was similar to that of birds.

More than a century after Seeley's (1887) initial classification, ornithischians are today recognized as an incredibly diverse group, which includes a whole range of herbivorous species, many of which are armored or ornamented, and which run the gamut from fleet-footed to plodding, dog-sized to larger than elephants (Sereno, 1997, 1999; Weishampel, 2004) (Fig. 1.21). Ornithischians are primarily united by two sets of characteristics: those relating to the "bird-like" pelvis morphology and others closely tied to herbivory. The pubis is retroverted, as originally noted by Seeley (1887), and there is a novel flange of bone, called the prepubic process, that projects forward at the articular surface between the ilium and pubis. Furthermore, there are additional sacral vertebrae to brace the pelvis, and the anterior wing of the pelvis (the preacetabular process) is long, thin, and strap-like (Sereno, 1997, 1999; Weishampel et al., 2004; Butler et al., 2007, 2008a; Irmis et al., 2007b; Butler, 2010). Many of these pelvic features relate to the reconfiguration of muscles and the accommodation of a larger gut, essential for herbivorous species

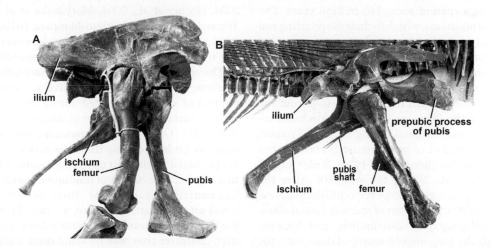

Figure 1.20 Saurischian and ornithischian pelves. (A) The pelvis of the theropod *Tyrannosaurus*, which exhibits the saurischian condition in which the pubis projects forward. (B) The pelvis of a hadrosaurid, which exhibits the ornithischian condition in which the pubic shaft projects backward (paralleling the ischium) and a novel prepubic process projects forward. Images courtesy of the American Museum of Natural History Library (image #35423, 4267).

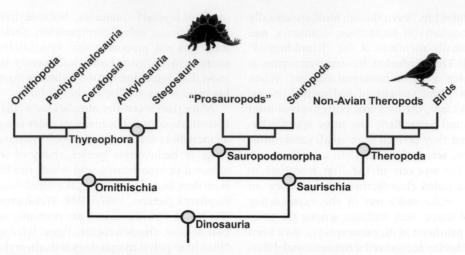

Figure 1.21 The general genealogical relationships of the major groups of dinosaurs.

that must ingest large quantities of plant matter. Other unique characters relating to herbivory include well-developed wear facets on leaf-shaped teeth, a large and strong dentary bone in the lower jaw, and the inferred presence of cheeks, which are vital for retaining chewed food in the mouth (Norman, 1984; Weishampel, 1984; Weishampel and Norman, 1989; Sereno, 1997, 1999; Butler et al., 2007).

The ornithischians that shared these characters persisted from the Late Triassic until the end of the Cretaceous, a span of some 165 million years. The oldest ornithischians, which include the puzzling and frustratingly poorly preserved *Pisanosaurus*, are known from the rock horizons in Argentina that preserve the world's oldest dinosaurs (Bonaparte, 1976; Sereno, 1991b; Butler et al., 2007, 2008a; Irmis et al., 2007b). Although they were present from the inception of the dinosaur clade, ornithischians remained rare and marginal during the Late Triassic, and only a handful of specimens are known from across the globe (Bonaparte, 1976; Butler et al., 2007; Irmis et al., 2007b; Butler, 2010). After the end-Triassic extinction, however, ornithischians exploded in diversity (number of species), faunal abundance, and geographic distribution, and became, along with the long-necked sauropod dinosaurs, the pre-eminent herbivores in most terrestrial ecosystems across the world (Butler et al., 2007, 2008a; Brusatte et al., 2008b). Many of the fundamental ornithischian subgroups arose and diversified during

the Jurassic: the plate-backed stegosaurs, the armored and tank-like ankylosaurs, and the beaked ornithopods (Galton and Upchurch, 2004a; Norman et al., 2004a; Vickaryous et al., 2004; Maidment et al., 2008; McDonald et al., 2010). One subgroup of the ornithopods, the duck-billed and fabulously crested hadrosaurids, flourished during the Cretaceous, along with two other fantastically ornamented ornithischian subgroups, the horned ceratopsians and dome-headed pachycephalosaurs (Dodson et al., 2004; Horner et al., 2004; Maryanska et al., 2004; You and Dodson, 2004; Prieto-Márquez, 2010a; Ryan et al., 2010a). No ornithischians, however, were able to endure the Cretaceous–Paleogene extinction.

The other major subgroup of dinosaurs, the saurischians, were also remarkably diverse, both in the number of species that lived during the Mesozoic and the variability in their size, anatomy, and diet (Fig. 1.21). Of course, saurischians also survive today as birds, meaning that this major subgroup has persisted for the past 65 million years in a world that is otherwise barren of dinosaurs. Seeley (1887) differentiated saurischians from ornithischians based on pelvic anatomy, but we now know that the "lizard-like" condition of saurischians is a primitive character that was retained from distant ancestors. After all, as its descriptive moniker implies, a lizard-like pelvis is present not only in saurischians but also in crocodiles and many other reptiles. This raises a problem. Because only shared, derived characters that are inherited from a common ancestor – in this case, the common ancestor of saurischians to the exclusion of all other reptiles – are useful in diagnosing a clade, scientists must recognize discrete evolutionary novelties of saurischians in order to retain this subgroup as a true division of Dinosauria. Otherwise, saurischians would simply be a nebulous assemblage of primitive dinosaurs, not their own unique group united by derived features. This is not a problem for ornithischians, because their highly peculiar pelvis is clearly modified from the primitive reptilian condition, and thus represents an undisputable evolutionary novelty. Yet, do saurischians possess any of their own novelties?

Fortunately, such characters do indeed exist, and the roster of saurischian novelties is being continuously refined as new fossils of Triassic dinosaurs emerge. All saurischians share derived features of the neck, hand, and feet (Sereno, 1997, 1999; Langer and Benton, 2006; Nesbitt et al., 2009b; Martinez et al., 2011). The neck is elongate, due to the increased length of individual vertebrae, and the epipophyses – the projections on the cervical vertebrae for muscle attachment that are present in all dinosaurs – are not limited to only the first few vertebrae as in ornithischians, but are present along the entire neck. The hand is long, nearly half the length of the arm, and the first finger is especially large and projects strongly medially relative to the remainder of the hand. All together, these features allowed the hand to function as a strong grasping organ, perfect for clasping prey. The metatarsals of the feet lie against each other in an overlapping, en echelon arrangement, and do not simply abut each other as in ornithischians and other archosaurs. Although some of these shared novelties have been revealed by new fossils, other discoveries have dismissed several previously held saurischian features as more widely distributed among dinosaurs. For instance, it was long considered that saurischians uniquely possessed a subnarial foramen between the premaxilla and maxilla, pneumatic openings and laminae on the vertebrae, and hyposphene–hypantrum articulations to reinforce the contact between vertebrae. These features, and many others, are now known to be present in some primitive ornithischians (Butler et al., 2007), and sometimes even in stem dinosaurs and crocodile-line archosaurs (Gower, 2001; Dzik, 2003; Nesbitt and Norell, 2006; Nesbitt, 2007; Weinbaum and Hungerbühler, 2007; Brusatte et al., 2010a).

The first saurischian fossils are also known from the same Late Triassic Argentine units that yield the first ornithischian fossils. In fact, the two major saurischian subgroups – the carnivorous theropods and long-necked herbivorous sauropodomorphs – are already present by this time (Sereno and Novas, 1992; Sereno et al., 1993; Langer et al., 1999; Bittencourt and Kellner, 2009; Martinez and Alcober, 2009; Ezcurra, 2010a; Ezcurra and Brusatte, 2011; Martinez et al., 2011). Unlike ornithischians, however, saurischians quickly diversified, became ecologically dominant, and spread across the globe soon after their origination. Theropods and sauropodomorphs are common fossils in Late Triassic rocks around the world, and in many ecosystems primitive sauropodomorphs were the most common, and the largest, herbivores (Benton, 1983).

Most Triassic sauropodomorphs were "prosauropods," an informal name for a nebulous grade of primitive species that did not comprise their own unique group distinguished by novel characters (Galton and Upchurch, 2004b; Upchurch et al., 2007; Yates, 2007; Pol et al., 2011). Prosauropods were the ancestors and closest relatives of the sauropods, the distinctive long-necked, small-headed, plant guzzlers typified by *Brachiosaurus* and *Diplodocus* (Upchurch, 1995, 1998; Wilson and Sereno, 1998; Wilson, 2002; Upchurch et al., 2004; Curry-Rogers and Wilson, 2005). These behemoths, which originated in the Late Triassic but reached their zenith in the Late Jurassic and Early Cretaceous, included the largest land animals to ever live. Compared to such giants, however, the Triassic and Early Jurassic "prosauropods" were much smaller and had shorter necks, and many species were likely omnivorous and could alternate between walking on two or four legs (Barrett, 2000; Barrett and Upchurch, 2007; Bonnan and Senter, 2007; Langer et al., 2007).

The earliest theropods were mostly small animals, dwarfed in comparison with their later, more familiar Jurassic and Cretaceous cousins such as *Tyrannosaurus* and *Allosaurus* (see Plates 1, 2, and 3). Most Triassic theropods belonged to a primitive grade of small species, the "coelophysoids," which like the "prosauropods" was not a unique group united by derived characters. These primitive theropods are exemplified by *Coelophysis*, a kangaroo-sized species overshadowed by much larger crocodile-line predators when it was alive (Colbert, 1989; Brusatte et al., 2008a). Some Triassic theropods, however, grew up to 5–6 m in length, and were surely apex predators in their ecosystems (Huene, 1934). A sudden and pronounced increase in theropod size is recorded across the Triassic–Jurassic boundary, immediately after the extinction of many crocodile-line archosaurs that filled top predator niches (Olsen et al., 2002), and throughout the Jurassic and Cretaceous theropods would diversify into a stupefying array of different subgroups. The most familiar of these are probably the tyrannosauroids, typified by the iconic *Tyrannosaurus rex*, and the dromaeosaurids, which include the sleek scythe-clawed predators *Deinonychus* and *Velociraptor* of *Jurassic Park* fame (see Plates 4–11). Regardless of their size or when they lived, theropods are united by a fairly conservative body plan: they are bipedal predators, most of which could run quickly and had a multitude of weapons, sharp teeth and claws included, to take down prey. Only some very aberrant, derived Late Jurassic and Cretaceous species, such as the beaked ornithomimosaurs, the toothless oviraptorosaurs, and the barrel-chested therizinosauroids, would deviate from this fast-running, predatory lifestyle; instead, these theropods were omnivorous or, in some cases, completely herbivorous (Kobayashi et al., 1999; Barrett, 2005; Zanno et al., 2009; Zanno and Makovicky, 2011).

Birds: Living Dinosaurs

The most atypical theropods, however, are undoubtedly the birds. One of the great revelations of dinosaur research, and perhaps the single most important fact ever discovered by dinosaur paleontologists, is that birds are descended from small carnivorous theropods. This idea was originally proposed by Thomas Henry Huxley, the acerbic 19th century advocate of evolution known as "Darwin's Bulldog," in the 1860s. This was a revolutionary decade in science. Darwin had published his *Origin of Species* in 1859, which persuasively and decisively laid out the evidence for evolution by natural selection. Rational thinkers had no

recourse: organisms evolved over great lengths of time, and shared characteristics were indicative of a close genealogical relationship. The public was a bit more skeptical, however, and pundits like Huxley were on the lookout for so-called "missing links" – transitional fossils that captured, like a freeze frame, the evolution of one group into another – that could viscerally demonstrate the reality of evolution to the masses.

It did not take very long for a convincing "missing link" to appear. In 1861, a mere two years after Darwin's groundbreaking publication, quarry workers in the Bavarian hillsides of Germany discovered the fossilized bones of a peculiar bird (Fig. 1.22A). This fossil had to be a bird: the fine preservation revealed an unmistakable halo of feathers around the body, it had a wishbone at the front of its chest, and the wrists and feet were almost identical to those of living birds. But something was amiss. This bird had teeth in its skull and a long bony tail, features that are not present in any living bird. And, even more puzzling, the skeleton of this bird, especially the form of its tail and skull, was eerily similar to another fossil discovered in the same lithographic limestone beds: a small predatory theropod called *Compsognathus*. To keen observers like Huxley this fossil was the Holy Grail: a "missing link" that possesses features of both dinosaurs and birds, and therefore captures an evolutionary transition between the two groups.

This fossil bird was named *Archaeopteryx*, and it immediately became a public sensation and still remains one of the most important and iconic fossils in the history of paleontology (Chambers, 2002). In a series of publications, and more important in a whirlwind sequence of public lectures, Huxley ebulliently argued that *Archaeopteryx* was proof positive that birds were descended from dinosaurs (Huxley, 1868, 1870a, 1870b). In an ironic twist, the strikingly half-bird, half-dinosaur skeleton of *Archaeopteryx* helped sway public perception in favor of evolution, but Huxley's specific idea that birds evolved from small carnivorous dinosaurs fell out of favor among scientists (Heilmann, 1926). It was not until the 1960s that a small, vocal group of paleontologists resurrected Huxley's ideas, buoyed largely by the discovery of spectacular fossils of the very bird-like dinosaur *Deinonychus* (Ostrom, 1969). Today, the hypothesis that birds evolved from theropod dinosaurs – nay, that

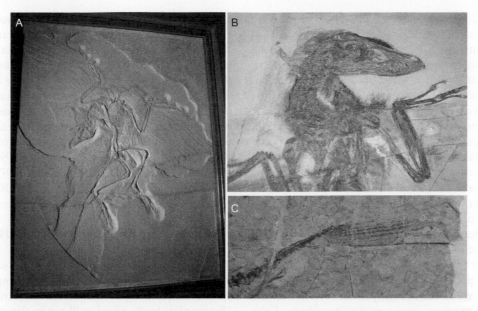

Figure 1.22 The evolutionary relationship between dinosaurs and birds. (A) The iconic Berlin specimen of *Archaeopteryx*, the oldest known bird; (B) the feathered non-bird dinosaur *Sinornithosaurus*, a dromaeosaurid closely related to *Velociraptor*; and (C) the tail of the feathered non-bird theropod dinosaur *Caudipteryx*. All photos by the author.

birds actually *are* theropod dinosaurs, since they are firmly nested within the group – enjoys nearly universal acceptance in the scientific community (Gauthier, 1986; Padian and Chiappe, 1998; Chiappe and Witmer, 2002; Chiappe, 2007).

The theory that birds descended from dinosaurs is so widely accepted because it is supported by a preponderance of evidence. Grand ideas demand strong evidence, and the dinosaur–bird link, which might be dismissed as too fanciful to be true by some critics, foots the bill. Living birds share hundreds of skeletal features with dinosaurs, and many characteristics that are unique to birds among living animals – a wishbone, a long S-shaped neck, a mesotarsal ankle joint, a wrist that enables the arm to fold against the body, and a retroverted pelvis being some of the most conspicuous – are also present in bona fide Mesozoic dinosaurs such as *Tyrannosaurus* and *Velociraptor* (see Padian and Chiappe, 1998; Shipman, 1998; Chiappe and Witmer, 2002; Chiappe 2007). Ditto for soft tissues: sinuses in the skulls and internal chambers in the vertebrae conclusively show that many dinosaurs had an extensive system of air sacs, a critical component of the bellows-like respiratory system un-

ique to modern birds, and proteins from a *Tyrannosaurus* skeleton that miraculously survived the rigors of 66 million years of fossilization share uncanny structural and molecular similarities with proteins of living birds (Britt, 1993; O'Connor and Claessens, 2005; Schweitzer et al., 2005a; Organ et al., 2008; Sereno et al., 2008; Benson et al., 2011). There is behavioral evidence as well: spectacularly preserved dinosaur fossils have been found in the characteristic sleeping and egg brooding postures of living birds (Norell et al., 1995; Xu and Norell, 2004), bone histology and texture indicate that dinosaurs grew rapidly like living birds (Padian et al., 2001; Erickson et al., 2007), and medullary bone – a novel tissue that provides calcium for the shelling of eggs – is known only among birds and dinosaurs (Schweitzer et al., 2005b).

Most extraordinary of all, thousands of spectacularly preserved dinosaur specimens, all of which have been discovered in northeastern China during the past 20 years, are unmistakably sheathed in a coat of feathers (Chen et al., 1998; Norell and Xu, 2005) (Fig. 1.22B,C; see Plates 5–7). Some of these feathers, with their central quill and radiating barbs, are identical to those of modern birds (Norell et al., 2002), and their preservation is so astounding that one could easily be fooled into thinking that they had just been plucked from a living, breathing species. Intensive molecular sampling demonstrates that many of these feathers preserve remnants of the melanosomes, the pigment-containing structures that give feathers their radiant (or in some cases drab) hues (Li et al., 2010; Zhang et al., 2010; Wogelius et al., 2011). And it wasn't only the closest relatives of birds that had feathers, but also much more distant cousins such as the herbivorous therizinosauroids (Xu et al., 1999) and, most incredible, the tyrannosauroids (Xu et al., 2004). Emerging evidence provocatively suggests that non-theropods had feathers, as a number of basal ornithischian specimens have also been found with a downy coat (Mayr et al., 2002; Zheng et al., 2009), and it is therefore possible that the common ancestor of dinosaurs was a feathered species.

Birds, therefore, are surviving members of the dinosaur clade. They are every bit as much a "dinosaur" as *Tyrannosaurus*, *Stegosaurus*, or *Brachiosaurus*, and their main distinction from other dinosaurs, aside from their novel flying lifestyle, is that they were able to survive the global meltdown at the Cretaceous–Paleogene boundary. Among dinosaurs, birds are most closely related to dromaeosaurids (*Velociraptor* and kin) and the troodontids, an intriguing subgroup of small, sleek, intelligent, and perhaps omnivorous theropods (Makovicky and Norell, 2004). These genealogical relationships – the nesting of birds within theropod dinosaurs and the particularly close relationships between birds, dromaeosaurids, and troodontids – are consistently recovered in phylogenetic analyses, and therefore these branches of the dinosaur family tree are on solid footing (Gauthier, 1986; Sereno, 1999; Norell et al., 2001a; Clark et al., 2002; Senter, 2007; Turner et al., 2007a; Csiki et al., 2010; Xu et al., 2011a).

The World of the Dinosaurs

It is difficult, if not impossible, to understand the biology and evolution of dinosaurs without an appreciation for the physical world they inhabited. Disregarding birds, which will not be the subject of this book, dinosaurs lived during the Mesozoic Era (from about 252 to 65 million years ago), an extraordinary time in earth history that witnessed the birth and death of a supercontinent and experienced some of the highest temperatures and sea levels in the geological record (Fig. 1.23). Indeed, as eloquently described by Sellwood and Valdes (2006), in an

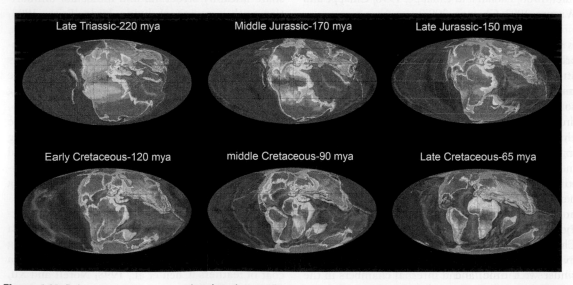

Figure 1.23 Paleogeographic maps, showing the configuration of the continents and oceans during six intervals throughout the Mesozoic history of dinosaurs. All images courtesy of Dr Ron Blakey (http://www2.nau.edu/rcb7/globaltext2.html).

important overview of Mesozoic climates that should be referred to for more specific information, "the Mesozoic earth was, by comparison with the present, an alien world."

Dinosaurs and their closest relatives originated in the Triassic Period (c.252–201 million years ago; see Sues and Fraser, 2010 for a general overview). The physical geography of this time was remarkable: most of the world's land was sutured together into a single landmass, the supercontinent Pangaea, which was centered on the equator and surrounded by a single giant ocean (Wegener, 1924; Smith et al., 1994; Scotese, 2004) (Fig. 1.23). Climates on this supercontinent were warm and arid, especially in the continental interior of Pangaea, which was far from the ameliorating effects of the coasts (Sellwood and Valdes, 2006). It is generally thought that temperatures became warmer and more arid throughout the Triassic, and by the time the first dinosaurs evolved much of the planet may have been covered in deserts (Tucker and Benton, 1982). Climate change during the Triassic may not have been gradual, however, as the rock record chronicles an abrupt transition from somewhat cooler and wetter conditions to substantially warmer and more arid climates during the Norian, the lengthy Triassic substage (c.228–208 million years ago) during which dinosaurs began their rise to dominance (Simms and Ruffell, 1990). Moreover, the most abrupt climatic change occurred at or near the Triassic–Jurassic boundary. Global temperatures and greenhouse gas levels spiked at this time, almost certainly the result of widespread volcanism associated with the initial rifting of Pangaea, and it is probably not coincidental that major extinctions in the terrestrial and marine realm occurred almost synchronously (McElwain et al., 1999; Whiteside et al., 2010).

Many of the extremes of Triassic geography and climate would dampen during the ensuing Jurassic Period (c.201–146 million years ago). Pangaea began to fragment during the Late Triassic, as an influx of heat from the deep interior of the earth tore open rift basins along what is currently the Atlantic coast of North America and Europe. These basins grew wider with time, and by the Middle Jurassic the nascent Atlantic Ocean separated Pangaea into northern and southern blocks (Fig. 1.23). The northern landmass, called Laurasia, contained North America, Asia, and Europe, the latter of which was flooded by high sea levels and reduced to a series of islands. The southern landmass, called Gondwana, was a still-giant block of crust that included South America, Africa, Australia, Antarctica, and India. The Jurassic was still a time of warm climates, but conditions were much wetter than during the arid Late Triassic and, as a result, a great diversity of plants (especially gymnosperms) were able to flourish at all latitudes (Rees et al., 2000; Sellwood and Valdes, 2006). The extreme peaks of Late Jurassic temperature are best illustrated by a simple comparison: geological evidence indicates that atmospheric carbon dioxide levels, a proxy for temperature, were up to four times higher in the Late Jurassic than in today's world (Berner, 2006; Fletcher et al., 2008).

The physical world continued to change, and to assume a more modern feel, during the final stanza of dinosaur evolution, the Cretaceous Period (c.145–65 million years ago). Laurasia and Gondwana, the two great remnants of Pangaea, further disintegrated during the Cretaceous, and by the end of the period the continents were positioned, more or less, in their current configuration (Smith et al., 1994; Scotese, 2004) (Fig. 1.23). Continental rifting was especially vigorous in Gondwana: what began as a single large landmass in the Late Jurassic fragmented into today's characteristic southern continents within a time frame of only a few tens of millions of years. Most remarkably, India began the Cretaceous as a wedge of crust between Africa and Antarctica, but steadily moved northeast until it had just begun colliding with Asia at the time the dinosaurs went extinct. This collision, of course, would be fully realized several million years later, with the Himalayas rising skyward as a consequence.

The Cretaceous world was still a hothouse, with high global temperatures and little evidence for polar ice caps, but temperatures fluctuated more wildly than during the Triassic and Jurassic (see Skelton et al., 2003). Temperatures were especially high throughout the middle Cretaceous, probably driven by intensive volcanism that belched large volumes of carbon dioxide into the atmosphere. High temperatures resulted in high sea levels, because little water was locked up in glaciers, and warm shallow seas lapped the continents. During the Late Cretaceous, for instance, North America was bisected by one such seaway, which stretched from the Gulf of Mexico to the Arctic. Global temperatures probably reached a peak approximately 100–120 million years ago, approximately at the same time as a middle Cretaceous extinction event thought to have been caused by rapid stagnation of the oceans (Jenkyns, 1980; Fletcher et al., 2008). Whether dinosaurs and other terrestrial organisms were affected by this brisk interval of climate change is uncertain. From this point on, however, atmospheric carbon dioxide levels, and thus global temperature, decreased throughout the remainder of the Cretaceous (Fletcher et al., 2008). Our modern climates, which are relatively cool compared with other intervals in earth history, come at the tail end of this long-term, 100-million-year decline. This being said, although current temperatures are cool compared with the Mesozoic, human-induced climate change is a pressing source of concern because of its rapid pace and its potential to alter physical environments that humans have become accustomed to.

Conclusions

Dinosaurs are an iconic group of archosaurian reptiles, whose living descendants include about 10 000 species of modern birds. Although the term "dinosaur" is part of the popular vocabulary, and is often used to denote anything that is old, huge, or frightening, the scientific concept of dinosaurs is precise: dinosaurs are defined as the clade on the family tree of life that encompasses *Triceratops*, the living sparrow (*Passer*), and all descendants of their most recent common ancestor. This group, formally referred to as Dinosauria, is diagnosed by several shared derived characters, or evolutionary novelties, that all dinosaurs inherited from their common ancestor. Most of these features relate to posture, locomotion, and musculature: dinosaurs stood upright, were likely faster than their closest relatives, and had a skeleton and associated musculature that were well adapted to endure a fast active lifestyle. Dinosaurs can be divided into two major subgroups, the "lizard-hipped" saurischians and "bird-hipped" ornithischians, each of which can be further subdivided more finely into other subgroups. The principal saurischian subgroups are the carnivorous theropods and long-necked sauropodomorphs, whereas familiar ornithischians include the armored ankylosaurs, plated stegosaurs, duck-billed hadrosaurs, horned ceratopsians, and dome-headed pachycephalosaurs. The evolutionary history of dinosaurs is a gripping narrative that took place against a backdrop of drifting continents, climate change, and mass extinction events. Dinosaurs originated in the Middle Triassic, gradually became more diverse and abundant during the Late Triassic, and were able to endure the end-Triassic mass extinction that decimated many competitor groups. During the Early Jurassic dinosaurs truly became dominant on a global scale, and for the remainder of the Mesozoic were the pre-eminent vertebrates, both carnivore and herbivore, at mid-to-large size in terrestrial ecosystems across the globe.

2 | Hard Tissues

When most people imagine a dinosaur, the first image that comes to mind is probably that of a skeleton. Dinosaur skeletons, especially the ominously athletic frame of *Tyrannosaurus* or the monument-like chassis of a long-necked sauropod, are always among the most popular exhibits in museums. As exhilarating as it may be to imagine the colors of dinosaurs, and the skin, feathers, and muscles that they may have sported, in most cases paleontologists are only left with bones and teeth (see Plates 1 and 4). Except in rare circumstances, only the hard mineralized tissues of dinosaurs are able to withstand the rigors of fossilization. It is no surprise, therefore, that most of our knowledge of dinosaur anatomy, and by inference our understanding of dinosaur biology and evolution, is based on careful consideration of bones and teeth. As frustrating as it may be for non-specialists to wade through the technical literature on dinosaurs, which is rife with descriptions of individual specimens and detailed comparisons of the bones of different species, this is the primary data upon which the entire science of dinosaur paleontology is built. This chapter will establish a framework of such primary data, by providing a general overview of bones and teeth, a summary of the dinosaur skeleton, and a more focused description of some of the most unusual, and sublime, dinosaur hard parts.

Dinosaur Paleobiology, First Edition. Stephen L. Brusatte.
© 2012 John Wiley & Sons, Ltd. Published 2012 by John Wiley & Sons, Ltd.

Bones and Teeth: A Vertebrate Innovation

Many organisms have hard parts, which are always the most common parts to fossilize, but the composition, development, and function of these tissues are remarkably variable. The spicules of sponges, shells of bivalves and gastropods, and carapaces of insects are but a handful of examples. Although these structures are all quite different, they do share several common functions: protection, support, and a firm anchor for muscles and other soft tissues.

Vertebrates have developed their own novel set of hard tissues – bone, enamel, and dentine – that together comprise the various hard structures of the skeleton (see Smith and Hall, 1990; Donoghue et al., 2002, 2006 for general reviews). Bone, which as its name implies is the primary component of bones, is a complex material comprising hardened mineral, fibrous connective tissue (collagen), and living cells (osteocytes) (Fig. 2.1). The mineral is hydroxyapatite, a compound of calcium and phosphate, which differs from the calcium carbonate mineralogy of the hard parts of most invertebrates such as bivalves, corals, and echinoderms. Enamel and dentine, which are the harder and softer components of teeth, respectively, are also formed

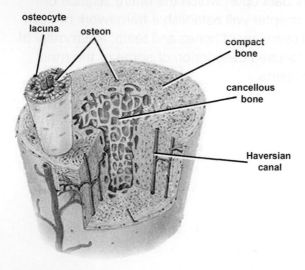

primarily from hydroxyapatite. The difference between the two is principally related to the amount of calcium phosphate that they contain: enamel contains a greater percentage of hydroxyapatite than dentine, and therefore it is harder, less brittle, and can be formed into a sharper cutting surface. As a result, enamel usually envelops the external surface of a tooth, whereas dentine comprises the inner, reinforcing core.

Although all bones have the same fundamental composition, there are two general types of bone in the vertebrate body – endochondral and dermal – which differ in how they develop and grow (see Romer, 1956 for a comprehensive overview of the vertebrate skeleton). Endochondral bones begin with a cartilage precursor, a scaffolding upon which bony tissue is deposited. Dermal bones, on the other hand, do not begin as cartilage, but rather develop directly in the skin or other membranous tissue. Because of their origin, these bones are often referred to as "membranous bones." The majority of the skeleton is composed of endochondral bone, including the vertebrae, limb bones, pelvis, and braincase (the solid box that surrounds the brain and primary sense organs of the skull). Much of the skull, however, is composed of dermal bone, including the major tooth-bearing elements of the jaws, the cheek region, and the skull roof. Other conspicuous dermal bones include armor plates, spikes, spines, and horns. Some bones, such as the scapula (shoulder blade), are a composite of dermal and endochondral ossifications.

Not only do different bones have different origins and developmental trajectories, but there is also a variety of bone textures (see Chinsamy-Turan, 2005 for an overview). The internal arrangement of the minerals and cells in a bone vary considerably, largely due to where in the bone they are located, how the bone grows, and how it responds to damage and applied loads. After all, bones are not static structures but growing tissues that dynamically respond to stress and which can be ravaged by disease or injury. The outer shell of a bone is usually composed of so-called compact bone (or cortical bone), which is hard, dense, and packed with hydroxyapatite, with only a minimum amount of empty space for osteocytes and blood vessels (Fig. 2.1). The interior of the bone, on the other hand, is filled with cancellous bone (also known as

Figure 2.1 Diagram of a generalized long bone (such as the femur or humerus) in cross-section. Schematic from Wikipedia Commons, modified by Anne Curthoys.

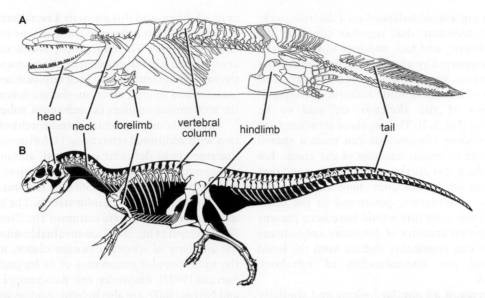

Figure 2.2 Skeletal reconstruction of an early tetrapod (*Ichthyostega*) and a dinosaur (the theropod *Allosaurus*) showing homologous parts of the skeleton. *Ichthyostega* image modified from Ahlberg et al. (2005) and *Allosaurus* image © Scott Hartman.

trabecular or spongy bone), a much lighter and more porous material through which blood vessels flow freely (Fig. 2.1). The great strength and rigidity of compact bone lend support to the skeleton, whereas the more permeable cancellous bone facilitates the free movement of blood, cells, and nutrients and therefore its primary role is not to provide support but to regulate metabolic activity and provide a substrate for the creation of new blood cells (in many animals, including humans, the cancellous bone is filled with marrow). Whereas compact and cancellous bone are functionally distinct, two additional bone textures are the result of differences in growth rate. Woven bone is composed of a chaotic arrangement of hydroxyapatite crystals and collagen, indicative of rapid growth, whereas the slower-growing lamellar bone has a more uniform, parallel microstructure.

The Vertebrate Skeleton

With an appreciation of the composition, microstructure, and growth of the primary tissues of the vertebrate skeleton, it is now possible to move on to the bones and teeth themselves. Just like the tissues

that comprise them, bones and teeth are innovations of vertebrates, or animals with a backbone (for a general review of vertebrate anatomy, see Romer, 1956; Carroll, 1988; Benton, 2005; Kardong, 2008). The first vertebrates, fish, were fully marine organisms and had a bony skeleton optimized for swimming. One subclade, the digit-bearing tetrapods, ventured onto the land during the Devonian (c.395 million years ago; Niedźwiedzki et al., 2010), and their skeletons changed dramatically as a consequence (Clack, 2002) (Fig. 2.2A). No longer passively supported by the buoyancy of water, tetrapods needed a more rigid skeleton to hoist their body weight, provide leverage for locomotion, and impart protection to vulnerable soft tissues and sense organs. Consequently, the limbs and girdles became more robust, the ribcage transformed into a rigid frame, a distinct neck developed to support the head, and the bewildering array of bones in the fish skull was reduced to a more countable number.

Dinosaurs, as well as humans and all animals with limbs and digits, are part of the great tetrapod radiation, and because they evolved from a singular common ancestor during the Devonian they all share the same general skeletal design (Fig. 2.2). The most conspicuous elements of the tetrapod body plan are a head balanced on a discrete neck; individual vertebrae that together comprise the neck, backbone, and tail; and four limbs, each of which is supported by a bony girdle and bears digits at its terminus. Because dinosaurs and other tetrapods evolved from a common ancestor, corresponding regions of the skeleton are said to be homologous (Fig. 2.2). That is, these structures are not independent objects, but can trace a shared origin to the common ancestor of the group. For instance, the upper arm bone of a dinosaur and that of a human are homologous: both are the exact same type of structure, positioned in the same place, and this structure would have been present in the common ancestor of dinosaurs and humans (which we can reasonably deduce from the fossil record and our understanding of vertebrate genealogy).

However, not all similar-looking and similarly positioned structures are homologous. The beak of a turtle and that of a parrot may appear similar and serve a similar function, but because the common ancestor of these two species – a carnivorous reptile – did not have a beak, then they cannot be homologous. Instead, such structures are called analogous. Conversely, not all homologous structures may appear similar to each other. The forelimb of a bird and that of a mole look nothing alike, but all of the individual bones were present in their common ancestor. These structures are indeed homologous, but have since been modified for different functional usages. This discussion underscores a major point of emphasis: the distinction between homology and anatomy is only possible if one knows something about the genealogy (phylogeny) of the organisms in question, so it can be assessed whether the common ancestor of two species would have possessed the potential homologous structure in question.

The Dinosaur Skeleton

The concept of homology greatly simplifies description of the vertebrate skeleton. Because all tetrapods have the same basic body plan, and share the same general sets of bones, all homologous bones inherited from the tetrapod common ancestor are given the same name. The femur in *Tyrannosaurus* is the same element as the femur in *Brachiosaurus*

or in a human, and this not only eases description, but also facilitates discrete comparisons between different species. Therefore, it is possible to give a brief overview of the dinosaur skeleton as a whole – the basic bones common to each dinosaur, and their location and general function – before delving into the anatomical nuances of each major subgroup.

The reader may wish to supplement this description with additional references. Unfortunately, the literature is so deficient in general, authoritative compendia of dinosaur skeletal anatomy that the textbook of Romer (1956) still remains one of the best single sources of information. The individual chapters in the second edition of *The Dinosauria* (Weishampel et al., 2004) are invaluable sources for the anatomy of specific dinosaur clades, whereas the more popular summaries of Holtz and Brett-Surman (1997), Fastovsky and Weishampel (2005), and Novas (2009) are also helpful. And, as with any topic relating to dinosaurs, it is instructive to understand the anatomy of birds and crocodiles: living dinosaurs and the closest extant dinosaur relatives, respectively. The standard reference for avian osteology is Baumel and Witmer (1993), whereas Iordansky (1973) and Langston (1973) are classic primers on crocodilian skull anatomy. Contemporary vertebrate anatomy textbooks, such as those by Hildebrand and Goslow (1998), Liem et al. (2008), Kardong (2008), and Pough et al. (2008), are also exceedingly useful.

The Dinosaur Skull

The skull is not a single bone but a complex of over 20 separate ossifications, some of which are dermal, others of which are endochondral, and some of which fuse together during growth (Figs 2.3–2.7). At first glance the skull may appear to be a befuddling amalgam of complexity, but there is an underlying pattern to the various bones and openings that comprise the dinosaur cranium. First, almost all the bones in the skull are paired; that is, there are corresponding elements on the left and right sides, in mirror-image positions. Second, individual bones meet at sutures which, unless fused, are usually discernible. Third, the bones of the skull can generally be divided into five complexes: (i) the upper jaw and cheek region, formed primarily of

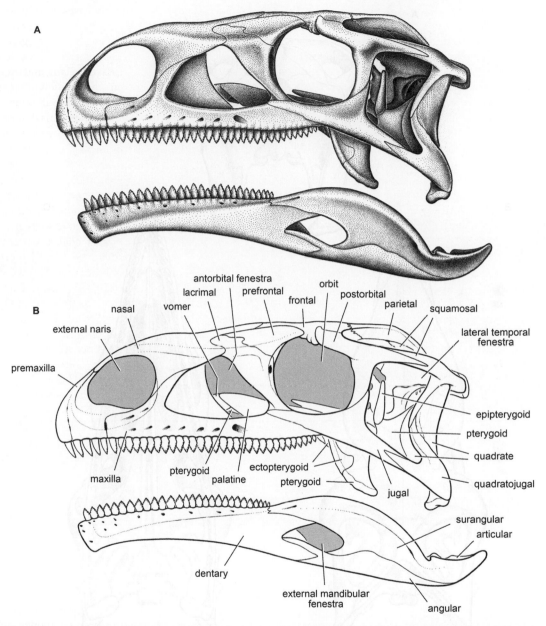

Figure 2.3 The skull of the basal sauropodomorph ("prosauropod") *Plateosaurus* in left lateral view, with major bones and cranial openings labeled. Scientific illustrations executed by Carol Abraczinskas and courtesy of Carol Abraczinskas and Dr Paul Sereno, University of Chicago.

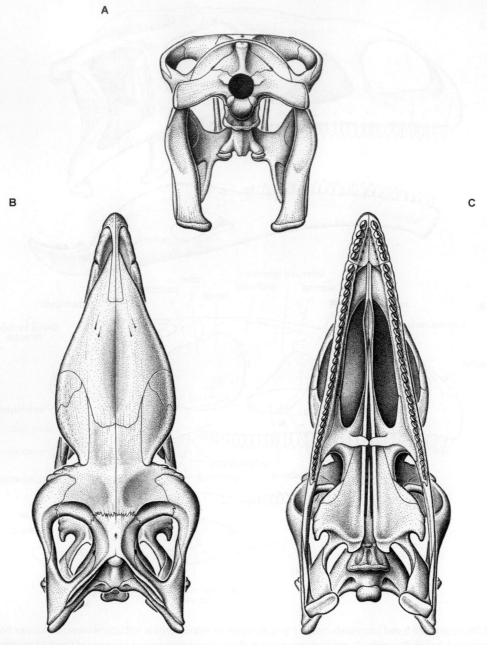

Figure 2.4 The skull of the basal sauropodomorph ("prosauropod") *Plateosaurus* in posterior (occipital) (A), dorsal (B), and ventral (palatal) (C) views. Major bones and cranial openings are labeled in the accompanying figure (Figure 2.5). Scientific illustrations executed by Carol Abraczinskas and courtesy of Carol Abraczinskas and Dr Paul Sereno, University of Chicago.

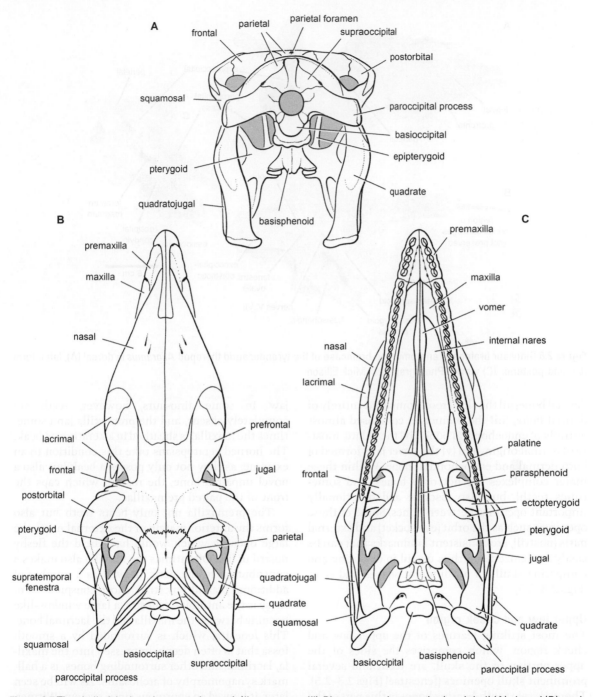

Figure 2.5 The skull of the basal sauropodomorph ("prosauropod") *Plateosaurus* in posterior (occipital) (A), dorsal (B), and ventral (palatal) (C) views, with major bones and cranial openings labeled. Scientific illustrations executed by Carol Abraczinskas and courtesy of Carol Abraczinskas and Dr Paul Sereno, University of Chicago.

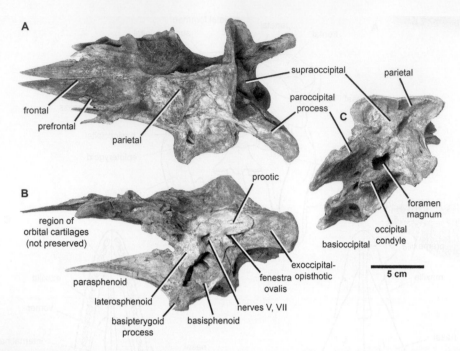

Figure 2.6 Dinosaur braincase anatomy. The braincase of the tyrannosaurid theropod *Alioramus* in dorsal (A), left lateral (B), and posterior (C) views. Photographs © Mick Ellison.

dermal bone; (ii) the skull roof, composed entirely of dermal bone; (iii) the braincase, composed almost entirely of endochondral bone; (iv) the palate, mostly of dermal origin; and (v) the lower jaw, formed of both dermal and endochondral bone. Within these major complexes are not only bones (and sometimes teeth), but also discrete and functionally important openings between bones. Some of these openings, such as the orbit (eye socket) and external naris (nostril), are consistent landmarks that can be easily recognized and thus useful in orienting and comparing skulls of all different shapes and sizes (Figs 2.3–2.5).

Upper jaw and cheek region

The most striking features of the upper jaw and cheek region, which comprises the sides of the upper portion of the skull, are teeth and several prominent skull openings (fenestrae) (Figs 2.3–2.5). There are two tooth-bearing bones in this part of the skull: the premaxilla at the front tip of the snout and the maxilla immediately posterior to it. Together, the teeth of both bones form a smoothly continuous tooth row, which occludes with that of the lower

jaw. In some dinosaurs, however, teeth are completely absent, and the premaxilla (and sometimes the maxilla) is sheathed in a keratinized beak. The horned ceratopsians take this condition to an extreme, as they not only posses a beak but also a novel unpaired bone, the rostral, which caps the front of the paired premaxillae.

The premaxilla not only bears teeth but also forms the anterior margin of the external naris, the large opening which would have held the fleshy nostril in life. Oftentimes the maxilla also makes a contribution to the naris, as well as a number of additional cranial fenestrae. Most conspicuous of these is the antorbital fenestra, a large window-like opening between the maxilla and the lacrimal bone. This fenestra, which is surrounded by a smooth fossa that is often deeply impressed into the maxilla, lacrimal, and other surrounding bones, is a hallmark synapomorphy of archosaurs. As will be seen later, it housed an enormous sinus system (Witmer, 1997a). The fenestra is usually large in saurischian dinosaurs, but is often reduced, and sometimes completely closed, in some ornithischians. In some derived theropod dinosaurs there are additional,

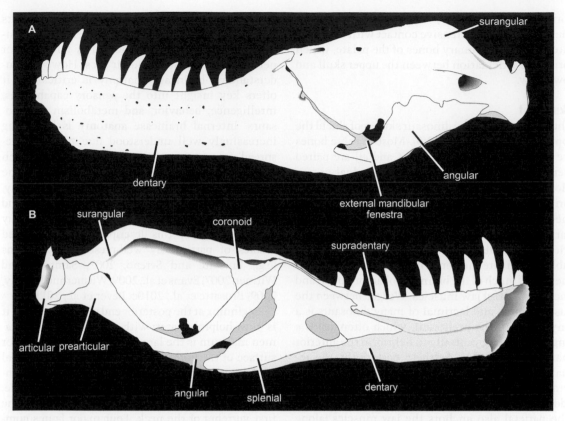

Figure 2.7 Dinosaur mandible (lower jaw) anatomy. The articulated lower jaw of the tyrannosaurid theropod *Tyranosaurus* in lateral (A) and medial (B) views. Modified from Brochu (2003) and reproduced with permission. Note that original image shows many pathologies, which have been removed for clarity here.

smaller openings associated with the antorbital fenestra. These orifices, the promaxillary and maxillary foramina, pierce the maxilla and are located within the fossa, anterior to the fenestra. They, too, were part of the same complex network of sinuses.

Posteriorly, the maxilla makes contact with two large bones, the lacrimal and the jugal. The lacrimal forms the bony strut between the antorbital fenestra and the orbit, and therefore comprises the anterior border of the eye socket itself. It houses the lacrimal ducts that deliver tears to the eyes, and in some dinosaurs sports a conical crest at its apex, where it rises above the orbit and the skull roof. The jugal bone is the cheek bone, and is usually a broad element that forms much of the lateral wall of the skull and borders the orbit ventrally. The majority of the posterior margin of the eye socket is made up of the postorbital, which in most species is

a T-shaped bone but in the ceratopsians is modified into the characteristic brow horn.

Together, the postorbital and jugal demarcate the anterior margin of yet another major skull opening, the lateral temporal fenestra (often referred to as the infratemporal fenestra). This expansive opening, along with another fenestra on the skull roof (the supratemporal fenestra), is one of two characteristic cranial windows that characterize the diapsid reptiles, a major tetrapod subclade that includes archosaurs, lizards, and snakes. The function of the lateral temporal fenestra is a matter of debate, but it is thought that it helped lighten the skull and provide increased attachment areas for the muscles that closed the jaws, which attached in this region (see review in Holliday, 2009). The posteroventral corner of the fenestra is formed by the quadratojugal bone, whereas its posterodorsal corner is surrounded by the squamosal. Both of these bones make extensive contact with the quadrate, one of the primary bones of the palate, which forms the articulation between the upper skull and lower jaw.

Skull roof

The eight bones of the dinosaur skull roof shield the top of the head (Figs 2.3–2.5). Most of these bones are only visible in dorsal view, and all are paired elements that often fuse as an individual grows. Most anteriorly, the elongate nasal forms the majority of the top of the snout, and defines much of the dorsal and posterior margins of the external naris. The nasals make broad contact posteriorly with the frontals, which usually form a wide skull table, demarcate the dorsal border of the orbit, form the roof of much of the brain and olfactory tract, and anchor powerful jaw muscles. Wedged between the nasal, frontal, and lacrimal of many dinosaurs is a tiny element, the prefrontal, which often sends a finger-like prong ventrally to help form the anterior margin of the orbit. Farthest posteriorly, the remainder of the skull roof is composed of the parietals, which overlie much of the braincase and lend additional protection to the brain and sense organs. The parietal also anchors the jaw muscles (along with the frontal), and located between the parietal, postorbital, and squamosal is the supratemporal fenestra. This fenestra is usually bordered anteriorly by a broad and deep fossa, which provided further attachment area for the jaw muscles. The parietal, along with the squamosal, is modified into the fantastic cranial frills of ceratopsians and skull domes of pachycephalosaurs.

Braincase

Surely the most complex portion of the skull, the braincase is almost entirely endochondral in origin, and therefore develops from a cartilage precursor early in development. More than 20 individual bones can constitute the braincase of dinosaurs, depending on the species, and most of these bones gradually fuse as the animal matures (Figs 2.4–2.6). Therefore, recognizing individual bones, much less the details of their shapes and articulations with other bones, is often challenging. Adding to this difficulty is the fact that bones of the braincase are geometrically complex: they

are not simple plates like many bones of the cheeks and skull roof, many are filled with internal sinuses, and they are often pierced by foramina for nerves, blood vessels, and air sacs. However, understanding braincase anatomy is critical, as it offers key insight into the sensory capabilities, intelligence, behavior, and metabolism of dinosaurs. Internal braincase anatomy is becoming increasingly well understood due to the widespread use of computed tomography (CT), which will be discussed later. For the time being, the focus here will be on external braincase anatomy, using the general descriptions of Romer (1956) and Currie (1997) as a guide. Adventurous readers may wish to follow up this description with published studies of particularly well-preserved specimens (e.g., Brusatte and Sereno, 2007; Sampson and Witmer, 2007; Evans et al., 2009; Witmer and Ridgely, 2009; Brusatte et al., 2010c; Bever et al., 2011).

Beginning at the posterior end of the braincase, it is most helpful to start with a landmark. The foramen magnum is the large opening on the posterior surface of the braincase, through which the spinal cord enters the skull (Fig. 2.6). It is bordered ventrally by a spherical knob, the occipital condyle, whose smooth articular surface contacts the first vertebra of the neck. Four major bones combine to form the posterior, or occipital, surface of the braincase. The occipital condyle is usually composed primarily of the basioccipital bone, a midline element, but its upper corners are often constructed from pedicle-like extensions of the exoccipital-opisthotics, of which there are paired left and right elements. As its name suggests, this bone is the fused composite of the exoccipital and opisthotic bones of more primitive reptiles, which are always completely joined into a single element, without any conspicuous sutures, in adult dinosaurs. Each exoccipital-opisthotic is widely exposed on the occipital surface of the braincase, where it anchors important muscles and ligaments of the neck, and its most prominent feature is the wing-like paroccipital process that expands laterally to contact the squamosal and the quadrate. Openings for three major cranial nerves (X, XI, XII) that control swallowing and digestion are enclosed within this bone. The final bone of the occipital surface is the midline supraoccipital, a single element that forms the dorsal border of the foramen magnum and rises dorsally to contact the parietals across a broad suture. Oftentimes the supraoccipital is bisected by a rugose midline ridge, from which arose neck muscles.

The lateral wall of the braincase is formed by three primary bones: the prootic, basisphenoid, and laterosphenoid (Fig. 2.6). In many species the exoccipital-opisthotic, which is primarily exposed in posterior view, also makes a large contribution to the lateral wall. The posteriorly and laterally facing portions of this bone are separated by a web-like strut, the crista tuberalis, which effectively divides the lateral and posterior walls of the braincase itself. Immediately anterior to this strut is the fenestra ovalis, which houses the stapes bone and leads into the inner ear. In front of this opening, the exoccipital-opisthotic makes broad contact with the prootic, which is pierced by foramina for the two fundamental cranial nerves (V and VII) that control the muscles of the face. Much of the prootic is hollowed by internal sinuses in many dinosaur species, and a rugose prominence on its lateral surface anchored muscles that opened the lower jaw. Beneath the prootic is the basisphenoid, whose most prominent feature is a pedicle-like process, the basipterygoid process, that articulates with the pterygoid bone of the palate. The basisphenoid is also invaded by an internal sinus in many dinosaurs, and accommodated the internal passage of the carotid artery. Finally, the laterosphenoid is a small bone, which is present as an ossified structure only in archosaurs, located immediately anterior to the prootic.

A few other bones also contribute to the braincase. The parasphenoid, the only dermal element of the braincase, is a long tapering wedge that is indistinguishably fused with the anterior part of the basisphenoid. It forms much of the ventral floor of the braincase, along with the basisphenoid and basioccipital, and extends anteriorly into the orbital region. Here, at the most anterior point of the braincase, it helped support a midline membrane that stretched to the skull roof bones dorsally to divide the orbital region into separate left and right chambers for the opposing eyeballs. Usually this membrane is fleshy but it can ossify into a structure called the interorbital septum. Finally, anterior to the laterosphenoid is a series of bones that develop from the cartilages of the orbital region. These

bones, which include the orbitosphenoid, sphenethmoid, and mesethmoid, are only sometimes ossified and often not preserved in fossil specimens (Ali et al., 2008). They help protect the anterior part of the brain and the cranial nerves that control vision and olfaction.

Palate

One of the most poorly understood and infrequently studied components of the dinosaur skeleton, the palate comprises a series of bones that lie within the skull, above the teeth of the upper jaw (Figs 2.4 and 2.5). Together, these bones effectively separate the oral cavity, and therefore ingested food, from the vulnerable soft tissues of the antorbital and nasal sinuses and the orbital region. Furthermore, they provide broad attachment sites for the jaw muscles that power mastication. Because of their location within the skull, and because of the delicate nature of most of these bones, it is no surprise that the palatal elements are rarely preserved as fossils, and in those rare cases that they are, are usually studied only superficially. However, the advent of CT scanning technology should permit more detailed understanding of the dinosaur palate in the near future.

The palate consists of six bones, five of which have a mirrored duplicate on the other side of the skull (Figs 2.4 and 2.5). The single unpaired bone, the vomer, is a midline element at the very front of the palate and forms a strut between the opposing maxillae. On either side of the vomer are the internal choanae – the interior nostrils that are continuous with the external naris – which are particularly large and elongate in dinosaurs. Posteriorly the vomer makes contact with the opposing palatines, each of which is usually composed of three or four discrete processes that articulate with the maxilla and jugal laterally and the vomer and pterygoid medially. The palatine defines the posterior margin of the internal choanae, as well as the anterior margin of a separate palatal opening, the suborbital fenestra, which is located between the palatine and the more posteriorly positioned ectopterygoid. This latter bone is a small, hooked element that makes broad contact with the jugal laterally to firmly brace the palate against the lateral wall of the skull. The posterior edge of the ectopterygoid forms the anterior corner of yet another palatal opening, the subtemporal fenestra, which is essentially the floor of the lateral temporal fenestra. Both the ectopterygoid and palatine articulate medially against the pterygoid, a complex bone with a bewildering three-dimensional geometry, due to numerous processes that project in many directions to articulate not only with other bones of the palate but also with the braincase. The two pterygoids meet on the midline, but diverge posteriorly to enclose a slit-like opening, the interpterygoid vacuity, between them.

The two remaining bones of the palate are endochondral in origin, both remnants of the upper jaw in primitive vertebrates that have become reduced, moved posteriorly, and changed their primary function in more derived tetrapods. The quadrate is the larger, and more functionally important, of these two bones. It comprises much of the posterior region of the skull, and lies lateral to the braincase. Ventrally it has smooth condyles for articulation with the articular bone of the lower jaw, and therefore is a primary component of the jaw joint in dinosaurs. Its dorsal head is a spherical structure that forms a rotary joint with the squamosal, permitting a wide range of motion in the posterior portion of the skull, likely useful in dissipating stress during feeding. Anteriorly, the quadrate has a broad thin flange that firmly articulates with the pterygoid. The second endochondral bone of the palate, the epipterygoid, is a small triangular plate that lies against the lateral wall of the braincase. It therefore contributes to the articulation between these two major portions of the dinosaur skull.

Lower jaw

Unlike the simple mandible of humans and other mammals, the lower jaw of dinosaurs is composed of a complex of bones, most of which are dermal in origin (Figs 2.3 and 2.7). Some of these bones are exposed only on the lateral surface of the lower jaw and others are visible only medially. The primary bone, and the only tooth-bearing element, is the dentary. The left and right dentaries meet at a symphysis at the very front of the jaw, but then diverge posteriorly. In ornithischian dinosaurs, but not saurischians, there is a novel unpaired bone – the predentary – that caps the dentary anteriorly. Behind the dentary, in all dinosaurs, are two additional dermal bones that are exposed laterally, the

surangular on top and the angular underneath, both of which are the site of jaw muscle attachment. Between the dentary, surangular, and angular of most species is a window-like opening, the external mandibular fenestra, that is a trademark character of archosaurs. At the far posterior end of the lower jaw, medial to the tips of the surangular and angular, is the articular, the sole endochondral element of the mandible. It has a smooth, saddle-shaped articular surface that receives the condyles of the quadrate, and therefore helps form the joint between the upper and lower jaws.

On the medial surface of the lower jaw, immediately in front of the articular, is a deep fossa on the medial surfaces of the dentary, angular, and surangular. This depression, the adductor fossa, houses many of the muscles that close the jaws. The fossa is actually a discrete chamber, not simply a depression, because it is bounded medially by a separate dermal bone that is only visible in medial view, the prearticular. Two additional dermal bones sheath the medial surface of the dentary anterior to the prearticular. The largest of these, the splenial, covers most of the medial dentary, and thus is often the most widely exposed bone of the lower jaw in medial view. Above the splenial is a long, thin, bow-shaped bone, the supradentary–coronoid complex, which shields the tooth roots medially. As its name suggests, this bone is formed by the fusion of what are two separate bones in more primitive reptiles: the elongate supradentary, which forms the majority of the fused element, and the triangular coronoid, a small region that braces the contact zone between the dentary and surangular medially.

Teeth

Many primitive reptiles have teeth that arise from not only the upper and lower jaws, but also many of the dermal bones of the palate. In all but the most primitive dinosaurs, however, only the premaxilla and maxilla of the upper jaw and the dentary of the lower jaw bear teeth. There is a tremendous difference in the size and shape of teeth across dinosaurs, mostly due to variation in diet. Regardless, no matter their diet, all dinosaur species had teeth that were composed of an outer sheath of enamel and an inner core of dentine, and which were continuously replaced during the lifetime of an individual.

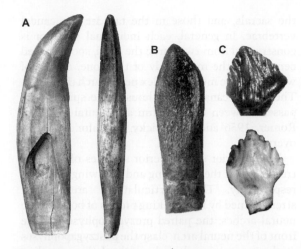

Figure 2.8 Dinosaur teeth. (A) A tooth of the carnivorous theropod *Alioramus* in lateral and distal (posterior) views, exhibiting the serrations and recurved shape typical of theropod teeth (photographs © Mick Ellison). (B) A tooth of the herbivorous sauropod *Bothriospondylus* in lateral view, exhibiting the blunt and spatulate crown shape that is typical of many sauropods (photo courtesy of Dr Jeffrey Wilson, Dr Mike D'Emic and the Muséum National d'Histoire Naturelle, Paris). (C) Teeth of a pachycephalosaurid (top) and basal ornithischian (bottom), exhibiting the characteristic leaf-shaped morphology of ornithischians (photos courtesy of Dr Thomas Williamson and Dr Richard Butler, respectively).

The carnivorous theropod dinosaurs have relatively simple blade-like teeth, ideal for cutting through the flesh of their prey (Fig. 2.8A). These teeth are usually thin in the mediolateral direction, recurved at their tip, and bear serrations. The serrations are located on discrete carinae, which are sharp enamel ridges located at the front (mesial edge) and back (distal edge) of the teeth. Most theropods with this type of tooth have a distinctive pattern of sculpturing on the external enamel, in which band-like "wrinkles" extend across the lateral and medial surface of the tooth, from carina to carina (Sereno et al., 1996; Brusatte et al., 2007). The function of these bands is uncertain, but they may have helped strengthen the otherwise thin and fragile tooth during feeding. Although this general description fits most theropod teeth, there is subtle variation in morphology depending on the theropod subgroup in question, and interested readers are

encouraged to consult the general reviews of Currie et al. (1990) and Smith et al. (2005) for more information. It is also important to note that several derived theropod subgroups lost teeth entirely, whereas the herbivorous therizinosauroids convergently developed teeth similar to those of other plant-eating dinosaurs.

The teeth of herbivorous species depart radically from the simpler morphology of carnivores (Fig. 2.8B,C). Most sauropodomorphs and ornithischians have teeth that generally resemble a leaf, as they are expanded apically away from the root and bear a set of coarse denticles at their apex, which were used to crop and grind plant matter. Some basal sauropodomorphs, however, have teeth that generally resemble those of living iguanas, which may have been well suited for an omnivorous diet (Barrett, 2000). On the other hand, some derived sauropods, such as *Diplodocus*, have much thinner pencil-like teeth, which seem to be suited for cropping softer vegetation such as ferns and shrubs (Fiorillo, 1998; Upchurch and Barrett, 2000; Sereno et al., 2007; Whitlock, 2011a). On the ornithischian side of the family tree, ceratopsians and ornithopods convergently developed an unusual condition in which enamel was limited to one side of the tooth only, whereas the opposing surface was covered in softer dentine (see Norman and Weishampel, 1985). This facilitated a built-in self-sharpening mechanism: as the dentine was worn away quicker than the harder enamel, a sharp cutting edge was formed and continuously sharpened. Other ornithischian subgroups evolved their own variations on the general leaf-shaped morphology of the group (see Galton and Upchurch, 2004a; Maryańska et al., 2004; Vickaryous et al., 2004).

Finally, new research indicates that the internal structure of the tooth enamel is also incredibly variable within dinosaurs. These differences, which relate to the shape and orientation of the individual crystals of enamel, may in some cases have functional significance, and in others may simply reflect morphology that was passively inherited from a common ancestor (Sander, 1999; Stokosa, 2005; Hwang, 2005, 2010, 2011). The identification of these microscopic differences, which in many cases are striking, was only recently made possible by the use of scanning electron microscopy (SEM), as more traditional polarizing light microscopy is inadequate to register anything but the simplest variation in enamel. Further SEM studies of dinosaur teeth, in the same vein as those cited above and a recent analysis of Mesozoic crocodylomorph enamel (Andrade et al., 2010), promise to shed new light on dinosaur feeding habits and transform the often-imprecise identification of fragmentary dinosaur teeth (Hwang, 2010).

The Dinosaur Axial Skeleton

All parts of the dinosaur skeleton aside from the skull are generally referred to as the "postcranial skeleton" (Fig. 2.9). The postcranium is further divided into two major complexes: the axial skeleton, which comprises the vertebrae and ribs, and the appendicular skeleton, which includes the girdles and limbs. The focus first is on the axial skeleton.

Vertebrae

The fundamental component of the axial skeleton is the vertebral column, which is composed of individual bones, the vertebrae, that stretch throughout the neck, trunk, and tail (Figs 2.10–2.12). Different regions of the vertebral column have different names: those elements in the neck are cervical vertebrae, those in the back dorsal vertebrae, the small set of vertebrae between the opposing pelvic girdles are

the sacrals, and those in the tail are the caudal vertebrae. In general, each individual vertebra is constructed from two parts: the solid, spool-shaped centrum is the main body of the bone, on top of which sits the more complex neural arch (Fig. 2.10). The neural canal, which housed the spinal cord, passes between the centrum and neural arch (see Romer, 1956 and Makovicky, 1997 for a general overview).

The anterior and posterior surfaces of the centrum contact the preceding and following vertebra, respectively. These articulations are further strengthened by interlocking prongs of bone on the neural arches: the paired prezygapophyses, at the front of the neural arch, clasp the postzygapophyses at the posterior end of the arch of the preceding vertebra (Fig. 2.10). The strength of these articulations, and their degree of interlocking, regulate how much motion is possible between individual vertebrae, and thus the extent to which the backbone can bend. The neural arches of the cervical, dorsal, and anterior caudal vertebrae sport a laterally directed, wing-like projection called the transverse process on each side. These thin flanges expand at their tip into a rugose diapophysis for articulation with one head of the rib. The other head of the rib usually makes contact with a stalk-like swelling of the centrum called the parapophysis. Finally, the neural arch is usually crowned with a single, thin, midline flange called the neural spine, which in

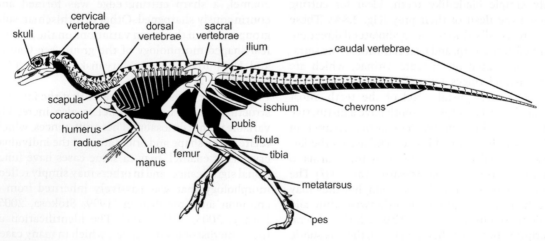

Figure 2.9 Dinosaur postcranial anatomy. A skeletal reconstruction of the Early Jurassic ornithischian *Eocursor*, with major regions of the skeleton indicated. Reconstruction © Scott Hartman.

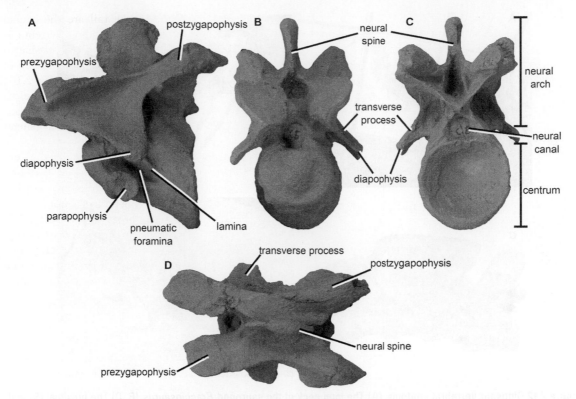

Figure 2.10 Dinosaur vertebral anatomy. A cervical vertebra of the large theropod *Aerosteon* in left lateral (A), anterior (B), posterior (C), and dorsal (D) views. Photos courtesy of Dr Roger Benson.

some dinosaurs is greatly expanded so that the back can support a hump or sail (Bailey, 1997).

Saurischian dinosaurs are immediately distinguished from ornithischian dinosaurs by several unusual features of the vertebral column, most of which are seen on the cervical and anterior dorsal vertebrae (Fig. 2.10). Many of the various processes of the neural arch and centrum are linked to each

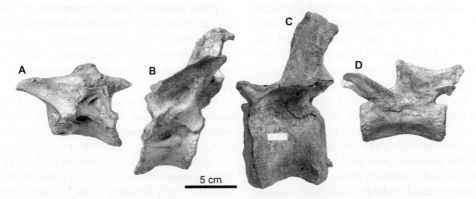

Figure 2.11 Dinosaur vertebral anatomy. A montage of different types of vertebrae from the same individual of the tyrannosaurid theropod *Alioramus*, including cervical (A), dorsal (B), anterior caudal (C), and posterior caudal (D) vertebrae, all shown to scale. Photographs © Mick Ellison.

Figure 2.12 Dinosaur vertebral anatomy. (A) The long neck of the sauropod *Brachiosaurus*. (B, D) The proximal (B) and middle (D) portions of the tail of the ornithomimosaurian theropod *Shenzhousaurus*. (C) The sacrum of the mid-sized theropod *Megalosaurus*.

other by bony struts called laminae (Wilson, 1999). These laminae, which often have astoundingly complex names and whose morphology is highly variable among species, probably relate to both an increased area for muscle attachment (Schwarz et al., 2007) and, more importantly, a pneumatic bird-like system of air sacs that invaded the vertebrae (Britt, 1993, 1997; O'Connor and Claessens, 2005; O'Connor, 2006; Sereno et al., 2008). These air sacs, which are described in more detail in Chapter 3, are clearly the cause of another distinguishing feature of saurischians: deep fossae and foramina, often referred to as "pleurocoels" or "pneumatic foramina," that pierce the lateral surfaces of the centrum and lead into a complex network of internal chambers (Britt, 1993). Provocatively, recent research indicates that both laminae and pneumatic foramina are present in some non-dinosaurian archosaurs (see below), which suggests that, like the saurischian pelvis, pneumaticity may

be a more primitive reptilian character that was simply retained in saurischians, whereas ornithischians are truly unique in losing air sacs and their concomitant bony indicators.

Most dinosaurs have approximately 9–10 cervical vertebrae, although some subgroups, most prominently the long-necked sauropods, increased their cervical count, both by adding new vertebrae and by converting dorsal vertebrae to cervicals (Upchurch, 1995; Makovicky, 1997; Wilson, 2002; Upchurch et al., 2004; Müller et al., 2010) (Fig. 2.12A). The ancestral dinosaur had a gracile S-shaped neck, which was enabled by the parallelogram-shaped morphology of the individual cervicals, in which the anterior face of the centrum is elevated relative to the posterior face (Gauthier, 1986; Sereno, 1991a; Benton, 1999; Brusatte et al., 2010a). Many dinosaurs retained this primitive condition, but some groups such as the stegosaurs and sauropods developed a straighter neck, without any offset of the cervical articular faces. In most dinosaurs the cervical neural arches are relatively modest in their morphology, and the zygapophyses, transverse processes, and neural spines are small. In sauropods, however, the cervical neural arches are extravagant in their complexity. The zygapophyses are large, the neural spines are often bifurcated, and a panoply of laminae weaves across the bone, making nearly every conceivable linkage between the array of bony processes on the arch.

The dorsal vertebrae provide strength and rigidity to the trunk of dinosaurs. The ancestral dinosaur probably had about 15 dorsals, and although there is variation in the dorsal counts of more derived dinosaur subgroups, most species do not deviate too far from the ancestral condition (see Makovicky, 1997 for general overview). Many gradual changes occur across the neck and back of most dinosaurs, defining a smooth transition from the cervical to the dorsal vertebrae (Fig. 2.11). The centrum becomes less elongate anteroposteriorly, the anterior and posterior articular faces lose their offset, the neural spines expand in height, the transverse processes become stouter and swing outward to become positioned fully laterally, and the parapophyses migrate from the centrum onto the neural arch. Along with this general pattern are morphologies specific to various dinosaur subgroups. Saurischian dinosaurs often have a set of accessory articulations below the zygapophyses, the hyposphene and hypantrum, which strengthen the union between adjoining vertebrae (Apesteguía, 2005). Sauropods retain elongate centra and their neural arches are reduced to a perplexing maze of laminae (Upchurch et al., 2004). Some sauropods, as well as some theropods, develop remarkably tall neural spines, probably to support a sail (Bailey, 1997). The dorsal vertebrae of stegosaurs are immediately recognized by their tall neural arches, which in anterior or posterior view resemble a normal dinosaur neural arch that has been stretched like taffy (Galton and Upchurch, 2004a).

The sacral vertebrae fulfill the crucial function of linking the backbone to the pelvis, and are therefore instrumental in supporting body weight and strengthening the trunk region (Fig. 2.12C). The common ancestor of dinosaurs had either two or three sacrals, and most primitive dinosaurs possessed three. However, several more derived dinosaur subgroups independently increased their sacral counts, usually through the addition of adjoining vertebrae from the back or tail (Makovicky, 1997). Most stegosaurs, ankylosaurs, sauropods, theropods, and primitive ornithopods have between four and seven sacrals, but some ceratopsians have up to 10 and some hadrosaurids up to 12 (Dodson et al., 2004; Horner et al., 2004). The functional significance of these drastically enlarged sacral counts is unclear, but at the very least they added stability to the pelvis and helped rigidify the backbone. In some species the sacrals are invariably fused together into a single elongate rod, whereas in others individual vertebral centra and arches are still discernible. Contact between the sacral vertebrae and pelvis is achieved by the transverse processes of the individual vertebrae and their associated sacral ribs, which usually are fused together into an amorphous mass.

Approximately 50 caudal vertebrae comprised the tail of the ancestral dinosaur, but more derived dinosaur groups achieved a remarkable variability in the number, size, and shape of individual caudal vertebrae, along with the length, range of motion, and morphology of the tail itself (Fig. 2.12B,D). There is a general decrease in caudal number throughout theropod evolution, which reaches its most extreme condition in the pygostyle of birds, a stout misshapen remnant of the tail formed by the fusion of the few remaining vertebrae. Although the tail of primitive dinosaurs was quite flexible, many groups developed a more rigid structure. Derived theropods such as dromaeosaurids evolved elongate prezygapophyses, which extend as reinforcing bony rods along the sides of 10–12 adjoining vertebrae (Ostrom, 1969). Many ornithischian dinosaurs ossified the tendons along their tails, providing a functionally similar, but anatomically distinct, condition to that of dromaeosaurids (Organ and Adams, 2005). Some sauropods, on the other hand, possess an even longer and more mobile tail. *Diplodocus* and its closest relatives have a greater number of tail vertebrae than other sauropods, and the most distal ones are modified to form a whip-like structure (Myhrvold and Currie, 1997). Other sauropods, and many ornithischians such as stegosaurs and ankylosaurs, sport spikes, armor, or clubs at the ends of their tails (Galton and Upchurch, 2004a; Upchurch et al., 2004; Vickaryous et al., 2004).

Ribs

One of the primary functions of the vertebral column is to support the ribcage, which provides a rigid trellis to protect the internal organs. Most dinosaurs have ribs emerging from the cervical, dorsal, and sacral vertebrae, the latter of which help form the articulation between the sacrals and the pelvis. However, caudal vertebrae rarely support ribs, but sauropodomorphs and ornithischians have short stout ribs on the first few caudals. Most dinosaur ribs are composed of a proximal "head" that is divided into two separate processes, the capitulum and tuberculum, for articulation with the parapophysis and diapophysis of the associated vertebra, respectively. Distal to the head, the shaft of the rib usually tapers to a thin point. Cervical ribs are always smaller and more gracile than those of the dorsal series, and in saurischians are extremely thin, delicate, and elongate, as they parallel the contours of the neck itself. Dorsal ribs are larger, longer, and more robustly built, and their shaft is usually outwardly convex to enclose a barrel-shaped chest cavity. In many saurischians air sacs extend into the ribs, and therefore they are internally hollow and pierced by pneumatic foramina on their surfaces. In some derived theropods, as well as some ornithischians, the shaft of the rib develops a plate-like flange, the uncinate process, which overlaps the adjacent rib (Fig. 2.13). These en echelon structures, which overlap in a series like shingles on a roof, lend further rigidity to the ribcage and may have been instrumental in supporting trunk muscles and powering an efficient, avian-like breathing apparatus (Codd et al., 2008; Codd 2010).

Chevrons

The bases of the caudal vertebrae sport a series of small bones that project downward to form the ventral surface of the tail itself (Fig. 2.12B,D). These are the chevrons, or haemal arches, each of which is composed of fused left and right laminae that enclose a large opening between them. When aligned together in articulation, the fenestrae of individual chevrons form a series along the tail, the haemal canal, which enclosed various nerves and blood vessels that innervated the tail muscles and controlled tail motion. The shapes of individual chevrons often vary across the length of the tail, as well as among different dinosaur subgroups. Some are simple spikes or plates that project ventrally, others have a kink that affords the overall bone an L-like shape, and still others are elongate in the anteroposterior direction and thus resemble a boat. Chevrons are usually absent in the distal portion of the tail, and in dromaeosaurids are thin and elongate, in the same style of the prezygapophyses, to lend strength and rigidity to the tail.

Gastralia

Perhaps the least studied and understood component of the axial skeleton is the gastral cuirass, an interlocking series of thin dermal bones – gastralia – along the ventral surface of the belly (Claessens, 2004) (Fig. 2.14). These bones are often called "dermal ribs" or "belly ribs," but are not technically ribs because they do not attach to the vertebrae. The presence or absence of gastralia is perplexingly variable across living and extinct tetrapods: they are present in most saurischian dinosaurs, as well as living crocodilians, but are absent in most living

Figure 2.13 Uncinate processes of the ribs. The non-avian theropods *Velociraptor* (A) and *Oviraptor* (B) with arrows pointing to uncinate processes. (C) is an image of an extant razorbill for comparison. Courtesy of Dr Jonathan Codd.

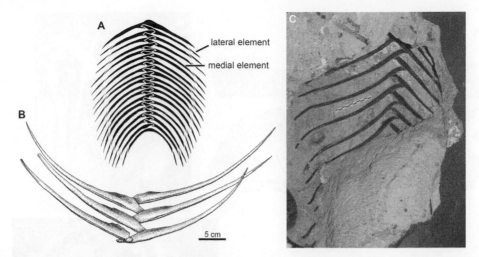

Figure 2.14 Dinosaur gastralia. (A) Schematic showing an articulated gastral cuirass of a theropod dinosaur in ventral view, with lateral and medial gastral elements labeled. (B) A portion of the gastral cuirass of the tyrannosaurid *Albertosaurus* in dorsal view showing the imbricating nature of the medial elements. (C) Photograph of the gastral cuirass of the theropod *Allosaurus* in dorsal view. Modified from images in Claessens (2004) and reproduced with permission.

reptiles and have never been discovered in ornithischians. Although at first glance the cuirass resembles a complicated web of interwoven slivers of bone, there is underlying order to the gastral skeleton (Fig. 2.14). The cuirass is divided into discrete rows, each of which contains four rod-like bones: two medial elements that contact each other on the midline, often in an overlapping fashion, and lateral elements that abut the outside edge of each medial bone. The conjoined left and right medial elements of each row usually overlap, in en echelon fashion, the medial elements of the two adjacent rows anteriorly and posteriorly. Therefore, all individual gastralia are connected to each other, permitting the cuirass to function as a single unit. In some dinosaurs, especially theropods, the gastralia may have been a primary control on the shape and size of the trunk cavity, and may have helped power the avian-like respiratory system (see Claessens, 2004 for a full review).

The Dinosaur Appendicular Skeleton

The final major division of the dinosaur skeleton is the appendicular portion (see Fig. 2.9). Included within the appendicular skeleton are the pectoral (shoulder) and pelvic (hip) girdles, as well as the limbs and digits that attach to them. Two additional bones of the pectoral region, the clavicle and sternum, are also listed among the roster of appendicular bones, although they are rarely preserved as fossils. The underlying geometry of the forelimbs and hindlimbs is identical – one proximal bone, followed by two more distal bones, and capped with a series of palmar bones and digits – but the form of the pectoral and pelvic girdles is dramatically distinct.

Pectoral girdle

The dinosaur shoulder girdle is composed primarily of two bones, the scapula and coracoid, with contributions from the clavicle and sternum, when present (Figs 2.15 and 2.16). The scapula and coracoid are endochondral bones, and together they contribute to the joint surface for the humerus of the upper arm (Fig. 2.15). The scapula is much larger than the coracoid and usually takes the shape of an elongate blade, which lies directly outside, and parallels the contours of, the ribcage. The coracoid, on the other hand, is a stouter and more plate-like element, usually shaped like a semicircle or square, that is appressed to the scapula anteriorly and

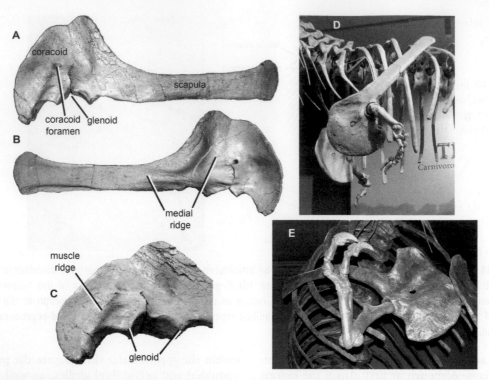

Figure 2.15 Dinosaur shoulder girdles: the scapula and coracoid. The conjoined coracoid and scapula of the large theropod *Aerosteon* in lateral (A) and medial (B) views, and a close-up of the glenoid region for articulation with the humerus in lateral oblique view (C). The articulated coracoid, scapula, and forelimb in skeletal mounts of *Carnotaurus* (D) and *Tyrannosaurus* (E). Images (A–C) courtesy of Dr Roger Benson.

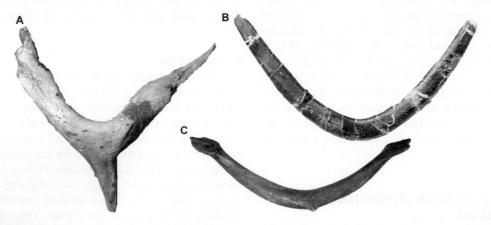

Figure 2.16 The dinosaur furcula. A montage of furculae in theropod dinosaurs, including the oviraptorosauroid *Citipati* (A), the dromaeosaurid *Bambiraptor* (B), and the tyrannosaurid *Daspletosaurus* (C). (A) and (C) courtesy of Dr Sterling Nesbitt; (B) © Mick Ellison.

ventrally. Along the ventral surface of where the two bones meet is the smooth, deeply impressed surface of the glenoid, which receives the head of the humerus. Various tuberosities of bone near the glenoid, on both the scapula and coracoid, supported muscles of the shoulder and forearm.

The sternum and clavicle are both dermal elements and are critical components of the shoulder girdle of living birds, which is highly modified to support the large muscles and complex tendons required for flight (Baumel and Witmer, 1993). Both elements are uncommonly encountered in fossilized dinosaur skeletons, but this likely reflects the fact that they were unossified or extremely fragile, not that they were truly absent. Some dinosaurs have been found with paired left and right clavicles, the normal condition in tetrapods (including humans). Theropods, however, have a unique morphology in which the two clavicles are fused into a single structure, the V-shaped furcula (Bryant and Russell, 1993; Nesbitt et al., 2009c; Vickaryous and Hall, 2010) (Fig. 2.16). A furcula is present in all living birds, in which it is thought to act as a spring that stores kinetic energy, and braces the left and

right shoulder girdles, during flight (Jenkins et al., 1988). In dinosaurs, the sternum is usually a thin platy element along the midline of the chest that is attached to the adjacent dorsal ribs by a series of cartilaginous extensions, the sternal ribs (Dodson and Madsen, 1981). Living birds exhibit a distinctive sternum: large, robust, and keeled ventrally. This morphology, which is not present in even the most bird-like dinosaurs, is intimately related to the mechanics of flight, because the principal flight muscles, which are enormous in birds, are attached to the sternum.

Forelimb

The forelimb is composed of a series of bones: the humerus proximally, followed by the radius and ulna, and then the carpal bones of the wrist, the metacarpals of the palm of the hand, and the individual manual digits themselves (see Christiansen 1997a, for an overview) (Figs 2.17 and 2.18). The humerus is usually the largest and most robust bone of the arm, and it anchors important muscles that control the majority of arm motion. The humerus meets the two forelimb bones, the radius and ulna, at

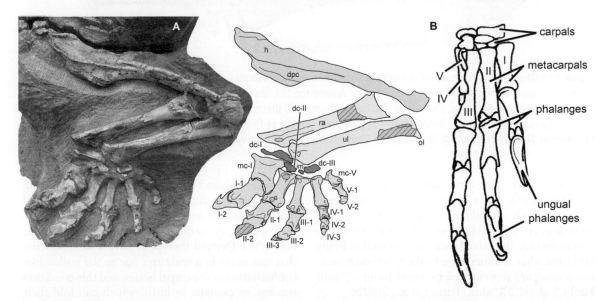

Figure 2.17 Dinosaur hand anatomy. (A) Photo and interpretive drawing of the forelimb and hand of the basal sauropodomorph ("prosauropod") dinosaur *Seitaad*. (B) Line drawing of the right hand of the primitive theropod *Herrerasaurus*. Images in (A) courtesy of Dr Mark Loewen and modified from Sertich and Loewen (2010); image in (B) drawn by Anne Curthoys based on illustration in Sereno (1994). For abbreviations in (A) refer to Sertich and Loewen (2010).

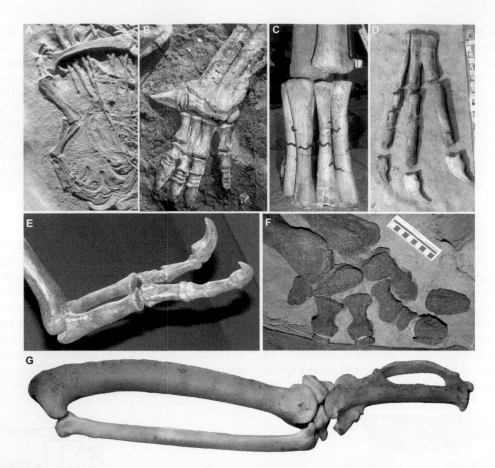

Figure 2.18 Dinosaur hand anatomy. A montage of different manual morphologies in dinosaurs, including the compsognathid theropod *Juravenator* (A), the ornithopod *Iguanodon* (B), the titanosaurian sauropod *Rapetosaurus* (C), the ornithomimosaurian theropod *Gallimimus* (D), the tyrannosaurid theropod *Tyrannosaurus* (E), the stegosaurid ornithischian *Stegosaurus* (F), and a living turkey (G). Images courtesy of Dr Jeffrey Wilson, Dr Mike D'Emic, and the Field Museum of Natural History (C), Dr Susannah Maidment (F), and Dr Corwin Sullivan (G).

the elbow joint. The ulna, which is usually larger than the radius, often has a pronounced olecranon process, which extends behind the elbow joint as a lever that attaches to large muscles that extend the arm outwards. In derived theropods, as well as living birds, the ulna anchors large feathers, whose attachment sites are represented by small bumps ("quill knobs") along the ulna (Turner et al., 2007b).

The articulation between the forelimb and the hand is complex: several individual carpal bones of the wrist, which are variable in number and size among species, connect these two important regions of the arm (Fig. 2.17). In large quadrupedal dinosaurs

like sauropods and ornithopods, the carpal bones are usually large, strong, and blocky. However, in bipedal dinosaurs that do not require the forelimb to support body weight, the carpals are usually small and often unossified. Derived theropods have a flexible wrist that can move in a wide arc, due to the pulley-like configuration of the carpal bones, and this condition reaches an extreme in birds, which can fold their wrists and hands against the body (Vazquez, 1992; Sullivan et al., 2010).

The hand, or manus, of dinosaurs is composed of a set of metacarpals and their corresponding digits, each of which is formed of multiple individual

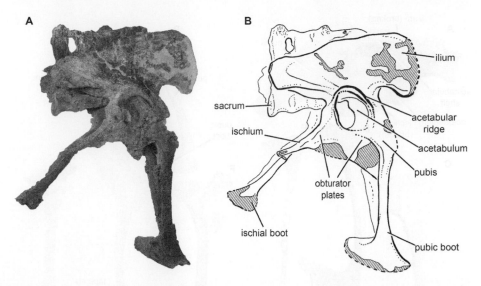

Figure 2.19 Dinosaur pelvic anatomy. Photo (A) and line drawing (B) of the pelvis of a theropod dinosaur from the middle Cretaceous of Africa. Photo and drawing by the author, under the guidance of Carol Abraczinskas. For more information see Sereno and Brusatte (2008).

phalanx bones (Figs 2.17 and 2.18). These are numbered I–V, with I the most medial (thumb) and V the most lateral (pinky). The form of the metacarpals and digits differs substantially depending on whether a species is bipedal or quadrupedal (Fig. 2.18). Bipedal dinosaurs such as theropods and some ornithischians have small gracile hands, which are often modified into grasping devices for gripping onto either prey or plant material, depending on diet. In theropods, there is a general reduction in digit number across phylogeny: basal species have four or five digits, whereas most theropods possess only three digits (Gauthier, 1986; Padian and Chiappe, 1998). Some aberrant species, such as the two-fingered tyrannosaurids and the bird-like alvarezsaurids with their single stout finger, exhibit even further reduction (Perle et al., 1993; Brochu, 2003; Xu et al., 2011b). Most theropod hands are capped with sharp claws, even the stubby arms of *Tyrannosaurus* and kin, which were clearly useless in prey capture (Lipkin and Carpenter, 2008). Quadrupedal dinosaurs, on the other hand, usually possess larger and more robust hands, whose terminal phalanges are sometimes modified into hooves. The individual metacarpals and digits are usually shorter, heftier, and arranged in a tight arc, all of which is useful in providing support for body weight

(Bonnan, 2003; Senter, 2010). In concert with differences in hand shape and size, quadrupedal dinosaurs generally have longer forelimbs than bipedal dinosaurs, although more derived theropods exhibit a trend of forelimb lengthening that reaches its apogee in the large wings of birds (Christiansen, 1997a; Middleton and Gatesy, 2000).

Pelvic girdle

The dinosaurian pelvis is composed of three different bones – the ilium, pubis, and ischium – which sometimes fuse with each other and whose relative sizes and shapes are highly variable among species (see Figs 1.14, 1.20, 2.19 and 2.20). The ilium is the largest bone of the pelvis and is located dorsally, above the pubis and ischium. The left and right ilia are separated on the midline by the sacral vertebrae, although in some species the two ilia make contact at their dorsal margin. In most species the ilium is an elongate blade-like bone, with two processes ventrally for articulation with the pubis and ischium. The most salient feature of the ilium is usually a robust curved ridge on the lateral surface, immediately above the acetabular joint (the ball-and-socket articulation between the pelvis and hindlimb). Here, the three bones of the pelvis come together to form a window-like socket for the femur, and the ridge on

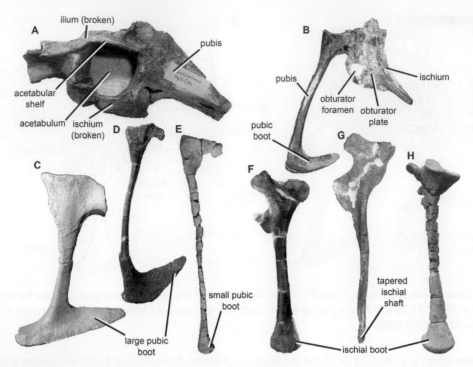

Figure 2.20 Dinosaur pelvic anatomy. (A) The acetabular region of the ornithomimosaurian theropod *Gallimimus* in right lateral view; (B) the articulated pubis and ischium of the small theropod *Miragaia* in left lateral view; (C) the pubis of the large theropod *Aerosteon* in left lateral view; (D) the pubis of the small theropod *Coelurus* in left lateral view; (E) the pubis of the primitive theropod *Liliensternus* in left lateral view; (F) the ischium of the large theropod *Allosaurus* in left lateral view; (G) the ischium of the tyrannosaurid theropod *Alioramus* in left lateral view; (H) the ischium of *Liliensternus* in left lateral view. Image (C) courtesy of Dr Roger Benson; image (D) taken by the author but copyright of Peabody Museum of Natural History; image (G) © Mick Ellison.

the ilium reinforces this region by providing a tighter fit for the femur and dissipating stress caused by locomotion.

The pubis and ischium are longer and more gracile than the femur, and generally resemble rods more than a blade (Fig. 2.20). The pubis is the anterior bone of the pelvis; in saurischians it projects both forward and downward from the ilium, whereas in ornithischians it is retroverted and projects backward and downward, paralleling the ischium. Both the pubis and ischium have an expanded plate-like region proximally, which articulates with the ilium. These regions, referred to as the obturator plates, also make broad contact with each other: the pubic obturator plate has a long contact with that of the ischium immediately below the acetabulum, further assisting in buttressing

the hip–limb joint. Distally the pubis and ischium taper into a long shaft. In some dinosaurs, most prominently large theropods like *Tyrannosaurus*, the very distal tip of one or both bones is expanded into a bulbous or foot-like swelling, referred to as the pubic, or ischial, boot. In ornithischians the pubis and ischium lay against each other due to the retroversion of the pubis, which also develops a novel flange, the prepubic process, projecting forward from the obturator plate immediately in front of the acetabulum.

Hindlimb
As with the forelimb, an orderly sequence of bones comprises the dinosaurian hindlimb. The femur, or thigh bone, is the most proximal element, followed by the tibia and fibula, an array of tarsals (including

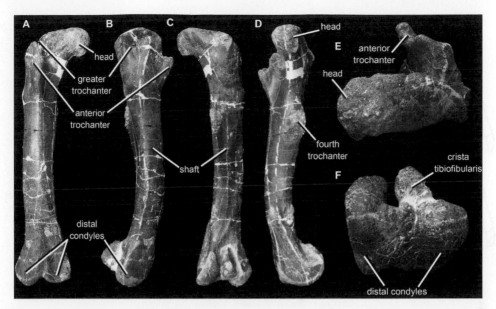

Figure 2.21 Dinosaur femur anatomy. The right femur of the large carcharodontosaurian theropod *Neovenator* in anterior (A), lateral (B), posterior (C), medial (D), proximal (E), and distal (F) views. Images courtesy of Dr Roger Benson.

the astragalus and calcaneum), metatarsals, and individual digits. The numbering of the digits, I being the most medial and V the most lateral, is identical to that of the hand. For a summary of dinosaur hindlimb anatomy and evolution, see Christiansen (1997b).

The femur is the largest, strongest, and most robust bone in the hindleg, although sometimes the tibia (shin bone) is longer in fast-running species (Fig. 2.21). The femur has a discrete head that projects medially relative to the shaft and is capped by a ball-like articular surface, which fits into the acetabulum of the pelvis. Several trochanters, or protuberances of bone for muscle attachment, are located immediately distal to the head, and anchored the most important muscles for hindlimb motion. The shaft of the femur is often bowed anteriorly in fleet-footed dinosaurs, but is straight and columnar to support the extreme body weight of sauropods and other colossal species. Distally, the femur joins the tibia and fibula at the knee joint. Dinosaurs, best as we know, did not have a separate kneecap bone like the patella of mammals. The tibia is much larger than the fibula in most dinosaurs, and this discrepancy reaches an extreme in derived theropods and birds, in which the fibula is reduced to a toothpick-like splint.

The tarsus of dinosaurs is composed of two different sets of bones: the larger and more functionally important proximal tarsals (astragalus and calcaneum) and a smaller series of flat disk-like distal tarsals that cap the proximal surface of the metatarsals (see Figs 1.17, 2.22 and 2.23). The astragalus is the most conspicuous bone of the tarsus: it is much larger than the calcaneum, forms the majority of the functional joint between the leg and foot, and has a tongue-like ascending process that is appressed to the anterior surface of the tibia. The calcaneum is a small crescent-shaped bone positioned lateral to the astragalus, and it primarily cups the distal end of the fibula. As described in Chapter 1, the astragalus and calcaneum are firmly united with the tibia and fibula, and the four bones together form a single complex. The functional ankle joint is between this complex and the foot itself, which is a second complex composed of the distal tarsals, metatarsals, and digits.

The size and morphology of the foot are highly variable among dinosaurs, but are correlated to general conditions of posture (bipedal vs. quadrupedal) and speed (fast vs. slow) (Fig. 2.23). Slower, heavier, quadrupedal dinosaurs have metatarsals and digits that are well suited for supporting the weight of the body. The individual metatarsals

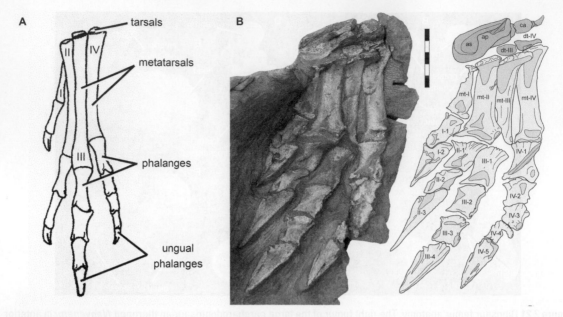

Figure 2.22 Dinosaur hindlimb and foot anatomy. (A) Line drawing of the metatarsus and foot of the primitive ornithischian *Lesothosaurus*. (B) Photo and interpretive drawing of the metatarsus and foot of the "prosauropod" *Seitaad*. Image (A) by Anne Curthoys, based on Thulborn (1972); image (B) courtesy of Dr Mark Loewen and modified from Sertich and Loewen (2010).

are short, robust, and arrayed in a wide arc, and often all five digits are large and weight-bearing. The terminal phalanges are frequently modified into hooves, as is seen in the hands, but in sauropods the medial two or three digits bear conical claws, whose function is debatable. Faster bipedal dinosaurs, such as most theropods, have extremely elongated metatarsals that are closely appressed to each other, which functionally comprise a third major long bone of the leg (along with the femur and tibia). Furthermore, in these species only the central three metatarsals and digits are large. Digit I is tiny, non-weight-bearing, and shifted to a medial position such that it does not contact the substrate during walking, whereas digit V is often reduced to nothing more than a small tapering metatarsal, without any digits.

Bizarre Structures

In addition to the general body plan described above, many dinosaurs possess so-called "bizarre

structures": sublime features of the skeleton that have captivated generations of museum-goers and long piqued the interest of scientists (I here use the term "bizarre structures" following its usage by Padian and Horner, 2011 in their influential review). The fantastic horns and frills of ceratopsians, extravagant cranial crests of hadrosaurids, and table-sized plates of stegosaurs are prime examples. Understanding the biological function of these structures, however, is fraught with difficulty, but is nonetheless a frequent subject of debate. The most common hypotheses hold that horns, frills, spikes, and other bizarre structures were weapons for defense against predators or ornaments used to attract mates, scare off rivals, or brand an individual as a member of its species. Testing these scenarios is difficult at best, and in most cases probably impossible. Many scientists make broad comparisons with living analogues, and often will argue that dinosaurs may have behaved like living species with similar aberrant structures, but it is always important to remember that no living species are perfect analogues for Mesozoic dinosaurs. More sophisticated analytical techniques, such as

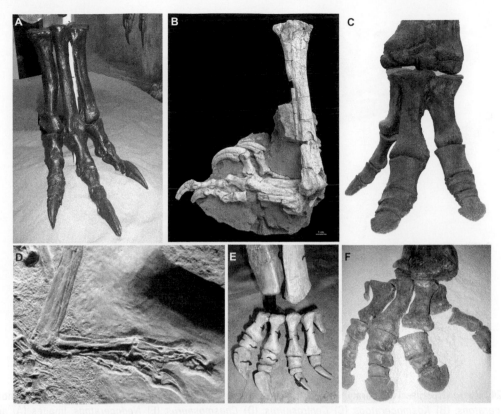

Figure 2.23 Dinosaur foot anatomy. Montage of different pedal morphologies in dinosaurs, including the carcharodontosaurian theropod *Neovenator* (A), the dromaeosaurid theropod *Balaur* (B), the hadrosaurid ornithischian *Edmontosaurus* (C), the compsognathid theropod *Juravenator* (D), the titanosaurian sauropod *Rapetosaurus* (E), and the ceratopsian ornithischian *Centrosaurus* (F). Images © Mick Ellison (B), and courtesy of Dr Albert Prieto-Márquez (C), Dr Jeffrey Wilson, Dr Mike D'Emic, and the Field Museum of Natural History (E), and Dr Thomas Williamson (F).

computer modeling, histological thin sectioning, and isotopic analysis, have been used to study the anatomy and composition of bizarre structures in detail, but can rarely pinpoint a specific functional usage with certainty. And of course, it is possible, and perhaps probable, that these structures performed many functions at the same time.

An astute review of bizarre features in dinosaurs was recently published by Padian and Horner (2011). They argue that species recognition may have been the primary function of many of these structures, but are candid in discussing the difficulties of testing functional hypotheses and the possibility that structures served multiple functions simultaneously. Indeed, they conclude, we may never know the exact reason why the horns of

ceratopsians evolved, or the exact function that stegosaur plates served, but this does not minimize the grandeur of these features or eliminate the need to study their anatomy. Here, in the concluding section of this chapter, I will describe several of the more salient bizarre structures of dinosaurs, and review some likely explanations for their function.

Ceratopsian horns and frills

The skulls of *Triceratops* and other ceratopsian dinosaurs are immediately recognizable due to their extraordinary size and fantastic array of cranial ornaments (Farlow and Dodson, 1975; Dodson et al., 2004; Ryan et al., 2010a) (Fig. 2.24). The skulls of derived Late Cretaceous ceratopsids could reach

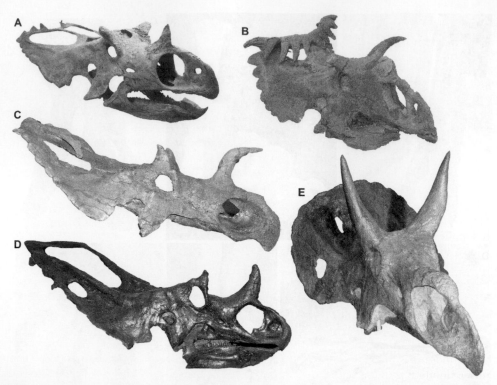

Figure 2.24 Bizarre structures: the ceratopsian skull. Montage of skulls showing a diversity of horn and frill morphologies: (A) *Utahceratops*; (B) *Kosmoceratops*; (C) *Centrosaurus*; (D) *Chasmosaurus*; (E) *Nedoceratops*. Images (A) and (B) courtesy of Dr Andrew Farke and modified from Sampson et al. (2010); images (C) and (D) courtesy of Dr Andrew Farke; image (E) courtesy of Dr Andrew Farke and modified from Farke (2011).

over 2 m in length, larger than in any other known terrestrial vertebrate, living or extinct. The back of the skull is expanded into a broad frill, composed of greatly enlarged parietal and squamosal bones, which varies from about 60% to slightly more than 100% of the length of the remainder of the skull. The peripheral edges of the frill are peppered with nubbins of dermal bone, the epoccipitals, which in some species are elaborated into a variety of horn-like projections, which are variably straight or curved. True horns, composed of a bony internal core that was covered with keratin in life, are usually present on the snout (on the nasal bone) and above the eyes (on the postorbital bones). The nasal horn, which is normally a single midline structure, reaches lengths of over 50 cm in some species. The postorbital horns, which are located on both sides of the skull, could be even larger, and in the gaudiest species stretched up to a meter in

length. Although most ceratopsids had horns, some species modified the nasal horn core into a gnarly rugose boss, which likely supported a thick pad of keratin-rich skin instead of a conical horn (Hieronymus et al., 2009).

Understandably, such ostentatious features have long generated wonder among both scientists and the public, and have attracted almost limitless speculation about their likely functions. Farlow and Dodson (1975), using ungulate mammals as a comparative modern analogue, suggested that the horns and frills of ceratopsians could have served a variety of functions, but were probably most useful in courtship and agonistic behavior toward rivals and predators. Noting how the frill and horns dramatically changed shape as an individual grew from a juvenile into an adult, Sampson et al. (1997) and Sampson (1999) argued that sexual display was a more probable function. Barrick et al. (1998)

performed a clever experiment by comparing the composition of oxygen isotopes (a proxy for body temperature) of bone in the frill and the core of the body. Their results suggested that the broad frill was something of a heat dump, which radiated excess body heat to the environment and thus helped to regulate body temperature, much like the floppy ears of elephants. Happ (2010) took this argument one step further, and suggested that the horns themselves were also used to radiate heat. However, his primary evidence was the presence of vascular surface texture, which he took as a sign of dense blood vessel networks that could bring excess heat to the surface of the horns. However, these vessels more likely supplied the growing kertain exterior of the horn, and would have had only a passing role in thermoregulation (see Horner and Marshall, 2002).

Perhaps the most common explanation for the bizarre skull of ceratopsians, especially the horns, is defense. Most ceratopsids lived in North America during the final 20 million years of the Late Cretaceous, a time and place swarming with bone-crunching tyrannosaurid predators. One of the most popular exhibits in many museums is a reconstructed "death match" between *Tyrannosaurus* and *Triceratops*, and indeed these two dinosaurs were among the most common species in the latest Cretaceous of North America. A fossilized *Triceratops* frill with bite marks closely matching the size and spacing of *Tyrannosaurus* teeth is unequivocal evidence that the two did, at least occasionally, engage in combat (Happ, 2008). Perhaps the horns and frill of ceratopsids often functioned as defensive weapons, used for protection from bites and to strike back at predators when attacked. Similarly, perhaps the horns were used as agonistic weapons to wrestle with conspecific competitors over mates or resources.

Several lines of evidence suggest that this is plausible. A computerized biomechanical study, using finite element techniques often employed by engineers to test the strength of roads and bridges, indicates that the broad and vaulted morphology of the frill imparted structural rigidity and strength, which may have provided reinforcement during battles with prey or conspecifics (Farke et al., 2010). A hands-on experiment using scale models of dueling *Triceratops* demonstrated that the size, shape, and position of the horns were ideal for

interlocking during intraspecies wrestling matches (Farke, 2004) (Fig. 2.25A). Finally, the empirical fact that *Triceratops* individuals did battle with each other is supported by abundant pathologies (lesions and rugose scars) on the frills of some fossils, in the same position as where the horns of a rival would be wedged during a tussle (Fig. 2.25B). Of course, although persuasive, this evidence does not "prove" that the horns and frills of ceratopsids functioned, either always or only, as weapons. Sexual display, species recognition, and thermoregulation may have also been important functions, and there are probably many other alternative functional explanations yet to be dreamed up by paleontologists.

Pachycephalosaur skull domes

The skulls of pachycephalosaurs, a peculiar subgroup of ornithischians closely related to the ceratopsians, are perhaps even more astonishing than the crania of their horned-and-frilled cousins (Fig. 2.26). In most species, the frontal and parietal bones are fused, thickened, and vaulted into an extravagant dome. Some species incorporate other surrounding bones, such as the postorbital and squamosal, into the dome, which is so enlarged and solid that the supratemporal fenestra is completely closed. All pachycephalosaurs sport a variety of bony excrescences along the margins of the dome, and in the largest and most derived species these are elaborated into a fantastic array of horns, spikes, and flanges (Goodwin et al., 1998; Maryańska et al., 2004).

The function of the pachycephalosaur skull has spawned much speculation and debate among researchers. No living animal possesses a domed skull remotely resembling that of pachycephalosaurs, but analogies to big horned sheep and other social mammals suggest that the vaulted and thickened cranium may have been used as a ramming device during intraspecific scuffles over mates or territory (Colbert, 1955; Galton, 1970). More quantitative biomechanical analyses indicated that the pachycephalosaur skull was ideal for such behavior (Sues, 1978), and debate has long focused on whether individuals battered each other head to head, or whether an individual would strike a rival on the flank (Alexander, 1989; Carpenter, 1997; Maryańska et al., 2004).

More recently, however, Goodwin and Horner (2004) examined the histological microstructure

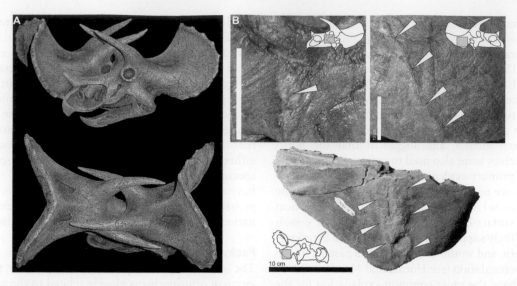

Figure 2.25 Bizarre structures: the ceratopsian skull and its possible functional usage. (A) Scale models of *Triceratops* showing how the horns may have interlocked during an intraspecific wrestling match. (B) Pathologies on specimens of *Triceratops*, placed at the regions where trauma would be expected if the horns of a rival *Triceratops* were the cause of damage. (A) Images modified from Farke (2004). © 2004 The Society of Vertebrate Paleontology. Reprinted and distributed with permission of the Society of Vertebrate Paleontology. (B) Images courtesy of Dr Andrew Farke (images modified from Farke et al. 2009).

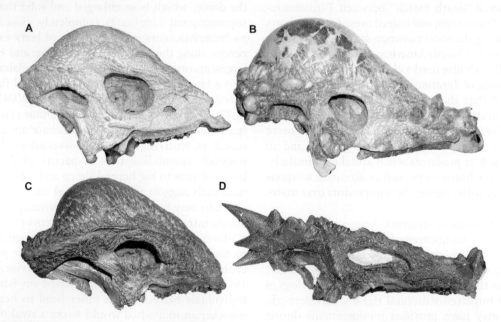

Figure 2.26 Bizarre structures: the pachycephalosaurian skull. Montage of skulls showing a diversity of dome and ornamentation morphologies: (A) *Prenocephale*; (B) *Pachycephalosaurus*; (C) *Stegoceras*; (D) *"Dracorex"* (considered by many to be a juvenile *Pachycephalosaurus*). Images (A) and (C) courtesy of Dr David Evans; images (B) and (D) courtesy of Dr Thomas Williamson.

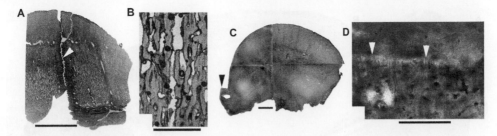

Figure 2.27 Bizarre structures: the pachycephalosaurian skull. Histology of a pachycephalosaurid skull dome. All images from Horner and Goodwin (2009), *Proceedings of the Royal Society B: Biological Sciences.*

of the pachycephalosaur dome and noted that its external surface is composed of brittle compact bone that is not well suited for enduring impact (Fig. 2.27). Heavily vascularized bone with radiating collagen fibers, which Galton (1970) and others argued would have helped dampen stress and cushion the brain during head butting, is merely fast-growing bone of juveniles, which becomes less spongy and stress-resistant as the individual matures. Moreover, the radiating pattern of fibers would likely have deflected stress into the brain, exactly the opposite of what would be required in a head-butting animal. Therefore, Goodwin and Horner (2004) argued that the fabulous dome of pachycephalosaurs was likely an ornament used for display, species recognition, or to attract mates.

But this study is not the final word. Motivated by Goodwin and Horner's (2004) histological work, Snively and Cox (2008) constructed finite element computer models that tested how skull domes with various shapes and histological microstructures would have performed during a hypothetical headbutt. Perhaps surprisingly, they found that the domes could withstand high impact forces, with more vaulted domes better suited for enduring stress. More recently, Snively and Theodor (2011) constructed additional computer models, based on CT scans of well-preserved skulls, which indicated that at least some pachycephalosaurs were as competent at head-butting as living head-rammers such as duikers and bighorn sheep. The function of pachycephalosaur domes therefore remains an open question, and it is possible that they served multiple functions, including both display and use in head-butting. More sophisticated computer modeling,

further histological examination, and a better appreciation of the soft tissues that surrounded the pachycephalosaur dome will surely refine our understanding of this bizarre structure.

Hadrosaurid cranial crests

Yet another group of plant-eating ornithischians, the "duck-billed" hadrosaurids, boast their own flavor of bizarre cranial anatomy (Fig. 2.28). The dermal bones of the face are modified into a crest, which is usually small and solid in hadrosaurines, one major hadrosaurid subgroup (also known as saurolophines), but elaborated into a gaudy, internally hollow extravagance in the other main subgroup, the lambeosaurines. In the latter taxa, the crest is composed primarily of the premaxilla and nasal, with contributions from surrounding bones, and it encloses an enlarged internal nasal cavity. Its shape varies dramatically among species, and ranges from the small hatchet-like quiff of *Lambeosaurus* to the bucket-shaped helmet of *Corythosaurus* and the ornate trombone-like tube of *Parasaurolophus* (Hopson, 1975; Horner et al., 2004).

The dizzying array of hadrosaurid crests has attracted an equally stupendous assortment of explanations for their function. Most of these were outlandish: suggestions of the crest being used for head-butting, snorkeling through water, or as a cranial reservoir for air storage were easily dismissed (Weishampel, 1981). Three more reasonable hypotheses, however, have been prominently discussed. These include a function in olfaction, sound production, or visual display. Ostrom (1962) argued that lambeosaurine crests housed a hypertrophied nasal capsule, which would have resulted in a greater sense of smell. Recent work, which largely relies on

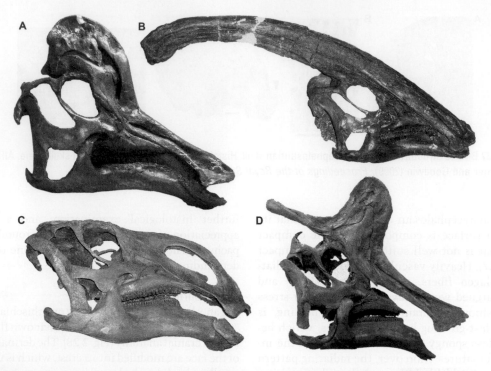

Figure 2.28 Bizarre structures: the hadrosaurian skull. Montage of skulls showing a diversity of crest morphologies: (A) *Corythosaurus*; (B) *Parasaurolophus*; (C) *Brachylophosaurus*; (D) *Lambeosaurus*. Images courtesy of Dr David Evans.

details of internal anatomy reconstructed by sophisticated CT scanning technology, sheds serious doubt on this hypothesis. Surprisingly, Evans (2006) illustrated that much of the nasal cavity of lambeosaurines was located outside the crest, indicating that the convoluted internal passages of the crest could not have improved the sense of smell. Moreover, Evans et al. (2009) demonstrated that the olfactory tracts and bulbs of the lambeosaurine brain were miniscule, utterly inconsistent with a heightened sense of smell.

The two other reasonable functions, vocal resonance and display, are well supported by anatomical evidence and currently favored by most paleontologists. Wiman (1931) was one of the first to draw parallels between the tubular crests of some lambeosaurines and the shapes of various musical instruments, and later authors have used mathematical and computer models to demonstrate that some hadrosaurids could indeed produce a throaty low-frequency call akin to that of a trombone (Weishampel, 1981; Sullivan and Williamson,

1999). Furthermore, Weishampel (1981) and Evans et al. (2009) have shown that the ear of lambeosaurines, studied in the latter case using CT technology, was capable of hearing such low-frequency sounds. Therefore, it is reasonable to hypothesize that lambeosaurines used their crests to produce sound that was heard by other members of their species, perhaps rivals or potential mates. The physical shape of the crest was probably also useful in attracting mates, intimidating rivals, or identifying its bearer as a member of a particular species. This idea was first seriously argued by Hopson (1975), who suggested that the crest had a dual role in sound production and display. More recent authors, such as Padian and Horner (2011), continue to lend favor to this hypothesis.

Scutes and osteoderms

Three major groups of dinosaurs, the stegosaurs, ankylosaurs, and sauropods, possess a variety of bony plates and scutes embedded in the skin (Figs 2.29 and 2.30). These dermal bones are called

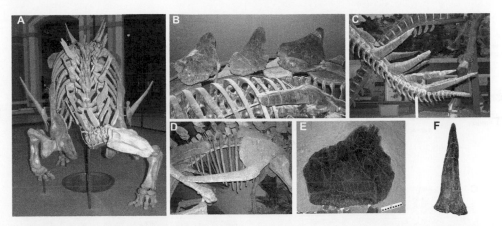

Figure 2.29 Bizarre structures: the plates and spikes of stegosaurids. Montage of stegosaur armor showing a diversity of morphologies: (A) the mounted skeleton of *Kentrosaurus* in anterior view; (B) the large plates over the back of *Stegosaurus*; (C) the spikes on the tail of *Stegosaurus*; (D) the flank spine of *Gigantospinosaurus*; (E) a plate of *Stegosaurus*; (F) a tail spike of *Stegosaurus*. Images (A) and (D–F) courtesy of Dr Susannah Maidment; photos (B) and (C) by the author but copyright of Peabody Museum of Natural History.

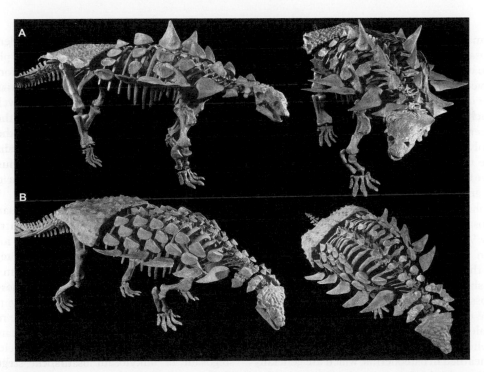

Figure 2.30 Bizarre structures: the osteoderms and armor of ankylosaurs. (A) *Gastonia* in lateral oblique and anterior views; (B) *Mymoorapelta* in lateral oblique and dorsal views. Skeletal reconstructions by Robert Gaston, photography by Francois Gohier (www.francoisgohier.com).

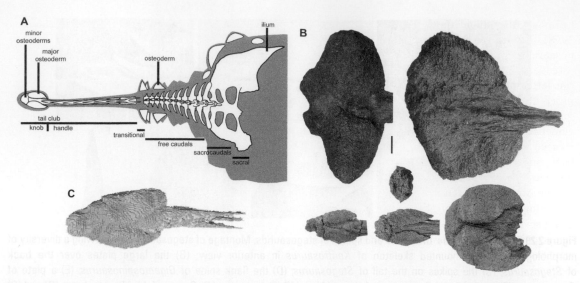

Figure 2.31 Bizarre structures: the tail club of ankylosaurs. (A) Schematic illustration of the tail and tail club of an ankylosaur; (B) a montage of different club morphologies in ankylosaur species; (C) a digital model of an ankylosaur tail club used in a computerized engineering analysis (see text). Images courtesy of Victoria Arbour; (A) and (B) modified from Arbour (2009) and (C) modified from Arbour and Snively (2009), and reproduced with permission.

osteoderms, and they are particularly prevalent in the stegosaurs and ankylosaurs, to the extent that extensive dermal armor is recognized as an important derived character uniting these two subgroups into their own larger clade (Thyreophora). In some cases, such as the triangular plates that line the back of stegosaurs, osteoderms can be modified into remarkably extravagant structures (see below).

Of the three armored dinosaur groups, ankylosaurs have the most extensive osteoderms across the skeleton and the greatest variety of osteoderm size and morphology (Fig. 2.30). Comprehensive reviews of ankylosaur osteoderms have been recently presented by Vickaryous et al. (2004) and Hayashi et al. (2010). In most ankylosaurs, osteroderms cover much of the skull, neck, back, and tail. One of the distinguishing features of ankylosaurs is that novel dermal ossifications, including rugose projections above the orbit and at the posterior corners of the skull, coalesce with the bones of the skull to form a single, solid cranium. Much of the postcranial skeleton is sheathed in osteoderms, which are generally arrayed in multiple elongate rows, arranged either anteroposteriorly or mediolaterally. These extend from the neck, across the back and pelvis, to the tip

end of the tail, and in some species osteoderms also envelop the belly. Oftentimes the large plate-like osteoderms of the neck fuse to each other, forming in some species a semicircular strap that partially cuffs the neck. Scutes along the back often develop into large conical spines, whereas those near the pelvic region fuse into a broad apron, called the sacral shield. However, the most remarkable modification is seen in the tail, where osteoderms fuse with vertebrae and ossified tendons to form a globular tail club (Fig. 2.31).

The osteoderms of stegosaurs and sauropods are more limited and less variable than those of ankylosaurs, and in the latter group only a handful of species actually possess them. Stegosaurs remarkably transform their dorsal osteoderms into a fabulous series of plates and spikes (Fig. 2.29). Only a very few stegosaurs retain any trace of osteoderms along their flanks, which are not only present in ankylosaurs but also in a handful of primitive thyreophorans that fall into neither the stegosaur or ankylosaur subclades (Galton and Upchurch, 2004a). Some stegosaurs, such as *Stegosaurus*, possess a complex of small interlocking ossicles embedded in the throat region, which forms a gular shield under the neck (Carpenter, 1998). In sauropods, members of the derived titanosaurian subclade possess two different kinds of osteoderms: large ovoid plates and smaller ossicles, the latter of which pack close together to cover much of the back and flank region (Bonaparte and Powell, 1980). There is limited, but provocative, evidence that some other sauropods may have sported keratinized spikes along their neck and back, but this remains to be validated by additional discoveries (Czerkas, 1992).

The growth and potential function of osteoderms, including ankylosaur tail spikes, has been studied using several techniques. The histology, or microstructural anatomy, of ankylosaur, stegosaur, and sauropod osteoderms has been analyzed in detail (Scheyer and Sander, 2004; Main et al., 2005; Cerda and Powell, 2010; Hayashi et al., 2010). Although osteoderms are often thin elements, their microstructure is optimized for strength. The outer cortex is composed of dense compact bone, whereas the more porous interior is strengthened by a network of collagen fibers that provides an internal scaffolding. In many ankylosaurs, the interior fibers are oriented in such a way that the osteoderm can endure large stresses from all possible directions (Scheyer and Sander, 2004). Therefore, osteoderms were akin to a coat of armor, ideal for protecting ankylosaurs, and in some cases sauropods and stegosaurs, from predatory attacks or other injuries to the soft tissues.

The tail clubs of ankylosaurs, on the other hand, were likely used as offensive weapons, akin to a mace (Fig. 2.31). Arbour (2009) and Arbour and Snively (2009) utilized finite element analysis and mathematical modeling techniques to assess two important qualities of any potential offensive armament: the amount of force that it could impart and the amount of impact force it could endure before breaking. By building a biomechanical model with parameters such as the estimated mass of the tail-swinging muscles, the flexibility of the tail, and the density of the caudal vertebrae, Arbour (2009) determined that the tail clubs of the largest ankylosaurs were powerful enough to shatter bone. By constructing three-dimensional finite element models of various tail clubs, which were subjected to a battery of stresses and strains in computer simulations, Arbour and Snively (2009) concluded that most clubs were strong enough to tolerate the high stresses of impact. Therefore, it is plausible that tail clubs were used as a retaliatory weapon to ward off predators. However, other functions are also possible, such as display.

Stegosaur plates and spikes

A special category of modified osteoderms, and probably their most fantastic manifestation among dinosaurs, are the plates and spikes of stegosaurs (Fig. 2.29). All stegosaurs possess two rows of plates and/or spines along the neck, back, and tail, which exhibit a bewildering variety in number, size, and morphology among species (Galton and Upchurch, 2004a; Maidment et al., 2008). Most stegosaurs have a series of triangular plates along the back and neck, transitioning into a more spike-like morphology across the posterior part of the back and tail. This condition is typified by *Kentrosaurus*, which has 14 or 15 plates in each row, along with a single elongate spike projecting backward from each shoulder. *Huayangosaurus*, perhaps the only stegosaur that retains the flank osteoderms of other thyreophorans, has a set of much smaller and shorter plates and spikes. The most extreme modifications are seen in the most familiar of the stegosaurs, *Stegosaurus*, which has a set of densely packed small plates along its neck, an imbricating array of table-sized plates along its back, and two sets of robust conical spikes at the tip of its tail (Carpenter, 1998; Galton and Upchurch, 2004a).

The function of these extraordinary structures has been debated since the first fossil discoveries of stegosaurs in the 1870s. Only with the advent of more modern quantitative techniques, however, has this debate veered away from unbridled speculation. Some of the first rigorous analyses suggested that stegosaur plates were likely used for thermoregulation, either to dump excess heat or to aid in warming the body, like solar panels (Farlow et al., 1976; Wheeler, 1978). A major step forward was taken by Buffrénil et al. (1986), who analyzed the histological microstructure of several specimens. They argued that the lightweight nature of the plates and their hollow internal structure made it very unlikely that they were used as armor or offensive weapons. However, extreme vascularization of the outer layer of bone suggested that the plates may have acted as thermoregulatory devices, by allowing blood to be either warmed or cooled as it coursed through the network of vessels and fanned out across the broad plate surface (see also Farlow et al., 2010).

Although a reasonable hypothesis, this idea was dismissed as unlikely by Main et al. (2005) based on more extensive histological research, buoyed by an additional 20 years of understanding of dinosaur bone growth and histology. These authors also noted the extreme vascular nature of the outer core of the plates, but found it more likely that a dense network of blood vessels was needed to supply an external keratin sheath, and to provide nutrients for rapid bone growth. Although they could not completely discount a thermoregulatory function for the plates, Main et al. (2005) suggested that a role in species recognition and display was more likely, and any physiological benefits were a mere byproduct. More recent histological work by Hayashi et al. (2009) confirms that the plates of stegosaurs grew remarkably fast, lending support to Main et al.'s (2005) functional interpretation of the vascularized surface texture.

The tail spikes of stegosaurs are quite different in shape than the plates, so perhaps they had a different function. It has often been speculated that these spikes – sometimes evocatively referred to as the "thagomizer" based on a joke that originated in a Gary Larson cartoon – were used for self-defense, to strike back at predators or other threatening species. A study by McWhinney et al. (2001) supports this hypothesis, as they identified several cases of trama-related pathologies on fossilized stegosaur tail spikes. Furthermore, a single spectacular fossil – a caudal vertebra of an *Allosaurus* punctured with a hole that matches in size and shape to the tail spike of a *Stegosaurus* – also lends evidence to this idea (Carpenter et al., 2005). Finally, a recent computer modeling study by Mallison (2011) showed that the tail spikes of *Kentrosaurus* were capable of both moving rapidly in a wide arc and delivering extreme force upon impact, lending evidence to the hypothesis that they were used as weapons.

Conclusions

Almost all known fossils of dinosaurs are bones and teeth, and therefore our understanding of dinosaur anatomy, biology, and behavior depends primarily on the interpretation of these hard tissues. The skeleton of dinosaurs follows the normal tetrapod bauplan, and consists of a skull composed of numerous individual bones, a vertebral column partitioned into different regions, and four sets of girdles and limbs. Dinosaurs exhibit great variety in their skeletal anatomy, but in general many of the differences between species are due to their diet (carnivory vs. herbivory), body size (large vs. small), and posture (quadrupedal vs. bipedal). Perhaps the most incredible features of the dinosaur skeleton are several co-called "bizarre structures," including the horns of ceratopsians and elaborate cranial crests of hadrosaurids, whose anatomy and function have long enamored paleontologists. Debates over the purpose of these sublime features are prime examples of how careful consideration of hard-part anatomy is used to infer the biology of long-extinct taxa.

3 Soft Tissues

The vast majority of dinosaur fossils are bones and teeth, but in order to understand how dinosaurs functioned as living animals it is necessary to have some appreciation for soft tissues as well. Like their living relatives, dinosaurs would have been covered in skin (and in many cases feathers), powered by muscles, and in possession of a suite of internal organs that regulated digestion, breathing, and metabolism. Understanding these soft parts, however, is no easy task. Occasionally, exceptionally preserved fossils may contain traces of skin, muscle, integument, or even the internal organs (Kellner, 1996; Briggs et al., 1997; Dal Sasso and Signore, 1998; Martill et al., 2000; Norell and Xu, 2005). Other specimens have been trumpeted as preserving proteins and other microscopic soft tissues (Schweitzer et al., 2005a, 2007a, 2007b; Schweitzer, 2011). Although provocative and often greatly informative, these remarkable specimens are not the norm. In most cases, the presence and morphology of soft tissues must be inferred by reference to both the anatomy of hard parts, which often anchor soft tissue, and the soft tissue anatomy of living relatives. While this method can be notoriously imprecise, a careful and conservative reconstruction of soft tissues can give critical insight into how dinosaurs moved and fed, and what they may have looked like as living, breathing creatures.

Dinosaur Paleobiology, First Edition. Stephen L. Brusatte.
© 2012 John Wiley & Sons, Ltd. Published 2012 by John Wiley & Sons, Ltd.

Inferring Soft Tissues: The Extant Phylogenetic Bracket

Bones are not simply objects, but components of a dynamic skeletal system that is integrated with other anatomical systems such as the muscles, viscera, and skin. Fossil bones therefore have the potential to provide information about soft tissues that are rarely themselves preserved as fossils. Muscles and tendons are fastened to bones, often in a predictable way, and the shape of the skeleton determines the size and position of organs. Even the integument can have a close correspondence with bone: some feathers anchor to certain bones, and skin and keratin often intimately overlie the bone surface, leaving a characteristic texture. This unlocks a powerful and tantalizing possibility: the long-decayed soft tissues of dinosaurs may be inferred by reference to their bones. However, this procedure must be exercised with caution. Soft tissues are difficult to reconstruct, and it is all too easy to fall into a habit of guessing where a muscle or organ may have been positioned based only on superficial consideration of bone anatomy (see discussion of the pitfalls of soft tissue reconstructions in Witmer, 1995a; Holliday, 2009). What is clearly

needed is an explicit, standard, and conservative rationale for inferring soft tissues by reference to the evidence at hand.

The extant phylogenetic bracket (EPB), independently outlined by Bryant and Russell (1992) and Witmer (1995a), has emerged as a clear framework for reconstructing soft tissues in extinct animals by reference to both hard tissues and knowledge of the anatomy of living relatives (Fig. 3.1). The EPB is not a foolproof method or absolute algorithm that tells a researcher what soft tissues were present in an extinct organism, but rather a means of making reasonable hypotheses and assessing the strength of competing inferences. The goal of the EPB is to identify structures on the bone, so-called osteological correlates, that are causally related to a certain type of soft tissue. For instance, a trochanter on the femur may be an attachment site for a hindlimb retractor muscle, or bony bumps on the ulna may be an anchor point for feathers. It is not enough to simply guess that such structures are intimately related to soft tissue; this must be demonstrated by reference to living organisms. But out of the entire panoply of life, which living organisms should be studied?

The EPB focuses on the two closest living relatives of the extinct species in question, or if these

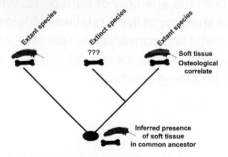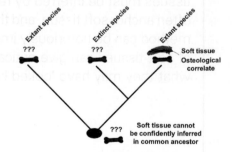

Figure 3.1 A schematic diagram illustrating how the extant phylogenetic bracket (EPB) operates in practice. The EPB approach uses features of hard-tissue anatomy (preserved in fossils and living animals), interpreted in a phylogenetic context, to make inferences about the soft tissues that may have been present in fossils. If a soft tissue is anchored to a specific hard-tissue structure in the two closest living relatives of a fossil species, and if the fossil species has the hard-tissue structure in question, then it can reasonably be inferred that the fossil species possessed the soft tissue. If, however, a hard-tissue structure is present in the living and extinct animals, but one or neither of the living relatives possesses the soft tissue in question, then that soft tissue is inferred with relatively less confidence in the fossil species. See text for more explanation.

are unknown or unavailable, then progressively more distant relatives. These relatives, or outgroups, "bracket" the extinct taxon on the phylogeny. If both living relatives have a certain soft tissue that is causally related to a discrete osteological correlate, then the presence of this soft tissue can be inferred in the extinct species (if it also has this bony correlate) with a high degree of confidence. Why is this the case? Because the soft tissue in the living organisms can be hypothesized to have evolved in their common ancestor, and since the extinct organism is bracketed by the living species then it also falls into the clade that originated with this ancestor. There are other cases in which an osteological correlate may be present in one outgroup taxon but absent in the other. In this situation, the associated soft tissue could very well be present in the extinct organism, but this hypothesis is less secure than an inference derived from two outgroup taxa, because it is not clear that the soft tissue in question originated in the common ancestor of the group. And what if a potential osteological correlate–soft tissue relationship is unknown in any outgroup? This does not disprove the possibility that dinosaurs possessed the soft tissue in question, but its inferred presence is weak. As can be seen, the EPB does not provide absolute answers – it doesn't demand that a certain tissue be present in a certain extinct organism – but explicitly outlines how much confidence can be placed in soft tissue reconstructions based on reference to living species.

The EPB method is straightforward, but may seem abstract at first. A few simple examples, outlined by Witmer (1995a, 1997a), should help clarify how to use the EPB in practice (also consult Fig. 3.1 for a visual explanation). Did *Tyrannosaurus* have eyeballs? It is reasonable to guess that it did, but it is better to make an explicit assessment based on tangible evidence. The closest living relatives of *Tyrannosaurus* are birds and crocodiles: these are the two outgroups that provide the phylogenetic bracket. In both birds and crocodiles numerous features of the bony skull are causally associated with supporting the eyeball, including a fenestra on the lateral side of the cranium surrounded by the postorbital and lacrimal bone (the orbit). *Tyrannosaurus* also possesses this cranial fenestra, and therefore we can hypothesize, with a high degree

of confidence, that it possessed an eyeball. The fenestra is not absolute proof of an eyeball in *Tyrannosaurus* – perhaps it sported a novel proboscis or enclosed an enlarged salt gland – but the most reasonable hypothesis is that it served the same function as in birds and crocodiles. What about another feature: could the mighty *Tyrannosaurus* have sported a beak at the front of its skull? Its premaxilla is peppered with numerous neurovascular foramina, which in birds are known to innervate the rapidly growing keratin beak. Crocodiles, however, do not have a beak, and therefore the inference of a beak in *Tyrannosaurus* is less secure. In other words, we place much more confidence in the inference of an eyeball than we do in the reconstruction of a beak.

In sum, the EPB is an explicit framework for anatomical comparison and hypothesis testing. Discrete comparisons are made with living relatives, causal relationships between hard and soft tissues are evaluated, and soft tissue reconstructions in extinct animals are assessed with varying degrees of confidence depending on whether two, one, or zero outgroups possess the hypothesized tissue in question. We will never know with certainty whether a dinosaur possessed a certain muscle in a certain location, but we can hypothesize that this muscle may be present based on a putative osteological correlate and comparison to living relatives. The EPB does have its limitations – imagine the difficulty of inferring soft tissues from human bones with only bats and whales as comparative outgroups – but it is a standard rationale for making reasonable soft tissue reconstructions.

Dinosaur Muscles

One of the most active areas of current dinosaur research is myology: the reconstruction and study of muscles (Fig. 3.2). Paleontologists have long been interested in dinosaur musculature, and muscle reconstructions were commonly published alongside descriptions of skeletons for much of the 20th century. Many of these reconstructions, most importantly those by legendary paleontologist Alfred Romer, were based on meticulous comparisons with the skeletons of living animals, which provided a guide for the reconstruction of muscles and

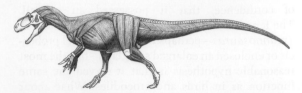

Figure 3.2 A reconstruction of the muscles of the large-bodied theropod dinosaur *Allosaurus*. Reconstruction © Scott Hartman.

other soft tissues in dinosaurs. More recently, with the advent of the EPB, muscle reconstruction has become more of a testable exercise, grounded in explicit consideration of osteological correlates and specific comparisons to living birds and crocodiles. Several important studies on dinosaur muscles, especially those of the skull and hindlimb, have been published in recent years, but most of these focus on theropods. One of the great opportunities for future research is to extend these studies to the remainder of the skeleton and to other groups of dinosaurs, especially aberrant species such as the

enormous sauropods and armored ankylosaurs. Here, I present a brief review of the major muscles of the dinosaur skeleton, with citations to important studies in the emerging field of dinosaur myology.

Skull and jaws

The various muscles of the skull are instrumental in feeding, and knowledge of the size, position, and morphology of these muscles is critical to understanding the dietary habits of dinosaurs. Fortunately, the bones of the dinosaur skull are rich in osteological correlates, which when interpreted in the EPB framework can indicate the presence, attachment areas, and size of the individual muscles (Holliday and Witmer, 2007) (Fig. 3.3). A comprehensive, well-illustrated summary of dinosaur jaw muscles was recently provided by Holliday (2009), and this should be consulted for further details. In general, the dinosaur skull musculature comprised three major groups (Fig. 3.4). First, the temporal muscles stretched from the skull roof to the lower

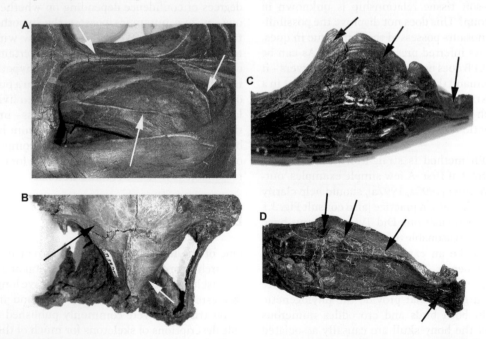

Figure 3.3 Osteological correlates of skull muscles in dinosaurs. Arrows point to various features on the skull roof of the hadrosaurid *Edmontosaurus* (A) and the bird-like theropod *Troodon* (B) and lower jaws of the small ornithischian *Thescelosaurus* (C) and the bird-like theropod *Dromaeosaurus* (D) that anchor temporal muscles (jaw adductor muscles). The identification of these attachment sites is based on comparisons with living birds and crocodiles, which form the extant phylogenetic bracket for dinosaurs. Images courtesy of Dr Casey Holliday.

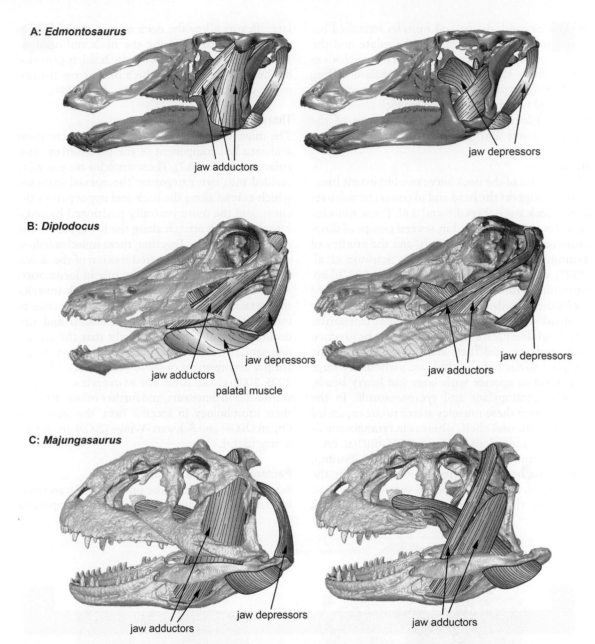

A: Edmontosaurus

jaw adductors

jaw depressors

B: Diplodocus

jaw depressors

jaw adductors

palatal muscle

jaw depressors

jaw adductors

C: Majungasaurus

jaw adductors

jaw depressors

jaw adductors

Figure 3.4 Dinosaur skull muscles. Reconstructed temporal and palatal muscles, which adduct (close) and depress the jaw, reconstructed on the skulls of the hadrosaurid *Edmontosaurus*, the sauropod *Diplodocus*, and the theropod *Majungasaurus*. Images on the left depict superficial muscles and those on the right deeper muscles that are obscured by the superficial muscles in lateral view. Modified from Holliday (2009), and reproduced with permission.

jaw, some branches passing through the lateral temporal fenestra and others extending across it laterally. These acted primarily to raise the lower jaw, and thus are often referred to as the "jaw adductor" muscles. These muscles were countered by the mandibular depressor muscle, which closed the jaw. Second, the palatal muscles stretched between the numerous bones of the palate and the lower jaw. Third, the orbitotemporal muscles extended from the braincase to the palate, filling much of the internal space of the orbital cavity and the lateral temporal fenestra. One such muscle, the pterygoid levator, likely enabled portions of the palate to move during feeding.

Neck

The muscles of the neck serve two important functions: to support the head and to power the motions of the neck itself (Figs 3.5 and 3.6). These muscles have been reconstructed in several groups of dinosaurs using the EPB approach, and the studies of Tsuihiji (2004, 2005, 2007, 2010), Schwarz et al. (2007), and Snively and Russell (2007a, 2007b) are particularly illuminating and should be consulted for further details. The neck muscles can generally be divided into five major functional categories. The head dorsiflexor muscles raise the head when they contract, and their attachment sites on the occipital surface of the skull are particularly large and broad in species with large and heavy heads, such as ceratopsians and tyrannosaurids. In the former group these muscles attach to the expanded parietosquamosal shelf, whereas in tyrannosaurids they are fastened to the expansive nuchal crest (Snively and Russell, 2007a, 2007b; Tsuihiji, 2010). The head ventroflexion muscles lower the head, the neck dorsiflexors and ventroflexors raise and lower the neck, respectively, and the neck

lateroflexors allow the neck to move from side to side. Assistance in raising the neck and head, as well as additional support for the head, is provided by the nuchal ligament, which runs across the tips of the cervical neural spines (Tsuihiji, 2004).

Thorax

The muscles of the trunk and thorax are the most understudied component of the dinosaurian muscular system (Fig. 3.7). These muscles are generally divided into two categories: the epaxial muscles, which extend along the back and upper part of the chest, and the more ventrally positioned hypaxial muscles, which stretch along the lower part of the chest and the belly. Together, these muscles helped support the trunk, facilitated motion of the dorsal vertebral column, and played a role in locomotion and breathing. Certain specialized trunk muscles that attach to the uncinate processes of the ribs in birds function in moving the sternum and ribs during breathing, and it is likely that the closest dinosaurian relatives of birds would have had a similar arrangement of these muscles (Codd et al., 2005, 2008; Codd, 2010). For an overview of thoracic muscles in dinosaurs, and further information on their morphology in specific taxa, the studies of Organ (2006) and Schwarz-Wings (2009) are particularly useful.

Pectoral girdle and forelimb

Several muscles sheath the pectoral girdle and forelimb of dinosaurs, and would have served important

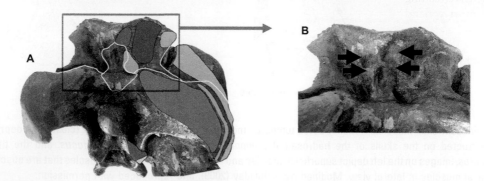

Figure 3.5 Dinosaur neck muscles (connecting to the back of the dinosaurian skull). The occipital surface of the skull of *Tyrannosaurus* shows the hypothesized attachment sites of several neck muscles, most of which function in raising and lowering the head (A). The insert (B) uses arrows to denote specific attachment sites (osteological correlates) where a particular muscle, m. spinalis capitis, attached. Modified from Tsuihiji (2010), and reproduced with permission.

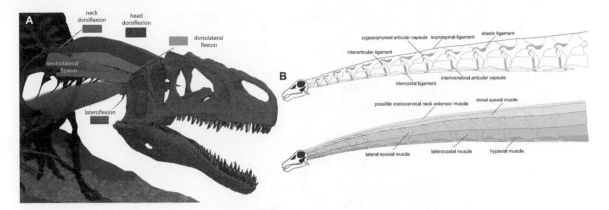

Figure 3.6 Dinosaur neck muscles. Reconstructed neck muscles in the theropod *Allosaurus* (A) and the sauropod *Diplodocus* (B). Image (A) courtesy of Dr Eric Snively and image (B) courtesy of Dr Daniela Schwarz-Wings and modified from Schwarz et al. (2007).

roles in moving the forearm, anchoring the pectoral girdle to the trunk, and assisting in moving the chest during breathing (Fig. 3.8). These muscles have been reconstructed for many individual dinosaur species, but often based on limited comparisons with living relatives and without a phylogenetic framework to guide inference testing (Ostrom, 1974; Coombs, 1978a; Norman, 1986; Dilkes, 2000; Carpenter and Smith, 2001). On the contrary, the more recent study of Jasinoski et al. (2006) explicitly uses the EPB to infer pectoral muscles in dromaeosaurid theropods (see also Langer et al., 2007). This paper, and an earlier but excellent offering by Nicholls and Russell (1985) with nascent echos of what would later be codified as the EPB, are both recommended for more comprehensive reading.

In the most general terms, the muscles of the shoulder and forelimb are divided into adductors, which bring the limb toward the body, and abductors, which extend the arm away from the body.

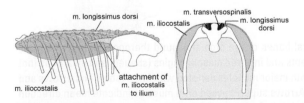

Figure 3.7 Dinosaur thorax muscles. Schematic showing major muscles of a sauropod thorax in left lateral (A) and anterior (B) views. Reproduced with permission from Schwarz-Wings (2009).

The main forelimb abductor is the deltoid, a large triangular muscle that extends from the deltopectoral crest of the proximal humerus to the lateral surface of the scapula. Within the forelimb, the flexion of the forearm relative to the humerus is powered by the biceps, which runs along the inside surface of the arm, from the humerus to the radius and ulna. Adduction of the arm occurs via contraction of the pectoralis, the dense mass of chest muscles which stretch from the deltopectoral crest of the humerus to the sternum (breast bone). These muscles are especially large in birds, as they provide the primary thrust for the flight stroke. The arm itself straightens due to the action of the triceps, the main intra-arm extensor muscle that extends from the humerus to the olecranon process on the ulna (the elbow joint).

One additional pectoral muscle deserves comment. The supracoracoideus was likely small in most dinosaurs, but in living birds is a large muscle that underlies the pectoralis and fans across the surface of the sternum. Perhaps counterintuitively, this muscle is the primary arm abductor in birds, as the deltoid muscle is greatly reduced. However, it makes little geometrical sense to have the arm abductor located directly on the underside of the chest, as a muscle in this position would normally adduct the forearm when it contracts. Indeed, this is exactly what occurs after contraction of the pectoralis. Therefore, a large tendon extends from the supracoracoideus muscle, through a bony canal between several bones of the pectoral girdle, and

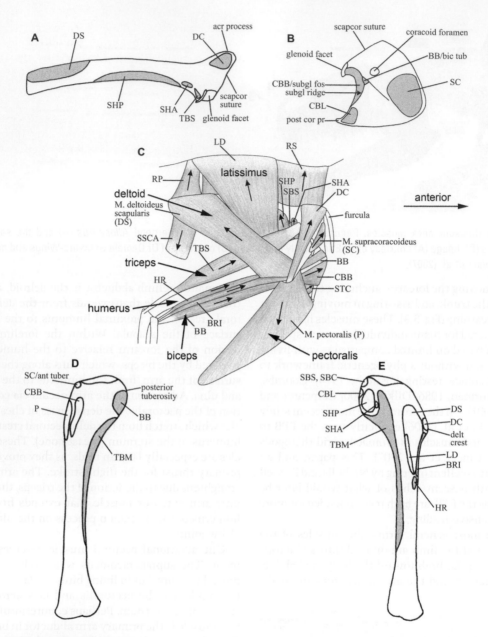

Figure 3.8 Dinosaur shoulder and forelimb muscles. Pectoral bones of the dromaeosaurid theropod *Saurornitholestes* (A, B, D, E), with osteological correlates for muscle attachments and inferred muscle origins (shaded) and insertions (not shaded). (C) Muscle reconstruction of the shoulder girdle, with major muscles denoted with large labels and arrows and detailed names of muscles following Jasinoski et al. (2006). Arrows superimposed onto muscles indicate mean direction of muscle action. Please consult Jasinoski et al. (2006) for details of the reconstruction and names of the individual muscles and osteological structures corresponding to the small labels. Modified from Jasinoski et al. (2006), and reproduced with permission.

attaches onto the head of the humerus. The contraction of the supracoracoideus is transmitted across this tendon, and as a result the arm is extended away from the body. This sequence of muscle and tendon activity powers the recovery stroke in flying birds.

Pelvic girdle and hindlimb

The pelvic and hindlimb muscles are usually the largest and most massive in the body, because they play an integral role in both body support and locomotion (Fig. 3.9). This is especially true in bipedal dinosaurs, in which the entire body is supported by the hindlegs. Large muscles require firm attachments to the bones that support them, and consequently the various tubercles, trochanters, and other muscle attachment scars on the pelvis and hindlimbs are exceptionally large and discrete. This allows for more confident and straightforward comparison to corresponding muscle attachment

sites in birds and crocodiles. It is no surprise, therefore, that the myology of the dinosaurian pelvis and hindlimb has been studied in great detail using the EPB framework (Gatesy, 1990, 2002; Hutchinson, 2001a, 2001b, 2002; Carrano and Hutchinson, 2002). Of these studies, Hutchinson's (2002) general overview of dinosaur hindlimb muscle structure, function, and evolution is particularly illuminating, and should be consulted for further details.

Although complex at first glance, the various muscles of the pelvis and hindlimb can be divided into a handful of functional categories. Most of the major muscles function in moving the hindlimb itself, and many of these extend from the broad lateral blade of the ilium and attach to the femur, the largest and strongest bone in the leg. The principal femoral protractors, which bring the leg forward, and abductors, which swing the leg outward, originate from the preacetabular blade of the ilium. They extend to the anterior and lateral surfaces of

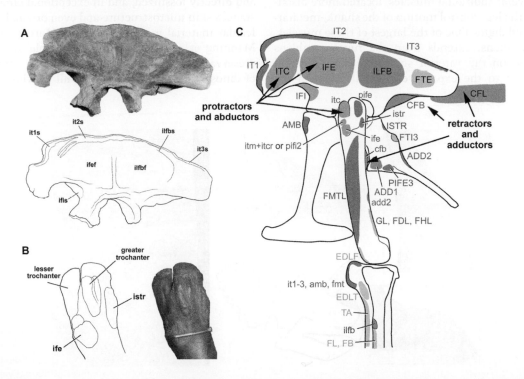

Figure 3.9 Dinosaur pelvic and hindlimb muscles. Photos and line drawings of the ilium (A, left lateral view) and femur (B, lateral view) of the theropod *Tyrannosaurus*, showing major muscle attachment sites. (C) Reconstructed muscle attachments to the pelvis and hindlimb of *Tyrannosaurus*, based on an EPB study. Modified from Carrano and Hutchinson (2002), and reproduced with permission.

the proximal femur, and attach to the prominent lesser trochanter and surrounding regions. Muscles controlling femoral adduction, or movement of the leg toward the body midline, are some of the smaller muscles of the pelvic region, and span from the ischium to the posterior and medial surfaces of the femoral shaft. Finally, the femoral retractors, which pull the leg backward, originate along the tail, pass through the broad and smooth brevis fossa on the ventral surface of the postacetabular ilium, and attach to the flange-like fourth trochanter on the posterior surface of the femur. As outlined by Gatesy (1990), these muscles, which are prominent in crocodiles and most dinosaurs, were reduced during the evolution of birds, in concert with the shortening and fusion of the tail. In birds, hindlimb retraction is usually powered by other muscles associated with knee flexion, because the tibia, not the small and largely immobile femur, is the most prominent bone in their hindlimb.

Several additional muscles, located more distally on the leg, control motion of the shank, metatarsus, and digits. One of the largest of these muscles, the ambiens, extends from the rugose ambiens process on the anterior surface of the pubis and attaches to the expansive cnemial crest on the

anterior surface of the tibia. Another prominent muscle, the iliofibularis, stretches all the way from the lateral surface of the ilium to a bulbous tubercle on the fibula. Furthest distally, a complex of small thin muscles regulated the motion of the individual digits of the foot.

Dinosaur Skin

Perhaps surprisingly, several fossils of dinosaur skin are known, including examples from most of the major dinosaur subgroups (Czerkas, 1997) (Fig. 3.10). The vast majority of these specimens are impressions of the skin imprinted into the sediment. Because this often occurs as an animal's foot impacts the ground, dinosaur footprints are a common source of skin impressions (Currie et al., 1991; Gatesy et al., 2005; Kim et al., 2010). In some rare cases, however, the skin itself is mineralized and directly fossilized, and in exceptional circumstances skin microstructure and even original molecular material may be preserved (Martill, 1991; Manning et al., 2009). This remarkable style of preservation is most famously seen in a number of dinosaur "mummies," almost all of which are

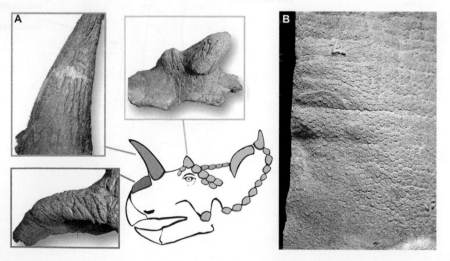

Figure 3.10 Dinosaur skin and integument. (A) Various skull bones of the ceratopsian *Centrosaurus* exhibit highly vascularized surface texture characteristic of a keratinous covering in life. (B) Skin impressions of the hadrosaurid *Edmontosaurus.* Image (A) modified from Hieronymus et al. (2009), and reproduced with permission; image (B) courtesy of Tyler Lyson.

specimens of hadrosaurids from the Late Cretaceous of North America (Czerkas, 1997; Manning et al., 2009).

The external patterning of dinosaur skin is variable but is broadly analogous to that of many living reptiles (Fig. 3.10B). The fabric of the skin is composed of individual scales, which are usually either round or polygonal in shape, that pack tightly together and often interlock like the individual leather panels that form the outer surface of a soccer ball. Two general categories of scales are present in most skin impressions: small densely packed scales that form a "ground pattern," within which are interspersed larger, more variably shaped scales, some of which are dome-shaped or even conical. The great variety of dinosaur skin patterning is reviewed by Czerkas (1997), Christiansen and Tschopp (2010), and Kim et al. (2010), and readers should consult these sources for further information.

Oftentimes dinosaur skin impressions and the texture and microstructure of the bones indicate the presence of keratinized integumentary tissue. Exceptionally preserved osteoderms of the basal thyreophoran *Scelidosaurus* and plates of a Late Jurassic stegosaurid have been discovered surrounded by a halo of soft tissue impressions, whose microstructure matches that of keratin (Martill et al., 2000; Christiansen and Tschopp, 2010). Similarly, two ornithomimosaur fossils from the Late Creataceous show evidence of a horny ramphotheca covering the edentulous tip of the snout, similar in shape and microstructure to the beak of many living vertebrates (Norell et al., 2001b; similar preservation is also seen in some hadrosaur fossils). Abundant evidence from external bone surface texture and interior histological microstructure indicates that ceratopsian frills and pachycephalosaur domes were sheathed in keratin (Horner and Marshall, 2002), and recently Hieronymus et al. (2009) used an EPB framework to identify a discrete set of osteological correlates that indicate the presence of keratin on bone (Fig. 3.10A). They concluded that the gnarly nasal bosses of some ceratopsians, such as *Pachyrhinosaurus*, were enveloped with a thick pad of cornified skin, and that the more traditional horns of other ceratopsians were almost surely encased in keratin as well.

Some examples of dinosaur skin show evidence of peculiar structures that would be difficult or impossible to identify if bones were the only sources of information. Czerkas (1992) described a series of triangular spikes along the caudal vertebrae of a remarkably preserved specimen of the sauropod *Diplodocus*, and these were composed of keratinized material without a bony core. Similar structures are known in some hadrosaurid specimens, and like many of the "bizarre structures" of the bony skeleton were likely used for display purposes (Czerkas, 1997). Fine preservation in the microbial mat deposits of Las Hoyas, Spain indicate the presence of a throat pouch, or dewlap, in the ornithomimosaur *Pelecanimimus* (Briggs et al., 1997). It is tantalizing to speculate about what other integumentary structures, completely unknowable by reference to bones only, are waiting to be revealed with new discoveries of fossilized skin.

Dinosaur Feathers

One of the greatest revelations in the history of dinosaur research is that some dinosaurs – perhaps all dinosaurs – sported feathers (see Plates 5–7). The hypothesis that birds evolved from theropod dinosaurs was a provocative and revolutionary idea when first proposed by Huxley in the 1860s. A wealth of corroborating evidence, most notably the strikingly bird-like skeletons of *Deinonychus* and *Velociraptor*, gradually won over a once-skeptical paleontological community, and by the 1980s it was widely recognized that birds were living dinosaurs. This genealogical hypothesis demanded a bold prediction: feathers most likely evolved before birds originated, and consequently some dinosaurs were probably feathered. Therefore, it was expected that, someday, a remarkably preserved dinosaur fossil would be found covered in feathers.

These specimens finally came to light in the late 1990s, with the discovery of two spectacular skeletons of the small theropod *Sinosauropteryx* from the Jehol Group, a sequence of Early Cretaceous rocks in Liaoning, China (Ji and Ji, 1996; Chen et al., 1998). These skeletons were encircled with a halo of thin hair-like filaments, preserved in fine-grained lithographic limestone deposited in lakes that were occasionally inundated with volcanic ash that immediately killed and buried all animals in the vicinity. In the nearly two decades since the discovery of *Sinosauropteryx*, hundreds of specimens of other feathered dinosaurs, ranging from primitive theropods to some of the closest relatives of birds, have been found in the tan, lacustrine, Jehol Group limestones of Liaoning (Zhou et al., 2003; Norell and Xu, 2005). This region is now regarded as one of the world's premier fossil sites, heralded by scientists as a unique (and very fortunate) window into the fine details of dinosaur soft tissue anatomy that are exceedingly difficult to preserve. Indeed, feathers and other soft tissues that are commonly retained in Jehol specimens, and fossils from a few other sites in China (Daohugou and Yanliao), are almost universally unknown in all other dinosaur fossils.

The auspicious "Feathered Dinosaurs of Liaoning," as they are commonly trumpeted in the popular press, have given paleontologists unprecedented insight into the origin, evolution, and development of feathers (Fig. 3.11). The feathers of living birds are complicated structures composed of keratin and exhibit a great amount of diversity. A typical feather is composed of a hollow calamus at its base, which usually attaches to skin or bone, a central rachis, and small barbs that project from the rachis to form the vanes (Stettenheim, 2000). Other feathers, such as the soft and fluffy down feathers, are much simpler but follow this same basic plan. Not all dinosaurs, however, possessed such complex structures. As might be expected, primitive theropods possessed simple feather-like integument and progressively more derived species – those that are closer and closer relatives to birds – developed more intricate feathers composed of a discrete calamus, rachis, and barbs. The variety of dinosaur feathers, and the inferred sequence of feather evolution on the line toward birds, is expertly summarized in a recent review by Xu and Guo (2009). Some of the more salient details of dinosaur feather morphology and evolution are described below, but this review paper, and the review of Prum and Brush (2002), should be consulted for further information.

It is likely that most, if not all, theropod dinosaurs were covered in feathers (Fig. 3.11). Nearly every well-preserved theropod specimen from the Jehol Group, as well as the Daohugou and Yanliao faunas, sports some type of feathery integument. The most primitive theropods in which feathers are unequivocally preserved are the basal coelurosaurs

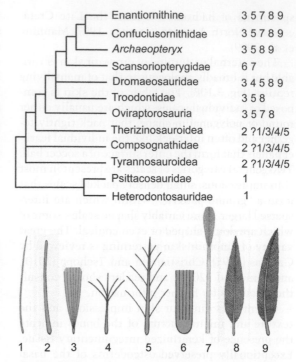

Figure 3.11 Dinosaur feathers. Generalized schematics of nine major morphotypes of feathers in dinosaurs (ranging from simple filaments to complex vaned feathers indistinguishable from those of living birds). The cladogram indicates the major feather morphotypes possessed by each major dinosaur group. Image courtesy of Dr Xu Xing and modified from Xu and Guo (2009).

Sinosauropteryx, a compsognathid, and *Dilong*, a tyrannosauroid (Chen et al., 1998; Xu et al., 2004). Their integument is simple in morphology and composed primarily of thin hollow filaments that range from a few millimeters to about 4 cm in length (Currie and Chen, 2001). More derived theropods, such as the therizinosauroid *Beipiaosaurus*, have larger and more complex filaments that were branched at their tips (Xu et al., 1999). Progressively more derived theropods, including oviraptorosaurs such as *Caudipteryx* and dromaeosaurids such as *Sinornithosaurus*, exhibit a greater degree of branching that is not limited solely to the tip and includes discrete branches arising along the length of a central filament. Finally, a handful of very close avian relatives, including some oviraptorosaurs, dromaeosaurids, and troodontids, possess truly pennaceous feathers that are differentiated into a central rachis and vanes composed of barbs (Norell et al., 2002; Xu et al., 2003, 2009a). Some of these feathers are basically identical in shape and size to those of modern birds, and are known to attach firmly to the bones of the forelimb just like in living avians (Turner et al., 2007b). Other feathers, however, exhibit their own unusual morphology, such as the elongate and branched "ribbons" of *Epidexipteryx* (Zhang et al., 2008), illustrating that there was great variety among the feathery integument of theropods.

Given that many theropods are now known to possess feathers of varying size, shape, and complexity, a new question beckons: how far back in dinosaur history did feathers originate? Emerging evidence intriguingly suggests that feathers were not a theropod innovation, but were also present in ornithischian dinosaurs, and perhaps even non-dinosaurs. This is a fascinating and active area of research, and new specimens are continuously pushing the origin of feathers further down the dinosaur family tree. A thin ridge on the ulna of the large allosauroid theropod *Concavenator* was recently interpreted as bony support for forearm feathers, akin to the quill knobs of living birds (Ortega et al., 2010). Although this identification is questionable, and the ridge may instead represent a muscle attachment site, if feathers were truly present in *Concavenator* then they must have evolved even earlier in theropod evolution than previously thought. More convincing are the thin, elongate, and presumably hollow filaments of the ornithischians *Psittacosaurus* (Mayr et al., 2002) and *Tianyulong* (Zheng et al., 2009). These structures are nearly identical to the filaments of *Sinosauropteryx* and *Dilong*, and the most reasonable conclusion is that these forms of integument are homologous. If this is the case, then feathers can be traced down the dinosaur family tree all the way to the common ancestor of Dinosauria, meaning that dinosaurs originated from a feathered species. But perhaps feathers can be traced even deeper in the phylogeny: filamentous integument has been reported in a few specimens of pterosaurs, the flying reptiles that are close relatives to dinosaurs (Bakhurina and Unwin, 1995; Kellner et al., 2009). It is unclear if these structures are homologous to the integument of ornithischians and theropods, but if so, then perhaps even the common ancestor of all archosaurs was feathered. There is no doubt that new discoveries will greatly clarify this picture in the near future.

The realization that many dinosaurs were feathered raises another important question: what function did feathers serve in these animals? Although the feathers of living birds are often assumed to be an adaptation for flight, the presence of feathers in obviously non-flying dinosaurs such as *Sinosauropteryx*, *Dilong*, and *Sinornithosaurus* is prime evidence that these structures must have evolved for another reason. It was not until the evolution of birds themselves, or perhaps their closest dinosaurian relatives, that feathers developed the more complex morphology and aerodynamic arrangement necessary for powered flight. It is likely that the simpler feathers of theropods, and likely ornithischians as well, functioned primarily in display and thermoregulation. Of these, display was probably more important. The haphazard distribution of filaments across the skeleton of many feathered theropods would not have provided an evenly distributed blanket to retain body heat. Instead, many of these filaments are concentrated around the head and along the back and tail, and are often modified into a fantastic array of shapes and sizes. It is likely, although not certain, that the diversity of feathery integument in early dinosaurs served the same purpose as the diversity in horn morphology among ceratopsians or crest shape in hadrosaurids: display, species recognition, or mate attraction.

Some of the most remarkable and fascinating research in contemporary dinosaur paleontology focuses on the color of extinct dinosaurs, inferred by microscopic examination of preserved feathers. Determining the color of dinosaurs, a long-standing subject of speculation and unbridled artistic license, was thought by most researchers to be impossible. Even the most delicately preserved dinosaur fossils were long ago stripped of their original soft tissues, leaving only the vaguest external traces of color such as dark and light bands that give little information on what the living animal would have actually looked like. The key to determining color, as it turns out, is in detailed microstructural analysis using the powerful technique of scanning electron microscopy (SEM). In living birds, microscopic structures in the feathers called melanosomes are packed with melanin, the pigment that grants feathers their exorbitant range of hues (Vinther et al., 2008, 2010). Melanosomes of different shapes and arrangements are diagnostic of certain types of melanin, and hence a certain color. Two research groups used SEM technology to analyze the feathers of several Mesozoic dinosaurs and were able to identify unequivocal melanosomes that are indicative of certain colors (Li et al., 2010; Zhang et al., 2010). For instance, the forelimb feathers of the troodontid *Anchiornis* were black and white and the tail of *Sinosauropteryx* was chestnut in color.

These discoveries are not mere novelties that will simply enable artists to draw dinosaurs in a more realistic fashion, but provide firm evidence that the feathers of dinosaurs had a fantastic variety of colors, lending support to the hypothesis that they were primarily display structures. Furthermore, the presence of unequivocal melanosomes, which are a very particular structure only found in certain types of tissues, provides incontrovertible proof that the integumentary structures of dinosaurs are true feathers, not simply degraded collagen fibers as some vocal critics had long been trumpeting (Lingham-Soliar, 2003; Lingham-Soliar et al., 2007).

Air Sacs

Postcranial air sacs

The fanciful vertebral laminae and fossae of saurischian dinosaurs can be so outlandish that it is difficult to imagine them serving any functional role. However, similar features are present in living birds, and are intimately related to a peculiar style of respiration (Duncker, 1971). Birds possess a "flow-through" lung in which the majority of air streams through the gas exchange tissues of the lung only in a posterior to anterior direction (Fig. 3.12). This is a fundamentally different system from the mammalian lung, through which air flows back and forth in a tidal cycle of inhalation and exhalation. Birds inhale and exhale, of course, but because air can flow in only one direction through most of the lung, it is necessary for inhaled air to be partitioned and stored within the body cavity. Some inhaled air immediately passes through the lungs during inhalation, and then flows into a series of

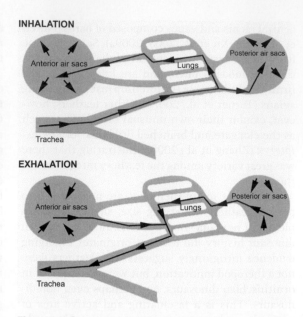

Figure 3.12 Schematic diagram illustrating the motion of air through the lungs and air sacs of a bird during inhalation and exhalation. Note that air moves across the gas-exchange tissues of the lungs only in an anterior direction, during both inhalation and exhalation. See text for further explanation. Original artwork modified from an image in the public domain.

balloon-like receptacles in front of the lungs, the anterior air sacs, before leaving the body during exhalation (Fig. 3.12). The remainder of the inhaled air flows first from the trachea into a series of posterior air sacs, which retain the air until exhalation, when it is passed through the lungs and back out through the trachea (Fig. 3.12). As a result of this complex system, oxygen-rich air flows through the lung during both inhalation (from the trachea) and exhalation (from the posterior air sacs). In essence, the series of air sacs, which are unique to birds and unnecessary in mammals and other animals with more traditional lungs, act as a set of bellows that partition and store air, help to maintain the rigidity of the thorax, and control airflow such that fresh air is able to continuously pass across the exchange tissues of the lungs (Fig. 3.13). It has long been thought that this intricate style of breathing is extremely efficient, because oxygen is captured throughout the breathing cycle, and therefore helps

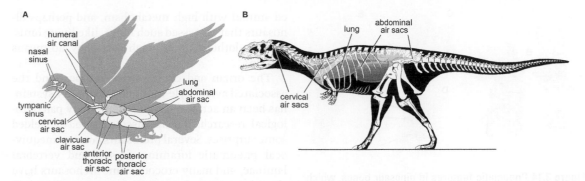

Figure 3.13 Diagrams showing the position and general morphology of pneumatic air sacs in a living bird (A) and the extinct large theropod *Majungasaurus* (B). Image (A) courtesy of Dr Paul Sereno and Carol Abraczinskas and modified from Sereno et al. (2008); image (B) courtesy of Dr Patrick O'Connor and modified from O'Connor and Claessens (2005).

power the metabolically expensive activity of flight (Perry, 1983; Brown et al., 1997; Maina, 2000).

The air sacs themselves are akin to balloons: they are soft, thin-walled, compliant structures that inflate and deflate during the ventilatory cycle. Often, small finger-like projections of the air sacs invade bone and occupy internal cavities in the vertebrae, shoulder girdle, forelimb, pelvis, and even the hindlimb. The reasons for this invasive process, called pneumatization, are poorly understood. The air sac projections that invade the skeleton are not pliant and do not help ventilate the lung, and it is thought that skeletal pneumatization may often be a means of lightening the skeleton and saving energy by replacing heavy and metabolically expensive bone with air. There are usually telltale signatures on the bone that unequivocally indicate the presence of air sacs. These include foramina on the external bone surface, which provide the entry for the air sac to invade the bone, the internal chambers within the bone itself, and fossae and laminae on the external surface that are indicative of an air sac nestling up against the outside of the bone. Important reviews of the morphology of air sacs and their bony correlates in living birds have been expertly provided by Duncker (1971) and O'Connor (2004, 2006, 2009), and these should be consulted for further information.

The presence of clear osteological correlates of air sacs in living birds enables the recognition of these soft tissue structures in extinct dinosaurs through identification of fossae, foramina, and laminae on the vertebrae (and, in some cases, other bones) (Figs 3.13 and 3.14). Only saurischians possess these bony signatures and, as far as current evidence suggests, they are completely absent in all ornithischians (Britt, 1997; Benson et al., 2011). Primitive saurischians, such as coelophysoid theropods and "prosauropods," have relatively simple pneumatic fossae and foramina that are usually restricted to the anterior cervical vertebrae (Britt, 1993, 1997; O'Connor and Claessens, 2005; Wedel, 2007; Benson et al., 2011). In living birds, foramina and fossae of similar size and position are associated with the cervical air sacs, which are among the anterior set of bellows as described above (O'Connor and Claessens, 2005). However, more derived theropods and sauropodomorphs develop increasingly complex pneumatic features across the entire cervical and dorsal column (Wedel, 2003a, 2003b; O'Connor and Claessens, 2005; O'Connor, 2006; Sereno et al., 2008; Benson et al., 2011). Pneumatic foramina and fossae are often larger, laminae are more intricate, and sometimes unequivocal foramina are present in the ribs, caudal vertebrae, pelvis, and shoulder girdle. Foramina and fossae within the posterior dorsal vertebrae and pelvic girdle of living birds are associated with the thoracic air sacs, some of the largest and most prominent air sacs in the posterior series (O'Connor and Claessens, 2005).

The inferred presence of air sacs in many dinosaurs, which is a strong deduction based on clear osteological correlates that can be examined in the EPB framework, has important implications for understanding dinosaur anatomy and physiology.

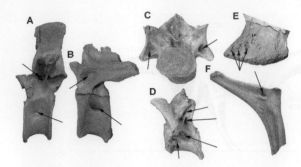

Figure 3.14 Pneumatic features in dinosaur bones, which indicate the presence of air sacs (either within or nestling against) these bones in life. Foramina and fossae related to air sacs are denoted by arrows. Vertebrae of the allosaur-oid theropod *Aerosteon* in left lateral view (A, B); a cervical vertebra of the tyrannosaurid theropod *Alioramus* in anterior (C) and left lateral (D) views; the pubic peduncle of the pelvis of *Aerosteon* (E); a dorsal rib of *Alioramus* (F). Images (A, B, E) courtesy of Dr Roger Benson and (C, D, F) © Mick Ellison.

At the simplest level, the presence of air sacs almost certainly implies the presence of an avian-style flow-through lung (Britt, 1997; O'Connor and Claessens, 2005; O'Connor, 2009; Benson et al., 2011). Other features of the dinosaur skeleton, especially the saurischian skeleton (but also the skeletons of ornithischians, which lack skeletal pneumaticity), are also consistent with this inference: the thorax is deep and robust, the thoracic vertebrae form a rigid column, the double-headed ribs are firmly joined to the vertebrae to produce an incompressible thorax, and the gastral basket provides an accessory pumping mechanism for changing the volume of the abdominal cavity (Schachner et al., 2009, 2011b). Taken together, these features provide the skeletal framework necessary for the unidirectional avian lung to function, because they provide precise volumetric control of the chest cavity (and hence air sacs) and a strong thorax to prevent lung collapse (O'Connor and Claessens, 2005). The presence of an avian-style lung in some, or all, dinosaurs may have physiological significance. The lungs of living birds are highly efficient, because oxygen is captured in both inhalation and exhalation and because they require less energy to operate (Perry, 1983; Brown et al., 1997; Maina, 2000). These are ideal for an energetic warm-blood-ed animal with high metabolism, and perhaps dinosaurs that possessed such lungs, like living birds, were endothermic (although this is by no means certain).

The origin and evolution of air sacs, and the associated avian-style lung and respiratory system, has been an active area of contemporary paleontological research. Recent discoveries have yielded some surprises. Several pterosaurs possess unequivocal pneumatic foramina, fossae, and vertebral laminae, and many crocodile-line archosaurs have clear laminae and possible pneumatic fossae on some vertebrae (Bonde and Christiansen, 2003; Nesbitt and Norell, 2006; Butler et al., 2009a; Claessens et al., 2009). Even some more primitive reptiles that fall outside the archosaur clade, such as the burly Triassic erythrosuchids, have a characteristic set of vertebral laminae that are nearly identical to those of dinosaurs and living birds (Gower, 2001). Most astonishingly, recent neontological research has demonstrated that living crocodiles, which have been dissected by anatomists and studied scientifically for several centuries, actually have a form of unidirectional breathing akin to that of birds, but without aid from air sacs (Farmer and Sanders, 2010). Therefore, it seems most likely that various elements of the avian-style breathing apparatus, including unidirectional lungs and air sacs, evolved piecemeal over hundreds of millions of years, beginning in reptiles that are only distantly related to birds. Surely our understanding of air sacs and the evolution of the avian style of respiration will continue to change with new fossil discoveries and research on living birds and crocodiles.

Cranial air sacs

One other type of air sac deserves comment. Many bones of the dinosaur skull are hollowed by internal sinuses, which were filled in life by fleshy air sacs that extended from the nasal cavity and tympanic system around the ear (Witmer, 1995b, 1997a, 1997b). These air sacs, which together comprise the paranasal and paratympanic sinus systems, were probably similar in structure to those of the postcranial skeleton, but they did not function in storing air and ventilating the unidirectional lungs. The most prominent sinus associated with the paranasal system was positioned adjacent to the antorbital fenestra, the large window-like opening between various bones of the snout that is a hallmark feature of archosaurs. Witmer (1997a) described in detail how this region, and many surrounding bones, were invaded by numerous extensions of the paranasal air sac, and dismissed previous ideas that the antorbital fossa housed enlarged jaw-closing muscles or a gland. The paratympanic system invades several bones of the braincase and the posterior upper and lower jaws (Witmer 1997b). The various hollow sinuses in this region have been extensively studied using computed tomography (CT) (Sampson and Witmer, 2007; Witmer and Ridgely, 2008, 2009), and they are clearly larger and spread throughout more bones in saurischians than in ornithischians. It is also likely that some bones of the skull were pneumatized by extensions of the pharyngeal sinus of the throat, although this is less certain (Witmer, 1997b).

Given that these air sacs did not function in respiratory gas exchange, it is possible that they may have played a role in thermoregulation (by dissipating heat) or served to reduce weight by replacing bone with air. Witmer (1997a, 1997b) persuasively argues, however, that many of these cranial air sacs, especially the complex and extremely variable sinuses of bird-like theropods, were likely the result of random opportunistic expansion of the sinuses. In other words, the nasal and tympanic sinuses of theropods were "pneumatizing machines" that advanced their way through the dinosaur skull in a capricious manner. Nevertheless, it is unclear why these sinuses are so extensive in derived theropods, but less elaborate in more primitive theropods and sauropodomorphs and often absent in ornithischians. Further research, especially CT scanning of ornithischian and basal saurischian skulls, should help determine the homologies of many of these sinuses, perhaps identify diminutive sinuses that were hitherto thought to be absent, and lead to a greater understanding of the function and evolution of the cranial air sacs.

Lungs, Hearts, and Internal Organs

Understandably the soft internal organs of dinosaurs are rarely fossilized, although a handful of spectacularly preserved specimens do retain impressions of organs such as the liver and intestines (Dal Sasso and Signore, 1998). One ornithischian specimen was trumpeted as preserving a three-dimensional fossilized heart, which apparently was divided into four chambers as in crocodiles and birds (Fisher et al., 2000), but careful examination of the specimen using CT scanning and histological thin sectioning reveals that it is more likely to be a non-biological concretion than a petrified heart (Rowe et al., 2001; Cleland et al., 2011). Some information on the organs of dinosaurs can be inferred by careful study of the skeleton and comparisons with living relatives. For instance, as outlined above, it is likely that all dinosaurs had a bird-like, rigid, unidirectional lung, as all dinosaurs possess several features of the skeleton instrumental in the functioning of such a lung, such as a rigid and incompressible ribcage, as well as air sacs (Schachner et al., 2009, 2011b).

Brain and Sense Organs

Because soft tissues readily decay after death, original fossil material of the brain, eye, ear, nose, and other organs involved in intelligence and sensory perception are wholly unavailable to paleontologists. Fortunately, many of these delicate organs were protected by bones of the skull, leaving a signature on the hard tissues that can be interpreted to give insight into the sensory capabilities of dinosaurs. Most importantly, the brain would have been enclosed within the bony braincase, and by examining the shape and size of the internal cavity that housed the brain scientists can gain valuable information on the anatomy of the brain itself. Pioneering paleoneurologists such as Osborn (1912) and Hopson (1977) would either have to slice open skulls to expose the brain region or rely on naturally preserved endocasts of the braincase cavity, usually formed when sediment washed into the brain cavity and hardened during the fossilization process. Recently, however, scientists have taken advantage of great advances in CT technology to digitally reconstruct endocasts (Fig. 3.15). This procedure has spread like wildfire through the field of dinosaur paleontology, and digital endocasts have been created and studied for many species of dinosaurs, including theropods (Rogers, 1998; Brochu,

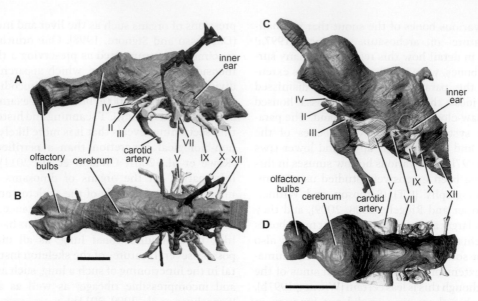

Figure 3.15 Dinosaur brains. Reconstructions of the brains of the theropod *Tyrannosaurus* (A, B) and the hadrosaurid *Hypacrosaurus* (C, D) based on CT scans. Images (A) and (C) in left lateral view, (B) and (D) in dorsal view. Roman numerals denote cranial nerves. Scale bars equal 5 cm. Modified from Witmer and Ridgely (2009) and Evans et al. (2009), and reproduced with permission.

2000, 2003; Franzosa and Rowe, 2005; Kundrát, 2007; Sampson and Witmer, 2007; Brusatte et al., 2009a; Witmer and Ridgely, 2009; Zelenitsky et al., 2009; Bever et al., 2011), sauropodomorphs (Sereno et al., 2007; Balanoff et al., 2010), and ornithischians (Zhou et al., 2007; Evans et al., 2009).

These studies have helped understand the morphology of the dinosaur brain, the intelligence and sensory capabilities of specific dinosaur taxa, and larger-scale patterns of brain and sense evolution over time. The brains of most dinosaurs probably did not completely fill the endocranial cavity, which is the case in living reptiles but not larger-brained extant birds (Jerison, 1973; Hopson, 1977). Cerebrospinal fluid, fatty tissue, and air sacs probably would have surrounded much of the dinosaur brain in life, perhaps rendering the preserved shape of the endocranial cavity a poor proxy for the shape of the brain as a whole. Those few regions of the brain that did fill the endocranial cavity, and thus nudged up against the bony braincase walls, imparted a complex vascular texture on the internal walls of the braincase (Osmólska, 2004). Recent research indicates that these heavily vascularized regions are actually quite common in dinosaur

braincases, suggesting that the dinosaur brain was larger than previously thought and, more importantly, that the size and shape of the endocast is probably a good approximation of the size and shape of the brain (Evans, 2005).

The shape of the dinosaur brain varied among groups: ornithischians, sauropodomorphs, and basal theropods generally possessed a brain that was similar in construction to that of living reptiles, whereas progressively more derived theropods exhibited increasingly more bird-like brains (Fig. 3.15). Regardless of the species in question, the most anterior region of the dinosaur brain comprised the olfactory bulbs and olfactory canals, which were confluent posteriorly with the cerebrum. In most dinosaurs the cerebrum was peaked, forming the highest point of the braincase, and the more posterior cerebellum and brainstem extended posteriorly and ventrally from this apex. Therefore, in many dinosaurs, the overall shape of the brain resembled an S or a Z. Derived theropods underwent a gradual change in shape, culminating in the more globular brain of living birds in which the olfactory tracts, cerebrum, and hindbrain are all on the same plane (Larsson et al., 2000; Domínguez Alonso et al.,

2004). The 12 major cranial nerves exited the brain laterally, anteriorly, and posteriorly, and their shapes and positions are often indicated by the presence of large stalks on the endocasts. The majority of these nerves, including those that controlled the facial muscles, viscera, and digestion, emerged from the hindbrain region. Also located in this region, and frequently preserved in endocasts, were the semicircular canals and other structures of the ear. The brain terminated posteriorly where the brainstem transitioned into the spinal cord and entered the neural canal of the vertebral column after emerging from the foramen magnum of the braincase.

The relationship between brain size and body size can give a rough indication of intelligence. Large brains do not simply indicate great intelligence, but body size must be taken into consideration: obviously, whales have physically larger brains than humans, but no scientist would argue that they are superior intellectually. Jerison (1973) outlined a simple equation that defines the "encephalization quotient" (EQ), which quantifies whether a species in question has a larger or smaller brain size than expected for its body size. Beginning with Hopson (1977), scientists have estimated the EQ of various dinosaurs, and the general conclusion is that relative brain size, and hence intelligence, became progressively larger in more derived theropods, culminating in the hefty brains and keen intelligence of birds (Hopson, 1977; Larsson et al., 2000). Much of this progressive enlargement occurred within the cerebrum, generally considered the seat of intelligence (Larsson et al., 2000). Sauropods had the lowest EQ among dinosaurs (Buchholtz, 1997), and many ornithischians had an EQ that was within the range of that of primitive theropods (Evans et al., 2009). The highest EQ calculated for a dinosaur belongs to the derived bird-like genus *Troodon*, whose relative brain size is within the range of that of modern birds (Currie, 1993).

The size of the olfactory bulbs relative to the remainder of the brain is a gauge of how acute the sense of smell was likely to have been. The olfactory bulbs of tyrannosaurids are enormous (Brochu, 2000; Witmer and Ridgely, 2009), and indeed a comparative quantitative study of olfactory bulb size among theropods found that tyrannosaurids

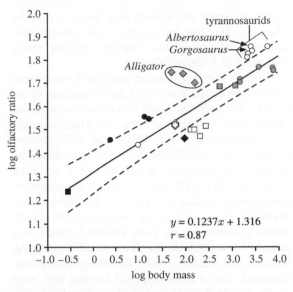

Figure 3.16 Dinosaur olfactory ability. A plot of olfactory bulb size relative to body size in theropod dinosaurs showing that tyrannosaurids and dromaeosaurids have larger olfactory bulbs (and therefore probably greater senses of smell) than other dinosaurs of similar size. From Zelenitsky et al. (2009).

and dromaeosaurids likely had keen senses of smell (Zelenitsky et al., 2009) (Fig. 3.16). Importantly, this study demonstrated that there was no progressive increase or decrease in olfactory acuity during most of theropod evolution, but rather that some groups had large olfactory bulbs, many had average senses of smell, and some groups had greatly atrophied olfactory bulbs (Fig. 3.16). Among the latter species are the ornithomimosaurs and oviraptorosaurs, two deviant theropod subgroups that developed an omnivorous or herbivorous diet, and therefore would not have required a sharp nose to detect prey or carrion. A follow-up study by Zelenitsky et al. (2011) showed that the relative size of the olfactory bulbs did increase during the evolution of the most derived non-avian theropods, culminating in the keen sense of smell of birds. Little information is available on olfactory acuity in other groups of dinosaurs, and comparative studies akin to that of Zelenitsky et al. (2009, 2011) are needed for sauropodomorphs and ornithischians. One bizarre feature of some ankylosaurs – a ludicrously complex three-dimensional looping of the internal nasal airway between the nostril and throat – may be related to olfaction, but its exact functional significance is debatable and requires further study (Witmer and Ridgely, 2008).

On the subject of hearing, studies of living birds and reptiles demonstrate that the dimensions of the bony cochlear ducts, which house the fleshy basilar papilla that controls hearing, correlate strongly with hearing sensitivity (Walsh et al., 2009; see also Gleich et al., 2005). These ducts can be isolated and measured in dinosaurs using micro-CT scanning technology (Fig. 3.17), and although comprehensive comparative studies like those of Zelenitsky et al. (2009, 2011) have yet to be conducted, some individual specimens have been analyzed. Based on a sample of several skulls that were CT scanned, Witmer and Ridgely (2009) illustrated that tyrannosaurids had a keen sense of hearing and were especially well equipped to hear low-frequency sounds. Evans et al. (2009) reported similar hearing capabilities for lambeosaurine hadrosaurs, and argued that these herbivores were ideally suited for perceiving the low-frequency sounds that were probably emitted by their ornate cranial crests (Weishampel, 1981; Sullivan and Williamson, 1999).

The orientation and development of the semicircular canals of the inner ear give insight into sense of balance, acuity of vision, and the alert head

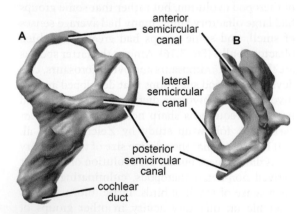

Figure 3.17 Dinosaur auditory anatomy. The semicircular canals of the inner ear of the hadrosaurid *Hypacrosaurus* in left lateral (A) and dorsal (B) views, as reconstructed by CT scans. Modified from Evans et al. (2009), and reproduced with permission.

posture an animal would employ (Fig. 3.18). When alert, living amniotes routinely orient their heads so that the lateral semicircular canal, one of three interconnected tubes that link to the cochlea, is horizontal (Hullar, 2006). Studies by Sampson and Witmer (2007) and Witmer and Ridgely (2009) report that most theropod dinosaurs held their heads at a slightly downturned angle, which would have been ideal for binocular vision, an important tool for an active predator. Many sauropods had a more pronounced downward orientation of their heads, perhaps related to their grazing lifestyle (Sereno et al., 2007; Balanoff et al., 2010). The most extreme condition is seen in *Nigersaurus*, whose head was held nearly perpendicularly to the neck, hypothesized to be an adaptation for browsing on low-lying vegetation (Sereno et al., 2007).

There is also a correlation between the size of the semicircular canals and visual acuity in living tetrapods, because the muscles that control visual focus as the animal moves are themselves controlled by the semicircular canals. Therefore, species in which the eyes are used to track predators or navigate complex environments (such as a forest canopy) usually possess robust and well-developed semicircular canals (Spoor et al., 2007). There has yet to be a broad comparative study of semicircular canal size, shape, and robusticity in dinosaurs, but it is worth noting that tyrannosauroids and other derived bird-like theropods have particularly elongate semicircular canals (Witmer and Ridgely, 2009). Perhaps these species had a particularly sharp sense of vision.

Information on visual acuity is also provided by the general shape of the skull and the size and shape of the sclerotic rings: the hoops of bone, composed of individual plates, which encircled and supported the eyeball of dinosaurs. Stevens (2006) examined the overall skull shape of theropod dinosaurs and found that all theropods were capable of some form of binocular vision, corroborating the evidence from semicircular canal orientation discussed above. However, he found that primitive theropods such as *Allosaurus* had a restricted form of binocular vision, similar to that of living crocodiles, in which there was only approximately a 20° overlap between the fields of vision of the individual eyes. More derived theropods, on the other hand, had wider fields of binocular vision (45–60° overlap)

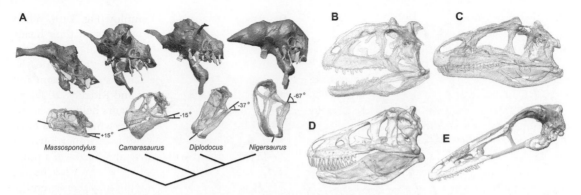

Figure 3.18 Head posture in dinosaurs. The hypothesized "alert" or "normal" head posture in sauropodomorph (A) and theropod (B–E) dinosaurs, which is the position the head assumes when the horizontal semicircular canal is held horizontally. Theropods include *Majungasaurus* (B), *Allosaurus* (C), *Tyrannosaurus* (D), and *Troodon* (E). Image (A) courtesy of Dr Paul Sereno and Carol Abraczinskas and modified from Sereno et al. (2007). Images (B–E) modified from Witmer and Ridgely (2009), and reproduced with permission.

similar to that in extant birds. Tyrannosauroids had particularly keen binocular vision, akin to that of a hawk, because the great increase in width of the posterior region of the skull allowed both eyeballs to face forward to a great degree. More recently, Longrich (2010) measured the sclerotic ring diameter in a handful of dinosaur specimens and compared the size of ring (relative to body size) in dinosaurs and living birds. One dinosaur in particular, *Protoceratops*, was found to have an enormous sclerotic ring, similar in relative size to that of extant noctural birds. It is possible therefore that *Protoceratops* may have been adapted for living in the dark. Most recently, Schmitz and Motani (2011) presented a broad comparative study of sclerotic ring dimensions in 19 non-avian dinosaurs and found that some species had the characteristic ring shapes of living diurnal (day-active) species, whereas others had the dimensions of nocturnal (night-active) or cathemeral (day and night-active) taxa. Dinosaurs therefore experimented with all the major types of daily activity strategies.

Microscopic Soft Tissues

Among the most astounding fossil discoveries of recent times has been the identification of numerous microscopic soft tissues – blood vessels, red blood cells, bone cells, and proteins – in a single spectacular specimen of *Tyrannosaurus rex* (Schweitzer et al., 2005a, 2007a, 2007b; Asara et al., 2007; Schweitzer, 2011; see also Embery et al., 2003 for similar tissues in a specimen of the ornithischian *Iguanodon*) (Fig. 3.19). Common wisdom has long held that soft tissues degrade during fossilization, and only in the most exceptional cases are they preserved by being replaced with hard mincrals (see Briggs, 2003 for a review of fossilization). What Mary Schweitzer and her colleagues discovered, however, was something quite different: non-mineralized, soft, still-pliable microscopic tissues that exhibit an uncanny resemblance to the vessels, cells, and proteins of living birds (Fig. 3.19). The putative blood vessels are not only soft and pliable, but also transparent and hollow, and they can be pulled and stretched just like the blood vessels of a living animal (Schweitzer et al., 2005a). The putative proteins were identified as collagen by the use of mass spectrometry, which found their chemical composition to be nearly identical to that of fresh collagen in birds and other living animals (Asara et al., 2007; Schweitzer et al., 2007a).

If these identifications are correct, they promise to offer radical new insight into the process of fossilization. Previous ideas about how vertebrate skeletal material is transformed into a fossil, and what tissues are actually preserved, may turn out to be incorrect. Furthermore, the identification of

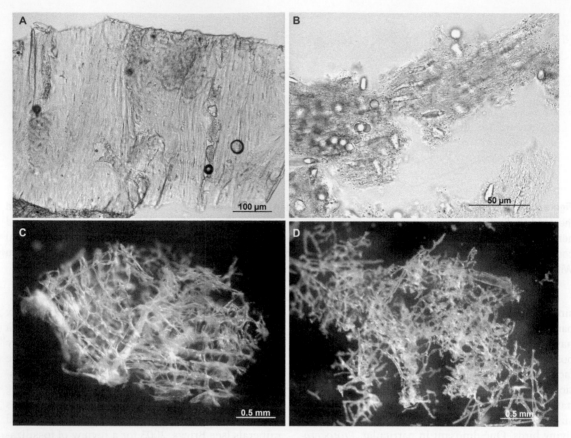

Figure 3.19 Microscopic soft tissues in a 66 million year old *Tyrannosaurus* (B, D) compared with similar tissues in an extant ostrich (A, C). (A, B) Soft tissues remaining after bone is demineralized, including osteocytes (transparent circles and ellipses); (C, D) interconnected and pliable blood vessels remaining after bone is demineralized. Images courtesy of Dr Mary Schweitzer and modified from Schweitzer et al. (2007b).

protein in dinosaur bones raises the exciting potential of conducting molecular phylogenetic analyses for extinct dinosaurs (Organ et al., 2008). In other words, the genealogy of dinosaurs could be examined not only by reference to skeletal anatomy, but also by analyzing molecular tissue, which is by far the most common method for studying living groups. However, as with any radical new scientific discovery that has the potential to shatter existing paradigms, caution is warranted when it comes to interpreting these soft tissue discoveries. Understandably, these findings have generated controversy, and it has been argued that some of the supposed soft tissue structures are simply remnants of bacteria (Kaye et al., 2008). It is essential that other laboratories and research groups confirm the find-

ings of Schweitzer and colleagues, and then take this research one step further by looking for similar tissues in a wide range of dinosaur species from across the Mesozoic, the globe, and the dinosaur family tree.

One of the greatest steps forward in identifying soft tissues and other original material of dinosaurs, studying the detailed microstructural anatomy of dinosaurs, and understanding the process of fossilization will probably come with advancements in synchrotron research. A synchrotron is a type of particle accelerator: a machine that uses electromagnetic fields to direct charged particles (such as protons and electrons) into beams. A simple example of a particle accelerator is a cathode ray tube in a televison set, but more sophisticated accelerators such as synchrotrons can be used as an ultra-powerful imaging device – akin to a supercharged X-ray machine or CT scanner – to study the detailed microstructure of materials, whether they are polymers for industrial use or dinosaur bones. Synchrotron X-ray tomography has already been used to examine the microstructure of a piece of ossified dinosaur tendon (Lonardelli et al., 2005), and was recently utilized to identify original chemical material in the fossilized feathers of *Archaeopteryx* (Bergmann et al., 2010) and the Mesozoic bird *Confuciusornis* (Wogelius et al., 2011). Synchrotron research is expensive, but as costs come down and access to the technology increases, it promises to be a rich new frontier in dinosaur research.

Conclusions

Although soft tissues are only infrequently preserved as fossils, the identification of osteological correlates (parts of the bony skeleton that anchor a certain soft tissue) and careful study of living dinosaur relatives enable scientists to understand the muscles, skin, feathers, air sacs, brains, and sense organs of dinosaurs in great detail. New technologies, especially CT scanning and SEM, have also proved to be extremely valuable, by allowing researchers to digitally peer into the bones of dinosaurs and identify the chemical composition and microstructure of tissues. The major skull and postcranial muscles have been reconstructed for many groups of theropod dinosaurs, but are comparatively more poorly understood in sauropodomorphs and ornithischians. Similarly, the brain morphology, inferred intelligence, and sensual acuity of a handful of dinosaur species have been examined in detail, but few large-scale comparative studies have assessed the development and evolution of intelligence and sense perception across the dinosaur family tree. It is now undeniable that many dinosaurs sported feathers, and even the colors of some species have been reconstructed by identifying diagnostic microscopic structures (melanosomes) that bear certain types of pigment. Reconstruction of dinosaur color, along with the identification of pliable microscopic soft tissues such as blood vessels and proteins, are currently some of the most exciting fields of dinosaur research.

4 Phylogeny

Building a family tree, or phylogeny, for dinosaurs is a major goal of current research. It seems like nearly every new paper on dinosaurs, especially the description of new species, includes a cladogram: a graphical representation of the genealogy of a group of organisms. Paleontologists have long been interested in reconstructing phylogeny for the same reason that many people are keen to piece together their own family genealogies. Having some concept of relationships is an essential framework for understanding large-scale patterns of evolution, whether it is the evolution of dinosaurs during the Mesozoic or the changes that occurred as a human family expanded, moved around, and merged with other families through marriages.

For those paleontologists interested in the evolution of dinosaurs, a phylogeny is necessary to answer many important questions. What type of animals did dinosaurs evolve from? Are there trends of increasing body size, more refined carnivorous habits, or more ornate cranial crests during the evolution of a certain subgroup? Did interesting and functionally important characters, such as feathers or leaf-shaped teeth for herbivory, evolve once or multiple times? Even one of the most supreme discoveries in the history of dinosaur paleontology – the evolution of birds from carnivorous dinosaurs – is fundamentally a genealogical hypothesis. Understanding phylogeny, therefore, is critically important for understanding the biology and evolution of dinosaurs. It is no surprise that dinosaur paleontologists are keenly interested, some might say obsessed, with phylogeny reconstruction. In this chapter I will review how paleontologists build family trees and present an outline of the major patterns of dinosaur genealogy.

Building the Dinosaur Family Tree: Reconstructing Phylogenies

Dinosaur paleontologists have always been interested in building and analyzing genealogical trees, but for most of the 20th century there was no standard, explicit, defensible method for reconstructing phylogeny. Researchers would usually draw a putative family tree by hand, connecting fossil species in ancestor–descendant sequences based on the temporal ordering of fossils in the geological record and some intuition about which species were most similar anatomically. Building phylogenies, and choosing between alternative family trees, was more of an exercise in art than science, because genealogies were simply asserted rather than constructed based on a repeatable methodology.

This practice came to a halt in the 1960s and 1970s, as a revolution in the philosophy and methodology of phylogeny reconstruction swept through the field of biology. Willi Hennig, a German entomologist, presented an explicit and standardized protocol for building genealogies. Now commonly referred to as "cladistics," Hennig's method (1965, 1966) holds that groups of species can be recognized as closely related by their possession of shared derived characters. Only derived characters, which represent an explicit evolutionary change from a primitive condition, are useful in grouping organisms. Cladistics therefore demands the identification of specific features of the anatomy (or other biological attributes) that vary (i.e., have two or more conditions, called "states") and the determination of which variable states are primitive and derived. This is done with reference to an outgroup, a species that does not belong to the group being studied but whose attributes are assumed to be primitive. For instance, dinosaurs can be regarded as a single group of organisms that evolved from a common ancestor – a "clade" in cladistic terminology – because they all possess a perforated acetabulum, which is a derived character that represents a discrete evolutionary change from the primitive solid acetabulum of other reptiles. The solid condition is assumed to be primitive because it is present in outgroup species, such as crocodiles and lizards.

Large family trees are built by compiling long lists of variable characters, using an outgroup to determine which variable states are primitive and derived, and then grouping species in hierarchically nested clades based on the presence of shared derived characters. Following from the above example, dinosaurs are recognized as a clade based on their perforated acetabulum, and within dinosaurs ornithischians are recognized as a clade based on their possession of a retroverted pubis, a derived condition that differs from the forward-pointed pubis of outgroup taxa such as crocodiles and lizards, as well as saurischian dinosaurs. By proceeding this way a branching cladogram is produced, illustrating the nested relationships between various dinosaur species. An example is shown in Fig. 4.1, which depicts the genealogical relationships of four species of dinosaur and one outgroup (a crocodile), based on a sample set of variable characters shown in Fig. 4.2.

On the cladogram, each species is denoted as a branch on the tree, and the most recent common ancestor of two species can be identified by tracing both branches to their common meeting point (called a "node"). In the example, the most recent common ancestor of *Tyrannosaurus* and *Brachiosaurus* is labeled "Node 1," whereas the most re-

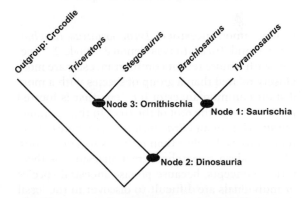

Figure 4.1 A simplified cladogram (genealogical tree) of dinosaurs showing the major clades. Saurischia and Ornithischia are sister taxa – each other's closest relatives – and together comprise Dinosauria. Dinosauria, in turn, is sister taxon to crocodiles in this diagram (as this is a simplified diagram, several species more closely related to dinosaurs than are crocodiles, such as pterosaurs and non-dinosaurian dinosauromorphs, are omitted).

Character–Taxon Matrix

Characters:
1) Acetabulum, presence of medial wall of bone: present, acetabulum closed (0); absent, acetabulum open (1).
2) Pubis, orientation of shaft: anteroventral (0); posteroventral, pubis is "retroverted" (1).
3) First digit of manus, orientation: in line with remaining digits (0); twisted and projects medially relative to other digits (1).

Characters as scored in various species:

Crocodile	000
Stegosaurus	110
Triceratops	110
Brachiosaurus	101
Tyrannosaurus	101

Figure 4.2 An example of a character–taxon matrix used in a phylogenetic analysis. The example shows three characters, each with a primitive state (denoted as "0") and a derived state (denoted as "1"). The "0" state is assumed to be primitive because it is present, in each case, in the outgroup (the crocodile). The derived state of character 1 is present in all the dinosaurs in the analysis, and therefore is a shared derived character (synapomorphy) of Dinosauria. The derived state of character 2 is present in the two ornithischians in the analysis, and is therefore a synapomorphy of Ornithischia. The derived state of character 3 is present in the two saurischians in the analysis, and is therefore a synapomorphy of Saurischia. A dataset such as this would be input into a computerized phylogenetic program, and a parsimony algorithm (in this case) would return the cladogram shown in Figure 4.1.

cent common ancestor of *Tyrannosaurus*, *Brachiosaurus*, and *Triceratops* is denoted "Node 2." Species with a more recent common ancestor are more closely related than a group of species with a more distant common ancestor (i.e., one that is located further toward the root of the tree). In this example, *Tyrannosaurus* and *Brachiosaurus* are more closely related to each other than either is to *Triceratops*. Most modern researchers treat ancestors as theoretical concepts, because precise ancestral species or individuals are difficult to discover in the fossil record. Ancestor–descendant lineages are therefore usually not depicted on cladograms, and instead the focus is on degree of common ancestry: which species are more closely related to each other than to other species. Finishing with the example, *Tyrannosaurus* and *Brachiosaurus* are each other's closest relatives, whereas *Triceratops* and *Stegosaurus* are more closely related to each other than

to any other species. All four dinosaurs, however, are more closely related to each other than any one of them is to the crocodile.

It is straightforward to reconstruct the phylogenetic relationships of five species, and to understand the major patterns expressed by their cladogram. However, how do researchers construct phylogenies for 10, 20, or even 100 species? And how do researchers choose between alternative phylogenetic trees? After all, a cladogram is not a revealed and absolute fact, but simply a hypothesis that may change with new discoveries and interpretations of fossils. How do researchers decide that one hypothesized tree is better than another? Both of these procedures – constructing the trees themselves and deciding between alternative trees – are performed with the help of computer algorithms. Comprehensive and helpful reviews of these algorithms, and the step-by-step procedures necessary to build cladograms, have been provided by Kitching et al. (1998), Felsenstein (2003), and Schuh and Brower (2009). These books should be consulted for further details, but the major steps in phylogeny reconstruction are outlined below.

First, it is necessary for the researcher to compile a list of characters that vary (have two or more states), identify an outgroup that determines (polarizes) which states are primitive and which are derived, and score each species for either the primitive or derived state for each character (Fig. 4.2). The end result of this exercise is a spreadsheet, or data matrix, that can be fed into a computer program such as PAUP or TNT (Swofford, 2000; Goloboff et al., 2008). These programs use search algorithms to build many possible cladograms and then decide which cladograms are best by employing an optimality criterion. The most common optimality criterion for dinosaur phylogenetic analyses is parsimony: those cladograms that minimize the total number of transitions from primitive to derived states among all characters are favored. Other optimality criteria are frequently employed in phylogenetic analyses of living taxa based on DNA sequence data, not anatomical characters as is usually the case with extinct species. These include maximum likelihood and Bayesian inference, both of which assume a model of evolution (some predetermined assumption about how fast and often characters change between primitive and derived states) and select cladograms that best fulfill the predictions of this model (Felsenstein, 2003). Bayesian analysis goes one step further and also takes into account so-called "prior probabilities" – information, assumed a priori, that certain cladograms are more likely than other cladograms – when selecting the most optimal trees (Rannala and Yang, 1996; Huelsenbeck et al., 2001; Archibald et al., 2003).

The most important fact to remember about cladograms is that they are not facts, but hypotheses. An optimal cladogram, identified as such because it best fulfills an optimality criterion such as parsimony or maximum likelihood, is not necessarily true. It is simply the best hypothesis available based on the evidence at hand. That evidence, however, is always open to change. The discovery of a single new fossil can reveal new characters that, in turn, can substantially alter the topology of the optimal phylogeny. Similarly, researchers are constantly studying and restudying specimens, and ideas about character variability and the identification of primitive and derived states frequently change as new eyes pour over the available evidence. Indeed, it has recently been shown that such disagreements about how to identify, polarize, and score characters lead to substantial differences in the reconstructed phylogenies of certain dinosaur groups (Brusatte and Sereno, 2008; Sereno and Brusatte, 2009). Researchers should always be explicit about the characters they are using, how the primitive and derived states are determined, and which computer algorithms and optimality criterion are used to derive their phylogeny.

The Closest Relatives to Dinosaurs

Dinosaurs are but one subgroup of Archosauria, the clade of "ruling reptiles" that arose soon after the Permo-Triassic mass extinction and survives today in the form of over 10,000 species of birds and crocodilians. The broad-scale phylogeny of Archosauria has been addressed by several recent phylogenetic analyses, including studies by Sereno (1991a), Juul (1994), Benton (1999, 2004), Nesbitt (2007, 2011), Irmis et al. (2007a), Brusatte et al. (2010a), and Nesbitt et al. (2010, 2011). These analyses are important because they identify the closest relatives of dinosaurs and give insight into what the immediate ancestors of dinosaurs would have looked like. Archosauria is divided into two major clades: Crurotarsi (the "crocodile line") and Avemetatarsalia (the "bird line") (see Fig. 1.5). Crurotarsi includes living crocodiles and all archosaurs more closely related to them than to birds, among which are several extinct Mesozoic subgroups such as the phytosaurs, aetosaurs, and rauisuchians. Avemetatarsalia, which is essentially an equivalent group to the more familiar clade Ornithodira (sometimes, because of historical legacy, the same clade has multiple names), includes all archosaurs more closely related to birds than crocodiles. Dinosaurs and pterosaurs fall into this group, along with a handful of close dinosaur relatives that are colloquially referred to as "basal dinosauromorphs."

The closest basal dinosauromorph relatives to true dinosaurs include taxa such as *Lagerpeton*,

Before moving forward, it is important to summarize several important terms that researchers use when describing phylogeny and cladograms. Remember that a clade is a group that comprises an ancestor and all of its descendants. Clades are said to be "natural groups," or "monophyletic," whereas paraphyletic grades are assemblages of species that do not include an ancestor and all descendants. Usually, such grades comprise a series of successively more closely related species to a more derived clade, but not that derived clade itself (as an example, *Triceratops*, *Stegosaurus*, and *Brachiosaurus* comprise a grade relative to *Tyrannosaurus* in Fig. 4.1). The distinction between clade and grade is visualized in Fig. 4.3. Sister taxa, or sister clades, are each other's closest relatives. In the above example, *Tyrannosaurus* and *Brachiosaurus* are sister taxa, and *Triceratops* and *Stegosaurus* are also sister taxa. Furthermore, the *Tyrannosaurus* + *Brachiosaurus* clade is the sister taxon to the *Triceratops* + *Stegosaurus* clade. An outgroup is a relative term, denoting a species or clade that is outside the designated group being studied. For instance, *Triceratops* and *Stegosaurus* are outgroups to the *Tyrannosaurus* + *Brachiosaurus* clade. *Tyrannosaurus* and *Brachiosaurus* in turn are outgroups to the *Triceratops* + *Stegosaurus* clade. Clades are identified by the possession of shared derived characters, which are properly referred to as synapomorphies. Primitive characters, which are not useful in designating clades, are called plesiomorphies.

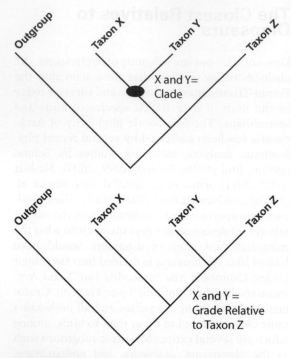

Figure 4.3 The difference between a clade and a grade. In the top cladogram Taxon X and Taxon Y form their own unique cluster – a monophyletic group – relative to all the other species. Therefore, X and Y form a clade. In the bottom cladogram X and Y do not form their own unique group, but are rather successive outgroups on the line to Taxon Z. Therefore, X and Y form a grade relative to Z.

Marasuchus, and *Silesaurus*, which are all small sleek animals that rarely reached sizes larger than a human (see Figs 1.6 and 4.4). Their phylogenetic relationships are shown in Fig. 1.11. The most

primitive dinosauromorphs are *Lagerpeton* and *Dromomeron*, two dimunitive genera that lived in South America and North America during the Middle and Late Triassic, respectively (Romer, 1971; Sereno and Arcucci, 1993; Irmis et al., 2007a). These basal dinosauromorphs were about the size of a house cat, and although no fossils of their forelimbs are currently known, the footprints of similar species indicate that they were quadrupedal (Haubold, 1999; Brusatte et al., 2011a) (see Fig. 1.7). More closely related to dinosaurs is another small, aberrant genus, *Marasuchus*, whose terrier-sized skeleton is well suited for a saltatorial, or hopping, style of locomotion akin to a rabbit (Romer, 1971; Sereno and Arcucci, 1994). The closest relatives of true dinosaurs comprise a speciose subclade of basal dinosauromorphs, Silesauridae, which includes the genera *Silesaurus*, *Sacisaurus*, *Asilisaurus*, *Lewisuchus*, and *Pseudolagosuchus* (Nesbitt et al., 2010) (see Figs 1.6 and 4.4). Recognition of this major subclade occurred only recently, with the discovery of *Silesaurus* (Dzik, 2003) and the realization that several other new specimens, as well as fragmentary fossils that had long confused scientists, belonged to close relatives. Most silesaurids were quadrupedal and herbivorous, with a small horny beak at the front of the jaws for cropping vegetation (Dzik, 2003; Ferigolo and Langer, 2007; Nesbitt et al., 2010).

The identification and appreciation of the closest relatives to dinosaurs gives insight into the origination and early evolution of the dinosaurian clade. All basal dinosauromorphs are small animals and many of them were quadrupedal. This suggests that the immediate ancestors of dinosaurs were also small species that walked on four legs, and that

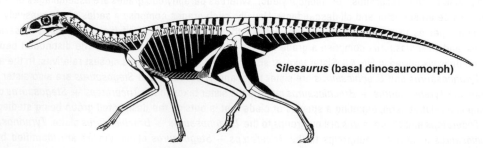

Figure 4.4 Skeletal reconstruction of the non-dinosaurian dinosauromorph *Silesaurus*, one of the closest relatives to dinosaurs. Reconstruction © Scott Hartman.

large body size and bipedality likely evolved later in dinosaur history, after the origination of true dinosaurs themselves (see Brusatte et al., 2011a). Furthermore, the herbivorous diet of most silesaurids, the closest relatives to dinosaurs, provocatively hints that the common ancestor of all dinosaurs may have fed on plants. Although it has long been assumed that the common ancestor of dinosaurs was a carnivore, a herbivorous diet would not be completely unexpected given that two of the major dinosaur subgroups, ornithischians and sauropodomorphs, were primarily herbivorous. Dietary evolution among early dinosaurs is currently an exciting, but unsettled, area of research, and new discoveries promise to help untangle this mystery (Barrett et al., 2011a).

The Earliest Dinosaurs: Saurischians, Ornithischians, and Uncertain Species

True dinosaurs are divided into two major subclades, Saurischia and Ornithischia, each of which is subdivided into numerous smaller subgroups, as was briefly reviewed in Chapter 1 (see Fig. 1.21). There are also a few species whose relationships have been the subject of much controversy, most prominently *Herrerasaurus* and *Eoraptor* from the Late Triassic of South America (Fig. 4.5). These genera, which are among the oldest dinosaurs yet discovered, were once thought to fall immediately outside of the Saurischia + Ornithischia clade, and thus represent the closest relatives to dinosaurs

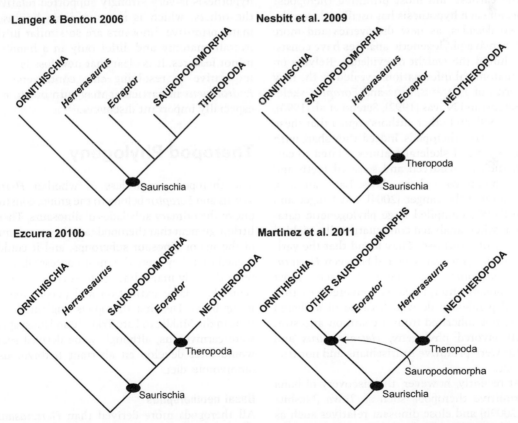

Figure 4.5 Four alternative hypotheses for the phylogenetic position of the Late Triassic dinosaurs *Herrerasaurus* and *Eoraptor*. Each hypothesis has been presented recently in the literature, but none is particularly well supported relative to the others. The genealogical position of these two dinosaurs will likely remain an active area of debate for some time.

rather than true dinosaurs by definition (Gauthier, 1986; Brinkman and Sues, 1987; Novas, 1992; Fraser et al., 2002). The discovery of important new specimens, including spectacularly preserved skeletons of both genera (Sereno and Novas, 1992; Sereno et al., 1993) and the fossils of several close relatives, have helped crystallize strong support for not only the dinosaurian affinities of *Herrerasaurus* and *Eoraptor*, but also their placement within Saurischia (Novas, 1996; Sereno, 1999; Langer, 2004; Langer and Benton, 2006; Irmis et al., 2007a; Nesbitt et al., 2009b, 2010; Ezcurra, 2010a; Nesbitt, 2011).

However, debate remains regarding the more precise phylogenetic placement of *Herrerasaurus* and *Eoraptor* (Fig. 4.5). Are they basal members of the saurischian clade that evolved before the split between theropods and sauropodomorphs, or are they the earliest and most primitive theropods? Support for each hypothesis has oscillated over the past two decades, as new discoveries and more comprehensive phylogenetic analyses have consistently shifted the weight of evidence. Relying on novel anatomical information revealed by the new discoveries of *Herrerasaurus* and *Eoraptor* skeletons, Sereno and Novas (1992), Sereno et al. (1993), Sereno (1999), and other authors argued that these genera were true theropods. Indeed, they share with theropods several skeletal features related to carnivory, such as recurved and serrated teeth and elongate hands with sharp claws. Later authors, most prominently Langer (2004) and Langer and Benton (2006), compiled large phylogenetic datasets that, when analyzed computationally, yielded something of a surprise. They found that the various carnivorous features shared between *Herrerasaurus*, *Eoraptor*, and theropods were most parsimoniously interpreted as convergences that were independently derived because of a shared lifestyle, not inherited from a common ancestor. In their favored phylogeny, *Herrerasaurus* and *Eoraptor* were primitive saurischians and not true theropods.

Most recently, however, the discovery of bona fide primitive theropods such as *Tawa* (Nesbitt et al., 2009b) and close dinosaur relatives such as silesaurids (Dzik, 2003; Nesbitt et al., 2010) have enabled the construction of even larger and more comprehensive phylogenetic datasets (Nesbitt

et al., 2009b, 2010; Ezcurra and Brusatte, 2011; Nesbitt, 2011). These analyses mostly agree that *Herrerasaurus* and *Eoraptor* are true theropods, as originally argued by Sereno and colleagues in the early 1990s. However, Ezcurra (2010a) reported a novel result in which *Eoraptor* was recovered as a true theropod but *Herrerasaurus* as a more basal saurischian. Perhaps even more surprising, Martinez et al. (2011) placed *Herrerasaurus* and a new carnivorous genus represented by a nearly complete skeleton, *Eodromaeus*, as basal theropods, but *Eoraptor* as the most primitive sauropodomorph (based on its possession of a large external naris, teeth with coarse denticles, and other features commonly seen in basal sauropodomorphs). It is premature, therefore, to consider this debate settled. Fluctuation between which hypothesis is favored is due to a simple fact: no hypothesis is very strongly supported relative to the others, which is not surprising because so many primitive dinosaurs are so similar in their overall anatomy and differ only in a handful of minor features. It is clear that new fossils will be imperative in resolving this controversy, and *Eodromaeus* in particular may turn out to be an especially important discovery.

Theropod Phylogeny

The theropods, regardless of whether *Herrerasaurus* and *Eoraptor* belong to the group, constitute one of the primary subclades of dinosaurs. There is little argument that theropods are the most familiar of the major dinosaur subgroups, and it could be argued that they were the most successful, as they were the only major lineage to survive the Cretaceous–Paleogene extinction and persist today as living birds. The first theropods originated sometime in the Middle to Late Triassic and most species were carnivorous, although some derived species would later develop an aberrant herbivorous or omnivorous diet.

Basal neotheropods

All theropods more derived than *Herrerasaurus*, *Eoraptor*, and *Eodromaeus*, if indeed they are theropods, constitute a subclade called the Neotheropoda (Fig. 4.6). Most primitive neotheropods can be

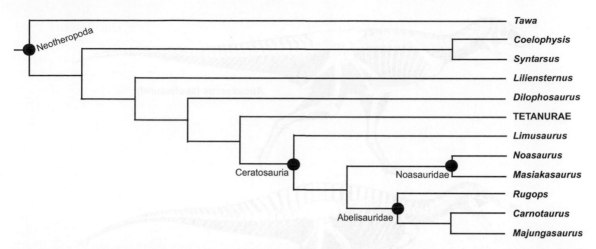

Figure 4.6 A generalized cladogram of Neotheropoda (the group of theropods exclusive of *Herrerasaurus* and/or *Eoraptor*). This is not a complete genealogy of primitive theropods, but shows a handful of representative species and denotes major subclades. Cladogram based on references discussed in text.

generally divided into two clusters, which may or may not represent actual clades that include an ancestor and all descendants. Several sleek, mostly small Late Triassic to Early Jurassic theropods are grouped together as "coelophysoids." The most recognizable coelophysoid is *Coelophysis*, a common discovery in the Late Triassic rocks of the western United States (Colbert, 1989; Rinehart et al., 2009), and other members of this cluster include the mid-sized *Liliensternus* from the Late Triassic of Germany, the double-crested *Dilophosaurus* from the Early Jurassic of North America, and *Syntarsus* from the Early Jurassic of South Africa. The other major cluster of basal neotheropods, the ceratosaurs, comprises the horned *Ceratosaurus* from the Late Jurassic, the svelte noasaurids that mostly lived during the Cretaceous, and the peculiar stubby-armed and bull-headed abelisaurids, the most common large predators on the southern continents during the Late Cretaceous (Fig. 4.7). Well-known abelisaurids include *Carnotaurus*, *Majungasaurus*, and *Rugops*, and the largest species reached up to 10 m in length.

The phylogenetic relationships of these basal neotheropods are currently in flux (Fig. 4.6). Some of the first cladistic analyses focusing on theropods found support for a monophyletic group uniting all coelophysoids and ceratosaurs, which was usually called Ceratosauria (Gauthier, 1986; Rowe and Gauthier, 1990; Holtz, 1994, 2000; Sereno, 1999; Tykoski and Rowe, 2004). More recent studies, however, have persuasively argued that ceratosaurs, which clearly do comprise their own monophyletic subgroup, are more closely related to derived theropods (such as birds) than they are to coelophysoids (Carrano et al., 2002; Rauhut, 2003; Wilson et al., 2003; Sereno et al., 2004; Carrano and Sampson, 2008; Xu et al., 2009b). In other words, ceratosaurs and coelophysoids do not form their own unique clade, but rather ceratosaurs are more derived and coelophysoids positioned closer to the base of the theropod cladogram. This fundamental change in the topology of the cladogram occurred because several derived features once thought to uniquely characterize coelophysoids and ceratosaurs, such as fusion in the bones of the lower leg, are now known to be more widely distributed among theropods.

Furthermore, it is uncertain whether coelophysoids constitute their own monophyletic group or, alternatively, whether they are simply a graded series of outgroup species on the line to the more derived group of ceratosaurs and other theropods. The most recent large-scale phylogenetic analysis sheds doubt on coelophysoid monophyly, and recovers *Tawa*, *Coelophysis*, *Liliensternus*, *Cryolophosaurus*, and *Dilophosaurus* as progressively more closely related to more derived theropods

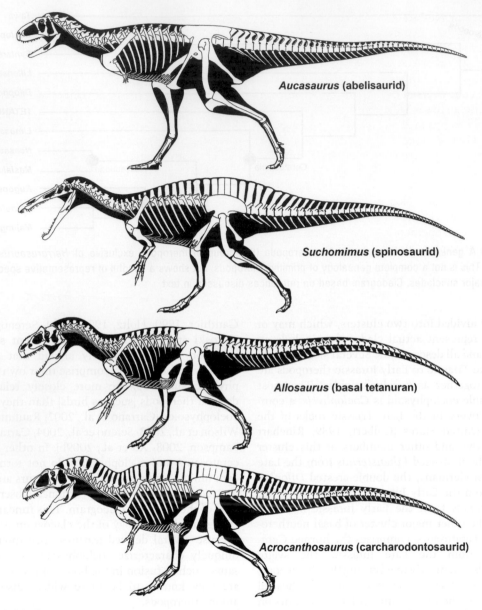

Figure 4.7 Skeletal reconstructions of four basal (non-coelurosaurian) theropod dinosaurs. Reconstructions © Scott Hartman.

(Nesbitt et al., 2009b). Similar to the case with Ceratosauria described above, this topological change resulted from the discovery that several supposed derived characters that united all coelophysoids, such as a gap between the teeth of the premaxilla and maxilla, are actually primitive features of theropods, and therefore not useful in grouping species into their own monophyletic subclade. A non-monophyletic coelophysoid group, however, is not a certainty. This result falls apart, and a more traditional monophyletic Coelophysoidea is recovered, with the addition of new species to the analysis of Nesbitt et al. (2009b) and subtle alterations to their character scores (Ezcurra and Brusatte, 2011).

Compared with the phylogeny of coelophysoids, the genealogical relationships of ceratosaurs are much less controversial (Fig. 4.6). Comprehensive phylogenetic analyses were recently presented by Carrano and Sampson (2008) and Xu et al. (2009b), and these broadly agree that the ceratosaur clade can be divided into three main clusters: a primitive assemblage of species, including *Elaphrosaurus* and *Limusaurus*, which form successive outgroups to two major subclades, Noasauridae and Abelisauridae. Noasauridae is one of the most poorly understood theropod subgroups, largely because only a few fossil specimens are known. This material, most notably the partial skeletons of the sleek and sickle-clawed *Noasaurus* and the buck-toothed *Masiakasaurus*, suggestively hint that noasaurids were strange small animals that filled a variety of ecological roles (Bonaparte and Powell, 1980; Sampson et al., 2001). Abelisaurids, on the other hand, are becoming increasingly better understood with the

discovery of new specimens, especially from the Late Cretaceous of South America (Novas, 2009) (Fig. 4.7). This group was first recognized less than three decades ago, with the discovery of the skull of *Abelisaurus* and the largely complete skeleton of *Carnotaurus* (Bonaparte, 1985; Bonaparte and Novas 1985). Today, more than 10 abelisaurid species have been discovered, including representatives from South America, Africa, India, Madagascar, and perhaps Europe. Some of these were apex predators and had short, deep, heavily ornamented skulls optimized for delivering strong bite forces (Mazzetta et al., 2009).

Basal tetanurans

All theropods more derived than coelophysoids and ceratosaurs constitute a subclade called Tetanurae (Gauthier, 1986) (Fig. 4.8). Tetanurans are further divided into three major subgroups: Megalosauroidea, Allosauroidea, and Coelurosauria. Allosauroids and coelurosaurs are sister taxa – each other's closest relatives – whereas Megalosauroidea is a more primitive clade. Regardless, because

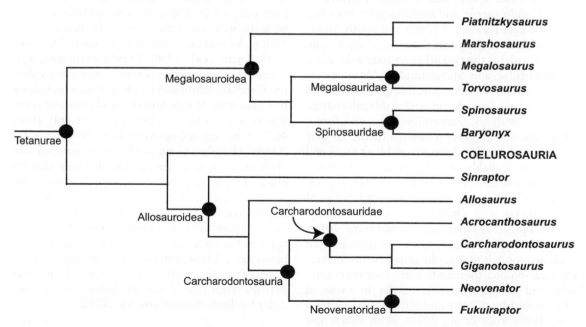

Figure 4.8 A generalized cladogram of Tetanurae (the group of theropods more derived than coelophysoids and ceratosaurs). This is not a complete genealogy of tetanuran theropods, but shows a handful of representative species and denotes major subclades. Cladogram based on references discussed in text.

coelurosaurs include birds and receive the brunt of attention from scientists who work on theropods, megalosauroids and allosauroids are collectively referred to, somewhat dismissively, as "basal tetanurans." Until very recently, little was known about the phylogeny of these basal taxa, which were largely ignored as theropod specialists focused preferentially on the relationships of the oldest and most primitive theropods or the most derived bird-like coelurosaurs. Over the past decade, however, these "middle children" of the theropod family tree have been studied in great detail, and understanding of their phylogenetic relationships has begun to stabilize.

The interrelationships of Megalosauroidea, which was often previously called Spinosauroidea in the literature, have been studied in detail by Benson (2010), whose analysis was later followed up by the slightly more inclusive study of Benson et al. (2010a). Buoyed largely by a detailed osteological redescription of the long-problematic and fragmentary genus *Megalosaurus*, Benson's (2010) phylogenetic analysis identified three principal megalosauroid subclades (Fig. 4.8). First, there is a cluster of basal forms that includes *Piatnitzkysaurus*, *Condorraptor*, and *Marshosaurus* from the Middle to Late Jurassic of North and South America. This clade is the sister taxon to a larger clade that encapsulates the two other primary subclades, Megalosauridae and Spinosauridae. Megalosauridae contains several once-puzzling genera from the Middle to Late Jurassic, including *Megalosaurus*, *Eustreptospondylus*, *Dubreuillosaurus*, and *Torvosaurus*. These are mostly mid-sized theropods that generally had slender skeletons and long arms and which were probably the apex predators in many Middle Jurassic ecosystems worldwide. Finally, Spinosauridae is one of the most peculiar subgroups of theropods (Fig. 4.7). Only four genera are known from well-preserved material – *Spinosaurus*, *Irritator*, *Suchomimus*, and *Baryonyx* – but these limited fossils are enough to describe spinosaurids as aberrant large-bodied theropods with enormous forelimbs and elongate snouts, similar to those of living crocodiles (Charig and Milner, 1997; Sereno et al., 1998; Sues et al., 2002). Skull shape, gut contents, and isotopic analysis indicate that spinosaurids were primarily piscivorous animals that lived along the shores of rivers and seas, an unusual lifestyle that is unknown in any other theropod subclade (Rayfield et al., 2007; Amiot et al., 2010). One spinosaurid, *Spinosaurus*, is perhaps the largest theropod ever discovered (Dal Sasso et al., 2005).

A flurry of recent work has also stabilized the phylogenetic tree of allosauroids (Fig. 4.8). The studies of Brusatte and Sereno (2008), Benson (2010), and Benson et al. (2010a) have analyzed the genealogy of allosauroids in great detail and include information gleaned from the discovery of several important new fossils over the past decade. Allosauroidea is divided into three major subclades: Sinraptoridae, Allosauridae, and Carcharodontosauria. The former two subclades comprise only a handful of species each, most of which lived during the Middle to Late Jurassic, including the familiar and eponymous *Sinraptor* and *Allosaurus*, respectively (Fig. 4.7). Carcharodontosauria, on the other hand, is a speciose and long-lived clade that spanned the entire Cretaceous and includes more than 15 species, which were distributed widely across the globe (Fig. 4.7). The most recognizable carcharodontosaurians are the enormous *Carcharodontosaurus* and *Giganotosaurus*, both of which may have eclipsed *Tyrannosaurus rex* in body size and were the apex predators on the southern continents during the middle Cretaceous (Coria and Salgado, 1995; Sereno et al., 1996). These mostly large-bodied and southern species comprise a subclade called the Carcharodontosauridae, which also includes a few taxa from North America and Asia and some genera, such as *Shaochilong*, that were only about 40% of the size of *Giganotosaurus* (Brusatte et al., 2009b). The other major carcharodontosaurian subclade is Neovenatoridae, a deviant group that includes genera such as *Neovenator*, *Fukuiraptor*, and *Australovenator*, many of which were much smaller, sleeker, and faster than other basal tetanurans and independently developed many bird-like features of their anatomy (Benson et al., 2010a). One neovenatorid, however, broke this mold: *Chilantaisaurus* from the middle Cretaceous of China was perhaps larger than *T. rex* and had a massive muscular forelimb (Benson and Xu, 2008).

Basal coelurosaurs

All theropods more derived than coelophysoids, ceratosaurs, and "basal tetanurans" comprise yet

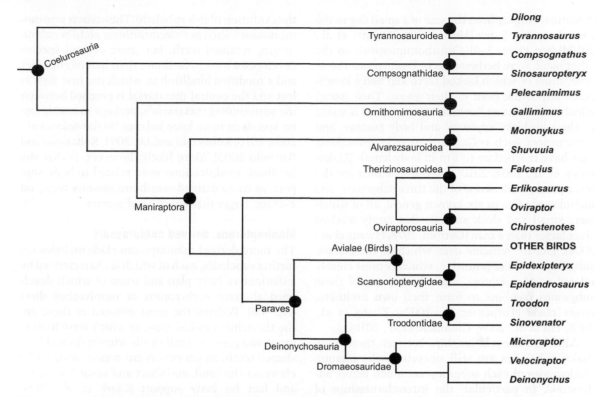

Figure 4.9 A generalized cladogram of Coelurosauria (the group of derived tetanuran theropods that includes birds and their closest relatives). This is not a complete genealogy of coelurosaurian theropods, but shows a handful of representative species and denotes major subclades. Cladogram based on references discussed in text.

another hierarchically nested major theropod subclade, Coelurosauria (Fig. 4.9). Among the most derived coelurosaurs are living birds, as well as their closest dinosaurian cousins. These bird-like coelurosaurs comprise a less inclusive subclade, Maniraptora, whereas more primitive coelurosaurs are lumped together under the umbrella term "basal coelurosaurs." Once again, as with "basal tetanurans," basal coelurosaurs do not encompass a monophyletic group, but rather are a graded series of progressively more derived outgroups on the line to Maniraptora.

Coelurosaur phylogeny has been the subject of fervent research over the past two decades, largely because paleontologists are interested in untangling the genealogy of birds and their closest relatives (Figs 4.9 and 4.10). Armed with a well-constrained family tree, researchers can confidently assess patterns of character change, rates of

evolution, and the timing of major morphological transitions as mid-sized ground-dwelling coelurosaurs evolved into small volant birds. Understandably, the primary focus has been on the phylogeny of birds and a handful of their closest maniraptoran relatives. Comparatively, the genealogy of basal coelurosaurs has received less attention and is understood with less certainty. Several major coelurosaur phylogenetic analyses have been published in recent years, some of which now include over 400 characters, and these should be consulted for further details (Norell et al., 2001a; Clark et al., 2002; Senter, 2007; Turner et al., 2007a; Zhang et al., 2008; Hu et al., 2009; Zanno et al., 2009; Choiniere et al., 2010; Csiki et al., 2010).

Three major subgroups comprise the basal coelurosaur grade: Tyrannosauroidea, Compsognathidae, and Ornithomimosauria (Fig. 4.11). Tyrannosauroids include approximately 20 species of bipedal predators, ranging from the size of a small dog to the 13-m frame of *T. rex* (Holtz, 2004; Brusatte et al., 2010d) (see Plates 1–3). Ornithomimosaurs, on the other hand, were herbivorous or omnivorous theropods, most of which lacked teeth and had a keratinized bill at the front of their snout. Their name, which translates as "ostrich mimic," pays homage to their bird-like anatomy and body posture, and some genera, such as *Gallimimus* and *Beishanlong*, may have reached up to 8 m in body length (Makovicky et al., 2004, 2010). Compsognathids are the most poorly understood of the three subgroups, and include only five or six known genera, all of which were small and sleek animals that rarely reached sizes larger than a man (Ostrom, 1978; Hwang et al., 2004). Debate remains over which of these three subgroups is most primitive, which is most closely related to Maniraptora, and whether any of these subgroups combine to form their own exclusive, larger clade (Turner et al., 2007a; Zhang et al., 2008; Hu et al., 2009; Choiniere et al., 2010).

Although the relationships between these three major subclades are still uncertain, the ingroup phylogenies of each subgroup are much better understood. In particular, the interrelationships of tyrannosauroids have generated substantial debate and revision over the past decade, in concert with the discovery of several new species of primitive tyrannosauroids that help reveal character and body size changes on the line toward colossal Late Cretaceous species such as *Tyrannosaurus* and *Albertosaurus* (Holtz, 2004; Carr et al., 2005; Brusatte et al., 2009a, 2010d; Carr and Williamson, 2010). The recent phylogenetic analysis of Brusatte et al. (2010d), which includes over 300 characters scored across nearly 20 tyrannosauroid species, helps to place the evolution of large tyrannosaurs in perspective (Fig. 4.10). Tyrannosauroidea is an ancient group that originated during the Middle Jurassic, at least 100 million years before *T. rex* lived (Rauhut et al., 2010). For the vast majority of their evolutionary history – approximately 80 million years – tyrannosauroids were mostly small-bodied animals, little bigger than a man, and gigantic size developed only during the final 20 million years of the Cretaceous, after the extinction of other large-bodied theropods such as the carcharodontosaurians.

A richer understanding of ornithomimosaur phylogeny also provides a framework for describing the evolution of this subclade. The earliest ornithomimosaurs, such as *Pelecanimimus* and *Shenzhousaurus*, retained teeth, but more derived species developed a beak to facilitate their herbivorous diet and a modified hindlimb in which the first digit is lost and the central metatarsal is pinched between the surrounding metatarsals, perhaps to enable faster speeds or more keen balance (Makovicky et al., 2004, 2010; Kobayashi and Lü, 2003; Kobayashi and Barsbold 2005). More likely, however, is that the hindlimb modifications were related to body support, as more derived ornithomimosaurs were, on average, larger than more basal species.

Maniraptorans: derived coelurosaurs

The more derived maniraptoran clade includes six further subclades, each of which is characterized by a distinctive body plan and some of which developed aberrant herbivorous or omnivorous diets (Fig. 4.9). Perhaps the most unusual of these are the therizinosauroids, most of which were herbivorous and possess small skulls with beaks and leaf-shaped teeth, an enormous gut region, scythe-like claws on the hand, and short and squat hindlimbs and feet for body support (Clark et al., 2004) (Fig. 4.12). The oviraptorosaurs also depart radically from the traditional theropod body plan, as most species have short, deep, and toothless skulls bearing elaborate crests and rugosities (Osmólska et al., 2004) (Fig. 4.12; see Plate 10). A third subgroup, the alvarezsauroids, are represented by only a handful of species, most of which are small, sleek, long-legged creatures with stubby forearms and a beaked skull (Perle et al., 1993). The dromaeosaurids and troodontids are, in many ways, more traditional theropods: they are all bipedal, had sharp teeth and claws, and were primarily carnivorous (Makovicky and Norell, 2004; Norell and Makovicky, 2004) (Fig. 4.13; see Plates 4–6, 8, 9, and 11). A signature feature of both subgroups is a hyperextensible claw on the second digit of the foot, the so-called "killer claw" that was likely used to slash or clamp onto prey (Manning et al., 2006). Finally, the sixth principal maniraptoran subclade are birds themselves, more properly referred to by their scientific name, Aves or Avialae (Fig. 4.14). All birds, even primitive species such as *Archaeopteryx* and *Confuciusornis*, were capable of at least weak flight, powered by their wings, large chest muscles, and asymmetrical

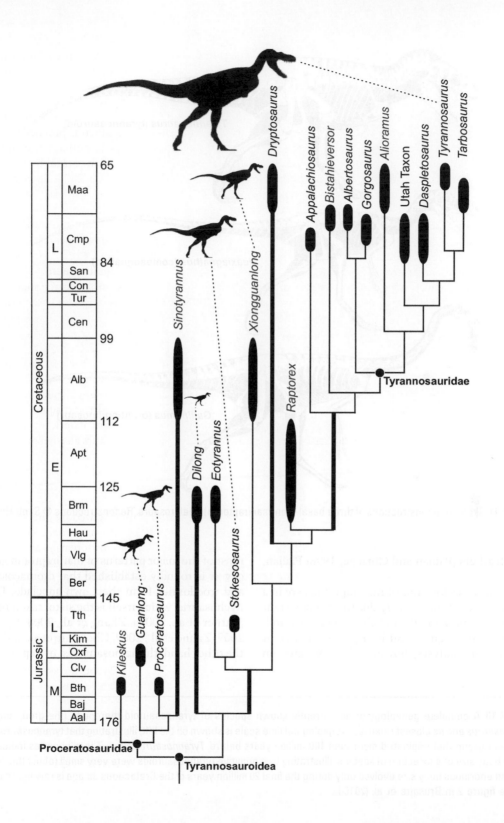

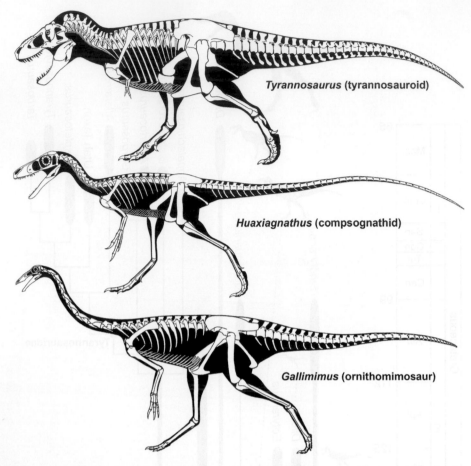

Tyrannosaurus (tyrannosauroid)

Huaxiagnathus (compsognathid)

Gallimimus (ornithomimosaur)

Figure 4.11 Skeletal reconstructions of three basal (non-maniraptoran) coelurosaurs. Reconstructions © Scott Hartman.

vaned feathers (Padian and Chiappe, 1998; Padian, 2004).

The interrelationships of maniraptorans are in a constant state of flux, largely due to the consistent discovery of new fossils and the intense research focus on the origin of birds (Fig. 4.9). Different phylogenetic analyses, however, largely agree on most of the major patterns of maniraptoran genealogy. It is roundly established that dromaeosaurids and troodontids form their own subclade, Deinonychosauria, which itself is the sister taxon to birds (Turner et al., 2007a; Zhang et al., 2008; Hu et al., 2009; Zanno et al., 2009; Choiniere et al., 2010). On the other hand, therizinosauroids, oviraptorosaurs,

Figure 4.10 A complete genealogy of all currently known species of tyrannosauroid theropods (the group including *Tyrannosaurus* and its closest relatives). A geological time scale is shown on the left, illustrating that tyrannosauroids are an ancient group that originated more than 100 million years before *Tyrannosaurus* lived. The silhouettes indicate the relative body size of a selection of species, illustrating that primitive tyrannosauroids were very small (about the size of a man) and enormous body size evolved only during the final 20 million years of the Cretaceous. Image is raw figure used to produce figure 2 in Brusatte et al. (2010d).

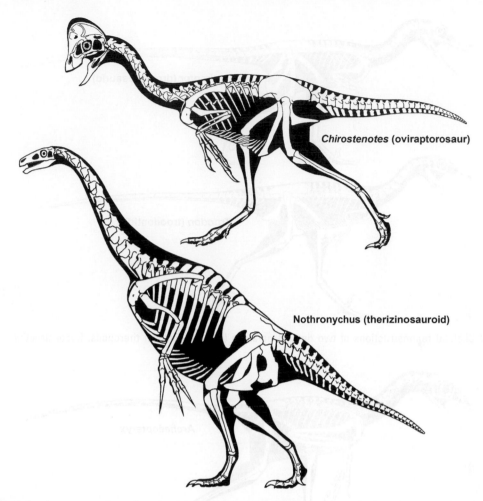

Figure 4.12 Skeletal reconstructions of two intermediate (maniraptoran, but non-paravian) coelurosaurs. Reconstructions © Scott Hartman.

and alvarezsauroids are more basal maniraptorans on the stem lineage toward birds and deinonychosaurs. However, the relative relationships of these three subclades are unsettled. Some studies find support for a clade of oviraptorosaurs and therizinosauroids (Turner et al., 2007a; Csiki et al., 2010), whereas other analyses place these subclades as successive outgroups on the line to birds (Zhang et al., 2008; Hu et al., 2009; Zanno et al., 2009; Choiniere et al., 2010). There is also debate over whether alvarezsauroids are the most basal maniraptorans (Turner et al., 2007a; Choiniere et al., 2010), are the closest relatives to the bird and deinonychosaur clade (Zanno et al., 2009), or are even non-maniraptorans that form a clade with ornithomimosaurs (Sereno, 2001). As with most phylogenetic debates, these uncertainties will surely be clarified with the discovery of new specimens.

The ingroup relationships of the six major maniraptoran subclades have also been the subject of keen interest and study. Many recent phylogenetic analyses demonstrate that the aberrant morphologies and dietary habits of many maniraptoran subgroups were not fully present in their most primitive members, which more closely resembled traditional theropods, but developed later in more

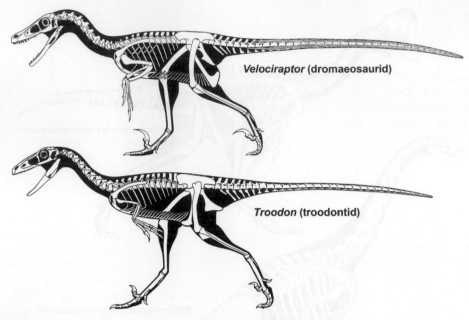

Velociraptor (dromaeosaurid)

Troodon (troodontid)

Figure 4.13 Skeletal reconstructions of two derived paravian (but non-avialan) theropods. Reconstructions © Scott Hartman.

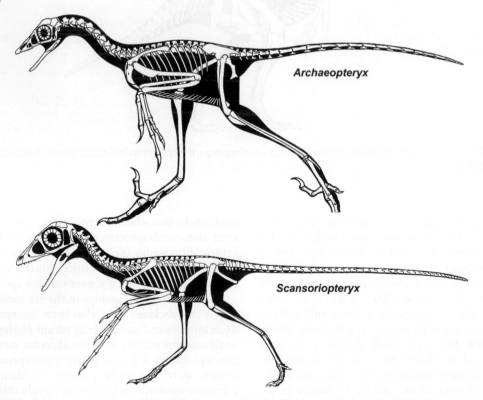

Archaeopteryx

Scansoriopteryx

Figure 4.14 Skeletal reconstructions of two basal avialans (birds). Reconstructions © Scott Hartman.

derived species. For instance, primitive therizinosauroids such as *Falcarius* and *Beipiaosaurus* had a larger head, smaller gut region, more diminutive claws, and less compact and weight-bearing hands and feet than more derived members of the subgroup (Zanno et al., 2009; Zanno, 2010). Similarly, the basal alvarezsauroid *Haplocheirus* had short legs and a more traditional theropod hand, not the elongate limbs and stubby hand – reduced to one digit – of more derived species (Choiniere et al., 2010).

Research on the interrelationships of dromaeosaurids and birds has also revealed some surprises. Dromaeosaurids are divided into several smaller subgroups, including a basal array of species that were remarkably small (Turner et al., 2007a) and perhaps capable of gliding (Xu et al., 2003; Chatterjee and Templin, 2007). Only more derived dromaeosaurids such as *Velociraptor* and *Deinonychus* were large, clearly terrestrial predators. This raises the question of whether flight may have evolved much earlier than previously thought, perhaps in the common ancestor of dromaeosaurids, troodontids, and birds (Makovicky et al., 2005). If this was the case, then *Velociraptor* and other "classic" dromaeosaurids evolved from flying ancestors but subsequently lost the ability to fly (Paul, 2002). Alternatively, it is possible that primitive dromaeosaurids, if they could glide, may have evolved their aerial capabilities separately from birds. The genealogy of these early birds has also been under scrutiny, and remarkable new fossil discoveries indicate that the most primitive birds were not animals like *Archaeopteryx*, but rather small pigeon-sized tree-dwellers like *Epidexipteryx* and *Epidendrosaurus*, which had elongate third digits and elaborate ribbon-like feathers (Zhang et al., 2002, 2008). These two species are grouped together into their own basal avialan subclade, Scansoriopterygidae, and future discoveries will likely expand the membership of this subgroup and help paleontologists understand the evolution of avian flight. There is even a recent suggestion, based on a cladistic analysis, that the iconic *Archaeopteryx* is not actually a member of the bird clade (Avialae), but is rather a deinonychosaur more closely related to *Velociraptor* and *Deinonychus* (Xu et al., 2011a). As always, further discoveries, and progressive refinement of cladistic datasets as new specimens are unearthed, should help test this intriguing hypothesis.

Sauropodomorph Phylogeny

The bulk of phylogenetic research has focused on theropods, and considerably fewer cladistic analyses have attempted to untangle the genealogy of sauropodomorphs and ornithischians. This situation is changing, however, as troves of new fossils of these dinosaurs are being found across the globe and scientists are becoming increasingly interested in their biology and evolution. Understanding the development of enormous body size in sauropods or the evolution and potential function of ornithischian plates, frills, and crests demands a phylogenetic framework for context, and recent reviews of each of these subjects have only been possible due to refinements in our understanding of sauropodomorph and ornithischian genealogy (Padian and Horner, 2011; Sander et al., 2011).

"Prosauropods"
Scientific understanding of sauropodomorph phylogeny has undergone a renaissance in recent years, with attention especially focused on the interrelationships of the smaller basal "prosauropods" of the Late Triassic to Early Jurassic and the genealogy of the colossal long-necked sauropods of the Late Jurassic (Figs 4.15 and 4.16). Many early cladistic studies found support for a bipartite division of Sauropodomorpha, composed of separate prosauropod and sauropod subclades (Galton, 1990; Sereno, 1999; Galton and Upchurch, 2004b). Prosauropods, including genera such as *Plateosaurus*, *Massospondylus*, *Lufengosaurus*, *Yunnanosaurus*, and *Riojasaurus*, were thought to be united by several derived characters, including a beak on the premaxilla, an inset first tooth of the lower jaw, an elongate first metacarpal that is offset relative to the other digits of the hand, a twisted first finger of the hand, and a broad weight-supporting foot (Sereno, 1999) (Fig. 4.17).

Over the past decade, however, the roster of known prosauropod and primitive sauropod species has skyrocketed. It is now clear that many supposed prosauropod synapomorphies are also present in the earliest sauropods, and therefore do not unite prosauropods as a discrete clade (Yates and Kitching 2003). Recent phylogenetic analyses instead posit that "prosauropods" form a primitive grade along

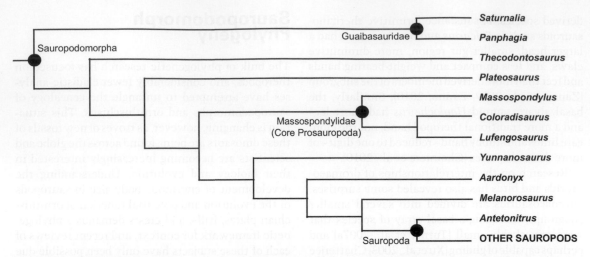

Figure 4.15 A generalized cladogram of basal sauropodomorphs ("prosauropods"). This is not a complete genealogy of basal sauropodomorphs, but shows a handful of representative species and denotes major subclades. Cladogram based on references discussed in text.

the line to Sauropoda (hence the designation as "prosauropods" within quotation marks, since they are not a clade; Yates, 2003, 2007; Smith and Pol, 2007; Upchurch et al., 2007; Ezcurra, 2010a; Yates et al., 2010; Pol et al., 2011) (Fig. 4.15). However, a

reduced subset of "prosauropods," informally termed "core prosauropods," probably does form its own clade, including *Coloradisaurus*, *Lufengo-saurus*, and *Massospondylus* (Fig. 4.15). Genera such as *Aardonyx*, *Melanorosaurus* and *Yunnanosaurus*

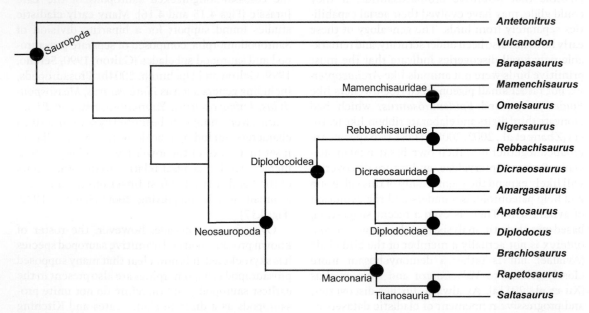

Figure 4.16 A generalized cladogram of sauropod dinosaurs. This is not a complete genealogy of sauropods, but shows a handful of representative species and denotes major subclades. Cladogram based on references discussed in text.

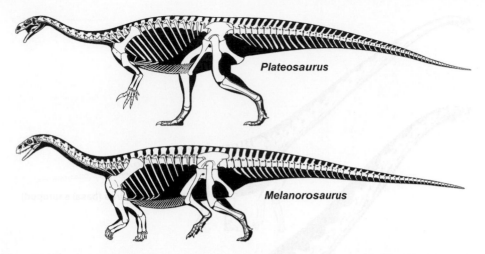

Figure 4.17 Skeletal reconstructions of two basal sauropodomorphs ("prosauropods"). Reconstructions © Scott Hartman.

are among the closest relatives to true sauropods. These "prosauropods" are clearly transitional between the smaller, sleeker, and facultatively bipedal primitive "prosauropods" and the larger, more massive, quadrupedal sauropods. *Aardonyx*, for instance, was a bulky animal with a massive femur to support its body weight, but it still moved bipedally most of the time (Yates et al., 2010). *Melanorosaurus*, however, was fully quadrupedal, but it lacked the columnar limbs and U-shaped weight-supporting hand of true sauropods (Bonnan and Yates, 2007) (Fig. 4.17). The first true sauropods, such as *Antetonitrus* and *Tazoudasaurus*, had the fully developed suite of sauropod synapomorphies, including columnar limbs, weight-bearing hands and feet with digits arrayed in a semicircular colonnade, a barrel-shaped chest, and an especially elongate neck (Yates and Kitching, 2003; Allain and Aquesbi, 2008) (Fig. 4.18). As can be seen, compared with what was known only one decade ago, the current understanding of saurodomorph genealogy now enables paleontologists to study the evolution of the classic sauropod body plan in great detail.

Some of the most important sauropodomorph discoveries of the past two decades have been new fossils of the oldest and most primitive known members of the clade. These fairly small, Late Triassic, omnivorous or herbivorous "prosauropods" include *Saturnalia* from Brazil (Langer et al., 1999) and *Panphagia* and *Chromogisaurus* from Argentina (Martinez and Alcober, 2009; Ezcurra, 2010a). These genera share several anatomical characteristics, including a long postacetabular process of the ilium and an acetabular wall that is only partially open, not fully window-like as in other dinosaurs (Ezcurra, 2010a). Many of these characters are also present in *Guaibasaurus*, a long-enigmatic and fragmentary Late Triassic genus that has usually been considered a primitive theropod or a basal saurischian outside the Theropoda + Sauropodomorpha group (Langer and Benton, 2006). On the other hand, the phylogenetic analysis of Ezcurra (2010a) finds support for a clade uniting all of these genera as their own discrete subgroup of primitive sauropodomorphs, termed Guaibasauridae. Because they are the earliest-branching members of the sauropodomorph lineage, it is no surprise that guaibasaurids share some characters with theropods.

Sauropods

The phylogeny of sauropods has become progressively better understood due to a series of cladistic analyses published over the past two decades (Upchurch, 1995, 1998; Salgado et al., 1997; Wilson and Sereno, 1998; Wilson, 2002; Upchurch et al., 2004; Harris, 2006; Allain and Aquesbi, 2008; Remes et al., 2009) (Fig. 4.16). Many of these studies have built on each other, often by adding new characters

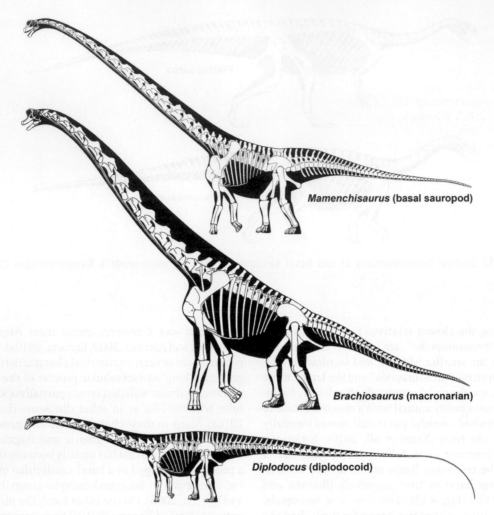

Mamenchisaurus (basal sauropod)

Brachiosaurus (macronarian)

Diplodocus (diplodocoid)

Figure 4.18 Skeletal reconstructions of three sauropod dinosaurs. Reconstructions © Scott Hartman.

and newly discovered species to previous datasets, and the most recent analyses include several hundred characters scored across more than 30 species.

The most primitive sauropods include genera such as *Antetonitrus*, *Vulcanodon*, *Kotasaurus*, *Shunosaurus*, *Tazoudasaurus*, *Spinophorosaurus*, and *Barapasaurus*, which comprise a grade along the line to more derived sauropods (Fig. 4.16). All these genera possess the classic characteristics of sauropods – the small head, long neck, columnar limbs, and enormous gut – but are generally smaller than more derived sauropods and have shorter necks, less delicately pneumatized cervical vertebrae, and somewhat less modified and foreshor-

tened skulls (Wilson, 2002; Upchurch et al., 2004) (Fig. 4.18). These primitive sauropods were widespread globally during the Late Triassic to Middle Jurassic, and were likely the principal herbivorous species in most terrestrial ecosystems. One peculiar subclade, Mamenchisauridae, includes Middle to Late Jurassic genera such as *Mamenchisaurus* and *Omeisaurus* that were restricted to Asia and which had some of the relatively, and absolutely, longest necks of any animal that ever lived (Fig. 4.18).

The larger, more derived sauropods comprise their own clade, Neosauropoda, which is divided into two primary subgroups: Diplodocoidea and

Macronaria (Fig. 4.16). Diplodocoid sauropods include such familiar genera as *Diplodocus* and *Apatosaurus* and are united by features such as a subrectangular snout, external nares that are fully retracted behind the eye socket, and a reduced supratemporal fenestra (Upchurch et al., 2004; Sereno et al., 2007; Whitlock, 2011b) (Fig. 4.18). Three primary subclades comprise Diplodocoidea (Fig. 4.16). The large-bodied Diplodocidae, which includes *Diplodocus* and *Apatosaurus*, were some of the largest and most common herbivores in North America during the Late Jurassic. The flamboyant Dicraeosauridae includes only a few known genera, among them *Dicraeosaurus*, *Amargasaurus*, and *Brachytrachelopan*, which have shorter necks than most other sauropods and greatly elongated neural spines on the vertebrae that may have supported a sail (Rauhut et al., 2005). Finally, the aberrant Rebbachisauridae constitutes one of the most unusual groups of sauropods. Some rebbachisaurids, most prominently *Nigersaurus*, have astonishingly delicate skulls and complex dental batteries, packed with hundreds of teeth that would literally wear down to dust before being replaced by a conveyer-belt of new teeth (Sereno et al., 2007).

The second major neosauropod subclade, Macronaria, groups together the familiar genera *Brachiosaurus* and *Camarasaurus* with a more derived subgroup, the Titanosauria (Fig. 4.16). All macronarians share an obscenely large external naris, which gives the group its name, as well as enlarged jaw muscles and a reduced number of teeth in the lower jaw (Wilson, 2002). Some macronarians, such as *Brachiosaurus* and a number of titanosaurs such as *Argentinosaurus* and *Paralititan*, were among the largest animals to ever live on land, perhaps reaching lengths of 25 m and masses of 60 tons (Smith et al., 2001) (Fig. 4.18). *Brachiosaurus* and close kin shared the Late Jurassic landscape with large diplococoids such as *Diplodocus*, whereas the titanosaurs were the sole major clade of large-bodied sauropods during most of the Cretaceous. They were especially common on the southern continents during the Late Cretaceous, and discoveries of new titanosaur species are occurring at a rapid pace as fossil sites in South America, Africa, Madagascar, and Australia are being explored with unprecedented vigor (Curry-Rogers, 2005). Research on titanosaur phylogeny has lagged behind the brisk pace of new titanosaur discoveries, and reconstructing the genealogy of the approximately 60 currently known titanosaur genera is a fundamental goal for future research.

Ornithischian Phylogeny

Ornithischian dinosaurs include a dazzling array of major subclades distinguished by signature body plans, such as the tank-like ankylosaurs, plated stegosaurs, horned-and-frilled ceratopsians, domeheaded pachycephalosaurs, and ornately crested hadrosaurids. The phylogenetic ingroup relationships of each of these major clades have been intensely studied, but considerably less work has focused on the higher-level phylogeny of Ornithischia as a whole (Fig. 4.19). In other words, there is considerable debate about how the various major subclades are related to each other and which ornithischians were the most primitive. The large-scale relationships of Ornithischia were addressed in one of the first cladistic analyses to focus on dinosaurs – the astute study of Sereno (1986) – but only more recently has a younger generation of researchers revisited this subject. Many of the major patterns of relationship elucidated by Sereno (1986) have been strongly corroborated by recent studies, but newer work has also revealed a few surprises.

Basal ornithischians

The most comprehensive analysis of large-scale ornithischian phylogeny was presented by Butler et al. (2008a), which followed two important earlier studies (Butler, 2005; Butler et al., 2007). These analyses identified a small subclade, Heterodontosauridae, as the most primitive group of ornithischians (Fig. 4.19). Heterodontosaurids include a number of mostly small and cursorial species, most of which were likely herbivorous or omnivorous and many of which possess pronounced canine-like teeth at the front of their snouts. The most familiar of these is the terriersized *Heterodontosaurus* from the Early Jurassic of South Africa, but recent discoveries have demonstrated that some species were feathered (Zheng et al., 2009) and others, such as *Fruitadens*, were truly miniscule and probably weighed less than 1 kg (Butler et al., 2010a).

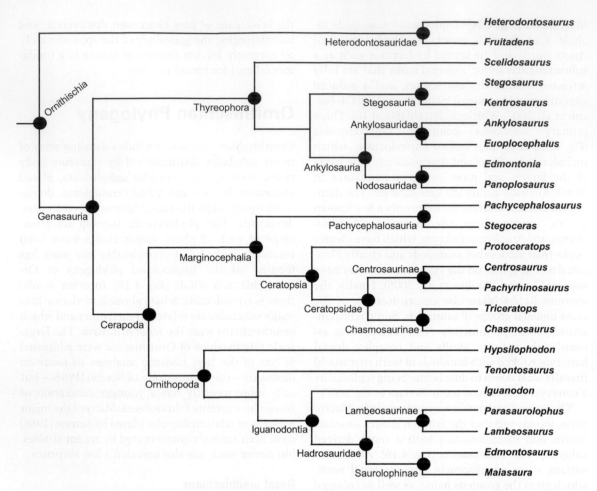

Figure 4.19 A generalized cladogram of ornithischian dinosaurs. This is not a complete genealogy of ornithischians, but shows a handful of representative species and denotes major subclades. Cladogram based on references discussed in text.

Sereno (1986, 1999) placed heterodontosaurids as more derived ornithischians, on the stem lineage toward the ornithopod clade that includes *Iguanodon* and the crested hadrosaurids. This placement was followed by other authors, including Norman et al. (2004a), whereas other workers suggested that heterodontosaurids may be closely related to the ceratopsians (Xu et al., 2006). A position at the base of the ornithischian tree, however, seems most plausible, largely for two reasons. First, heterodontosaurids share many characters, such as an enlarged grasping hand with prominent claws and elongate phalanges, with primitive saurischian dinosaurs. These are most parsimoniously interpreted as

characters that evolved in the common ancestor of dinosaurs, but were lost in ornithischians more derived than heterodontosaurids. Second, all ornithischians other than heterodontosaurids share many features, such as a broad symphysis between the opposing dentaries and teeth that are expanded into a leaf shape above the root (Butler et al., 2008a). Therefore, there is evidence that heterodontosaurids retain many primitive dinosaur features and that all other ornithischians share their own set of derived features, making it most plausible that heterodontosaurids are basal ornithischians. As with any genealogical hypothesis, however, this finding is subject to change based on the discovery of new

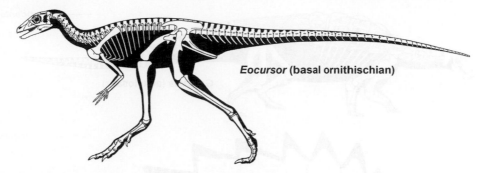

Eocursor (basal ornithischian)

Figure 4.20 Skeletal reconstruction of the basal ornithischian *Eocursor*. Reconstruction © Scott Hartman.

specimens, and, in particular, support could emerge for a heterodontosaurid + ceratopsian group as outlined by Xu et al. (2006).

A handful of genera, some of which may or may not belong to the heterodontosaurid subclade, are also among the most primitive ornithischians (Fig. 4.20). These include *Pisanosaurus*, known from a scrappy fossil from the Late Triassic of Argentina that is the oldest unequivocal ornithischian specimen yet discovered, and *Lesothosaurus* and *Eocursor* from the Early Jurassic of South Africa (Sereno, 1991b; Butler et al., 2007; Irmis et al., 2007b). These are all diminutive animals, smaller than a human and dwarfed in comparison to most derived ornithischians, and show adaptations for a fast-running and agile bipedal posture.

Thyreophora
All ornithischians more derived than heterodontosaurids and other basal taxa comprise a subclade called Genasauria (Fig. 4.19). This subgroup is divided into two of its own principal subgroups, Thyreophora and Cerapoda. The thyreophoran dinosaurs include the ankylosaurs, stegosaurs, and a small subset of singleton genera such as *Scelidosaurus*, *Scutellosaurus*, and *Emausaurus* (Sereno, 1986, 1999; Norman et al., 2004b; Butler et al., 2008a) All thyreophorans share one unequivocal and instantly recognizable feature – bony body armor – which gives the group its name, which translates as "shield bearers". This armor is modified into the bony osteoderms, shoulder spikes, and tail clubs of ankylosaurs and the expansive plates and tail spikes of stegosaurs (Fig. 4.21).

The ingroup phylogeny of Thyreophora has been addressed by several cladistic analyses, most of which focus on either the stegosaurs or the ankylosaurs (Fig. 4.19). *Scutellosaurus*, *Emausaurus*, and *Scelidosaurus* likely comprise a series of successive outgroups on the line to the stegosaur + ankylosaur clade (Butler et al., 2008a). These genera lived during the earliest Jurassic, were not much larger than pigs, and had less elaborate suits of armor than the more derived stegosaurs and ankylosaurs. Stegosaur phylogeny was recently addressed by the thorough cladistic analysis of Maidment et al. (2008), which greatly improved the number of analyzed taxa and characters relative to previous studies (Sereno and Dong, 1992; Sereno, 1999; Carpenter et al., 2001; Galton and Upchurch, 2004a). Similarly, the phylogenetic analyses of ankylosaurs presented by Vickaryous et al. (2004) and Thompson et al. (2011) comprehensively built upon previous work (Sereno, 1999; Carpenter, 2001; Hill et al., 2003). Both stegosaurs and ankylosaurs exhibit overall trends of increasingly ornate armor in more derived species. Ankylosaurs are divided into two major subgroups, Ankylosauridae and Nodosauridae, which are distinguished based on the form of their armor (Fig. 4.19). Ankylosaurids generally possess globular tail clubs and heavily rugose skulls, whereas nodosaurids lack tail clubs but have a strongly ornamented snout and a bulbous protuberance above the eye (Vickaryous et al., 2004).

Marginocephalia
The remaining ornithischians – ornithopods, ceratopsians, and pachycephalosaurs – comprise the

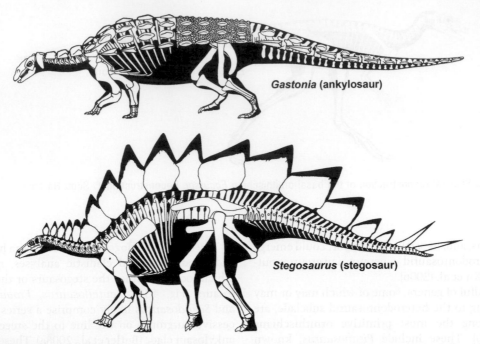

Gastonia (ankylosaur)

Stegosaurus (stegosaur)

Figure 4.21 Skeletal reconstructions of two thyreophoran ornithischian dinosaurs. Reconstructions © Scott Hartman.

major subclade Cerapoda, which is often referred to as Neornithischia (Fig. 4.19). Of these major cerapodan subgroups, ceratopsians and pachycephalosaurs are sister taxa and together comprise the clade Marginocephalia. This group takes it name from one of its most distinguishing features: the presence of a shelf, composed of the parietal and squamosal bones, that extends posteriorly from the margins of the skull (Sereno, 1986). This shelf is modified into the flamboyant frill of ceratopsians (Figs 4.22 and 4.23) and forms much of the cranial dome of pachycephalosaurs (Fig. 4.24). Other unique features of the group include a reduction in the number of premaxillary teeth and an atrophied pubis.

A great deal of recent work has focused on the ingroup genealogy of ceratopsians, motivated by a flood of new fossil discoveries over the past decade. Higher-level cladistic studies elucidating the broadscale shape of the ceratopsian family tree have been provided by Sereno (1999, 2000), Makovicky (2001), Xu et al. (2002a, 2006), You and Dodson (2004), Makovicky and Norell (2006), and Chinnery and Horner (2007). Although there are many subtle differences between the phylogenies proposed by

these studies, all analyses agree that Ceratopsia comprises a primitive grade of smaller, less bulky, and less ornately ornamented species that form successive outgroups to the derived, elaborately frilled Ceratopsidae (Fig. 4.19). Among the smaller and less gaudy "basal ceratopsians" is *Psittacosaurus*, a genus of mostly housecat-sized, beaked herbivores that ranged across Asia for much of the Early Cretaceous (Sereno, 1990, 2010) (Fig. 4.22). Other primitive ceratopsians include an array of species such as *Protoceratops*, *Bagaceratops*, *Leptoceratops*, *Yamaceratops*, and *Montanoceratops* that lived across Asia, North America, and Europe during much of the Cretaceous (You and Dodson, 2004; Makovicky and Norell, 2006; Ősi et al., 2010). Some of these genera may combine to form small subclades, such as Protoceratopsidae (*Protoceratops*, *Bagaceratops*) and Leptoceratopsidae (*Leptoceratops*, *Montanoceratops*).

The family tree of the largest, gaudiest, and most derived ceratopsians, the elaborately horned Ceratopsidae, is also becoming further refined with new discoveries (Figs 4.19 and 4.23). The most comprehensive cladistic analysis for all ceratopsids was

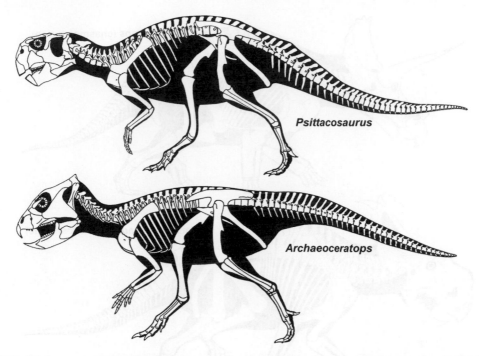

Psittacosaurus

Archaeoceratops

Figure 4.22 Skeletal reconstructions of two basal ceratopsian ornithischian dinosaurs. Reconstructions © Scott Hartman.

presented by Dodson et al. (2004) and augmented with data gleaned from new fossils by Wu et al. (2007). These studies, along with many pre-cladistic classifications of the horned dinosaurs, support the subdivision of Ceratopsidae into separate subclades, Chasmosaurinae and Centrosaurinae (Fig. 4.19). Among the most familiar chasmosaurines are the three-horned *Triceratops*, the five-horned *Pentaceratops*, the broad-frilled *Chasmosaurus*, and the recently discovered spiky-headed *Kosmoceratops*. The interrelationships of these species were recently addressed by the in-depth analysis of Sampson et al. (2010), which should be consulted for further details. Centrosaurines, on the other hand, are a slightly less diverse group that is centered on *Centrosaurus*, a common species in the Late Cretaceous ecosystems of western North America that is distinguished by its enormous erect nasal horn. Both McDonald and Horner (2010) and Ryan et al. (2010b) recently presented focused analyses of centrosaurine phylogeny, and find evidence for a specialized subclade of aberrantly horned taxa comprising *Pachyrhinosaurus* and *Achelousaurus*,

which have replaced their nasal horn with a gnarly rectangular boss, and *Einiosaurus*, whose nasal horn droops forward like a limp noodle. Until recently it was thought that ceratopsids only lived in North America during the Late Cretaceous, but the discovery of primitive centrosaurine *Sinoceratops* in China reveals that the clade was more widespread (Xu et al., 2010).

The genealogy of the second primary marginocephalian subgroup, the dome-headed Pachycephalosauria, has been addressed by only a few cladistic analyses, including those of Sereno (1986, 1999, 2000), Williamson and Carr (2002), Sullivan (2003), Maryańska et al. (2004), Schott et al. (2009), and Longrich et al. (2010a). The most recent and comprehensive study finds evidence for separate Asian and North American subclades, the latter of which includes some of the best-known members of the group: the large-bodied, spiky-skulled, and strongly domed *Pachycephalosaurus*, *Stygimoloch*, and *Stegoceras*, all of which lived during the latest Cretaceous (Longrich et al., 2010a) (Fig. 4.24).

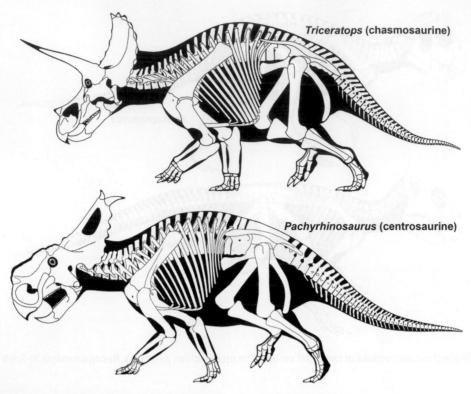

Triceratops (chasmosaurine)

Pachyrhinosaurus (centrosaurine)

Figure 4.23 Skeletal reconstructions of two derived ceratopsian ornithischian dinosaurs. Reconstructions © Scott Hartman.

Ornithopoda

The final major ornithischian subclade is Ornithopoda, which groups together a number of different herbivorous species, including the fabulously crested hadrosaurids, the iguanodontids, and a range of smaller and fleet-footed genera such as *Dryosaurus*, *Thescelosaurus*, and *Hypsilophodon* (Fig. 4.25).

Broad-scale phylogenies, such as that of Butler et al. (2008a), indicate that *Hypsilophodon*, *Thescelosaurus*, *Orodromeus*, *Parksosaurus*, and other svelte genera form a series of successive outgroups on the line to more derived ornithopods such as iguanodontids and hadrosaurs (Fig. 4.19). Many of these smaller primitive taxa could probably run

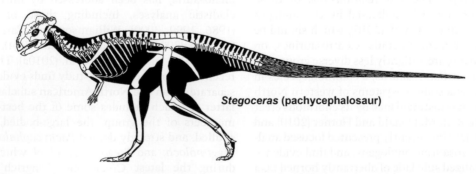

Stegoceras (pachycephalosaur)

Figure 4.24 Skeletal reconstruction of a pachycephalosaurid ornithischian dinosaur. Reconstruction © Scott Hartman.

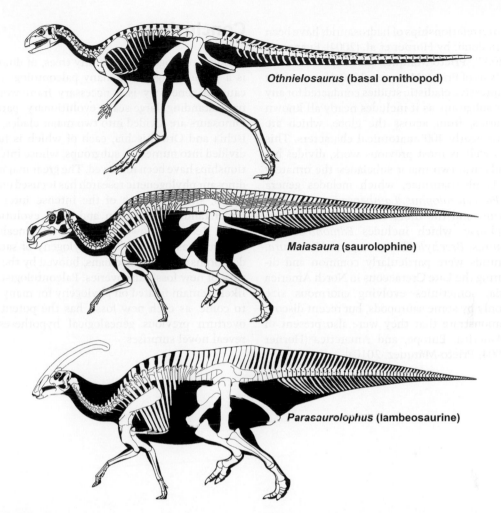

Othnielosaurus (basal ornithopod)

Maiasaura (saurolophine)

Parasaurolophus (lambeosaurine)

Figure 4.25 Skeletal reconstruction of three ornithopod ornithischian dinosaurs. Reconstructions © Scott Hartman.

bipedally, cropped vegetation with the aid of a sharp beak, and some were capable of burrowing (Varricchio et al., 2007). There is a general trend of increasing body size across ornithopod phylogeny. "Intermediate" ornithopods such as *Tenontosaurus* and the small subclade Rhabdodontidae are larger and bulkier than the primitive grade taxa, but not quite as large and lumbering as the iguanodontids and hadrosaurids, most of which probably walked quadrupedally most of the time and possess blunt hooves on their hands and feet to support their body weight (Fig. 4.25).

The phylogeny of more derived ornithopods – the iguanodontids, hadrosaurids, and their closest relatives – has been assessed by a handful of recent studies (Fig. 4.19). This derived subclade is referred to as Iguanodontia, and comprises a series of more primitive taxa forming a grade relative to the more derived Hadrosauridae. The interrelationships of these gradational stem taxa have been assessed by the recent cladistic studies of Norman (2004) and McDonald et al. (2010), which illustrate that the familiar genera *Dryosaurus*, *Camptosaurus*, and *Iguanodon* are successively closer relatives to the hadrosaurids. The study by McDonald et al. (2010) is particularly comprehensive and illuminating, as it examines the relationships of more than 40 species of these "stem hadrosaurids," many of which have come to light only over the past decade.

The interrelationships of hadrosaurids have been studied in detail by Horner et al. (2004), Evans and Reisz (2007), Prieto-Márquez (2010a), and others. The analysis of Prieto-Márquez (2010a) is one of the more impressive cladistic studies conducted for any dinosaur subgroup, as it includes nearly all known hadrosaurids, from across the globe, which are scored for nearly 400 anatomical characters. This study, as well as most previous work, divides hadrosaurids into two major subclades: the ornately crested Lambeosaurinae, which includes genera such as *Parasaurolophus*, *Lambeosaurus*, and *Corythosaurus*, and the more modestly ornamented Saurolophinae, which includes *Edmontosaurus*, *Gryposaurus*, *Brachylophosaurus*, and *Maiasaura*. Hadrosaurids were particularly common and diverse during the Late Cretaceous in North America and Asia, sometimes evolving enormous size rivaled only by some sauropods, but recent discoveries demonstrate that they were also present in South America, Europe, and Antarctica (Horner et al., 2004; Prieto-Márquez, 2010a).

Conclusions

Building phylogenies, or family trees, of dinosaurs is a particular goal for many paleontologists, because a genealogy is a necessary framework for understanding large-scale evolutionary patterns. Dinosaurs are divided into two major clades, Saurischia and Ornithischia, each of which is further divided into numerous subgroups, whose interrelationships have been reviewed. The great majority of dinosaur phylogenetic research has focused on theropods, surely because of the intense interest in understanding the origin and early evolution of birds. Recently, however, a flurry of genealogical research has addressed the phylogeny of sauropodomorphs and ornithischians, buoyed by the rapid pace of new fossil discoveries. Paleontologists will likely remain fixated on phylogeny for many years to come, as each new fossil has the potential to overturn previous genealogical hypotheses and reveal novel surprises.

5 | Form

The individual hard parts and soft tissues of dinosaurs, which were the subject of detailed focus in earlier chapters, together comprise the overall "form" or "body plan" of a living species. Intense fixation on one or a few bones is all too common in the dinosaur literature. Sometimes a handful of poorly preserved bones is all that is known from a certain species, and in these cases obsession with pedantic features of the anatomy is the only recourse for interpreting the material at hand. In general, however, it is more illuminating to step back from the minutiae of subtle ridges and depressions on bones and focus more broadly on form. What are the overall body plans that distinguish different groups of dinosaurs, and what may these indicate about locomotion, feeding, or other aspects of biology? What sort of variability in size and shape is present in the skeletons of dinosaurs, and what can this tell us about how individual species grew, how multiple species may have partitioned ecological resources, and what differences may have characterized males and females? How did the individual bones articulate and function together, in concert with muscles and other soft tissues, to power a living, breathing, feeding organism?

Understanding form, especially being able to quantify and explicitly study it, underpins almost all research in dinosaur paleontology. It is impossible to study locomotion, feeding, growth, reproduction, and ecology without having a firm appreciation of form. This chapter will introduce several general explicit methods for studying form, which will continue to be used throughout this book in later discussions of dinosaur feeding, locomotion, and behavior. Some of these methods have briefly been touched upon earlier, but this chapter will describe them more comprehensively and outline several case studies in the literature where they have been employed. The overall goal of each of these methods is to analyze form – body shape, size, and variability – in an explicit, rigorous, and quantitative way. This ever-expanding toolkit of analytical techniques is fueling a revolution in contemporary paleontology. Vague superficial statements that were

Dinosaur Paleobiology, First Edition. Stephen L. Brusatte.
© 2012 John Wiley & Sons, Ltd. Published 2012 by John Wiley & Sons, Ltd.

all too common in the older literature – "species X was probably about the size of an elephant" or "species Y had a hindlimb that probably allowed it to run fast" – are now obsolete and must be eliminated. The emerging use of explicit methodologies, and the folly of imprecise generalizations, is illustrated at the end of the chapter with a look at two focused questions: How big were dinosaurs? Can we recognize the difference between male and female dinosaurs?

Methods for Studying Form

Computed tomographic scanning

Perhaps the single most valuable technological advance in recent paleontological research is the use of X-ray computed tomography, known colloquially as CT scanning (Fig. 5.1; see Plate 12). CT technology is widely used in the field of medicine, as it provides a quick, powerful, and non-invasive means to view internal structures such as organs and tumors, which could only otherwise be accessed via surgery. When applied to fossils, CT scanning allows researchers to observe details of internal anatomy that are concealed within the specimen and virtually impossible to study with the naked eye alone. In essence, CT opens up a new realm of potential anatomical observations and analysis, by allowing paleontologists to digitally peer into the brain cavities, cranial sinuses, marrow cavities, and other internal structures of dinosaurs. It is no surprise therefore that CT imaging has rapidly swept through the field of paleontology, and this trend shows no signs of slowing because scanning technology is continuously becoming cheaper and more widely available. Paleontologists will frequently scan specimens at local hospitals and some especially well-funded researchers now possess CT scanners in their own labs.

Two general reviews of the potential for CT technology in paleontology have been presented by Carlson et al. (2003) and Witmer et al. (2008), and many of the promising potential applications they outlined have already been realized by a flurry of research (Fig. 5.1). In particular, CT scanning is frequently used to study the brain cavities, cranial nerves and vessels, and skull sinuses of dinosaurs (Rogers, 1998; Brochu, 2000, 2003; Franzosa and Rowe, 2005; Sanders and Smith, 2005; Kundrát, 2007; Sampson and Witmer, 2007; Sereno et al., 2007; Zhou et al., 2007; Balanoff et al., 2009, 2010; Brusatte et al., 2009a; Evans et al., 2009; Norell et al., 2009; Witmer and Ridgely, 2009; Zelenitsky et al., 2009). Other workers have used CT to map extensive pneumatic sinuses in vertebrae and appendicular bones (Wedel, 2003a; Sereno et al., 2008), isolate and digitally reconstruct the fragile bones of dinosaur embryos (Balanoff et al., 2008), and build computerized models of certain bones or entire skeletons that can be subjected to biomechanical analyses to help study locomotion, feeding, and other functions (Rayfield et al., 2001; Snively and Russell, 2002; Arbour and Snively, 2009; Bates et al., 2009a; Mallison, 2010a, 2010b).

The primary utility of CT scanning is that it can provide a detailed view of internal anatomy, which was usually intractable by previous methods of studying fossils, without destroying the specimen or demanding several months of meticulous and often risky mechanical preparation. High-end CT scanners are powerful enough to capture resolution at the micrometer scale, detail that even the most highly trained and delicate preparators cannot match with mechanical tools. The series of images captured by a CT scanner – basically a flip-book of slices, a few microns or millimeters apart, stretching across the length of a specimen – can be digitally manipulated using accessible computer software and used to generate a three-dimensional digital model of an internal space such as a brain cavity or sinus. These digital datasets, unlike the often fragmentary and singular specimens themselves, can be shared with colleagues or sent to other researchers with the click of a mouse. As this entire process – scanning specimens, digitally manipulating the scan images, creating virtual models, and sharing them with colleagues – becomes quicker, cheaper, and easier, CT scanning promises to remain at the forefront of methods that paleontologists use to study form.

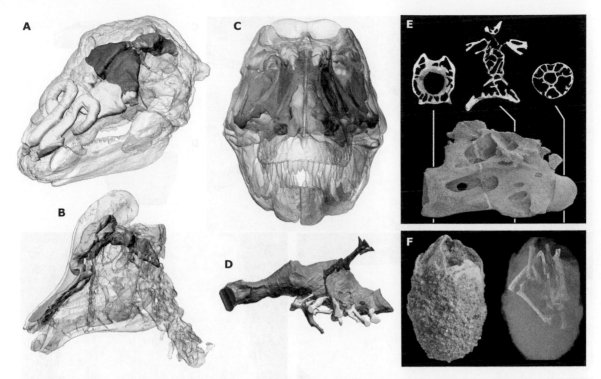

Figure 5.1 The use of CT scanning in dinosaur research. (A) Digital reconstruction of an ankylosaurid skull showing the looping nasal passage at the front of the snout. (B) Digital reconstruction of a hadrosaurid skull showing the internal morphology of the cranial crest and the nasal passage at the front of the snout. (C) Digital reconstruction of the skull of *Tyrannosaurus* showing various cranial sinuses that would have housed air sacs. (D) Digital reconstruction of the endocast of *Tyrannosaurus* showing the general shape of the brain, various cranial nerves, and the semicircular canals of the inner ear. (E) Cervical vertebra of *Apatosaurus*, with CT scans of transverse sections denoted by lines. (F) A ceratopsian egg and a digital image of the small embryonic bones within the egg, which are too fragile to physically prepare from the specimen. Images (A–D) courtesy of WitmerLab at Ohio University and modified from Witmer and Ridgely (2008, 2009) and Evans et al. (2009); image (E) adapted from Wedel (2003b), and reproduced with permission; image (F) courtesy of Amy Balanoff and modified from Balanoff et al. (2008).

Laser surface scanning

CT scanning is invaluable in studying the internal anatomy of fossils. A related technique, laser surface scanning, can capture detailed information on three-dimensional external morphology, which is especially useful for building digital models of dinosaur skeletons (Figs 5.2–5.4). As discussed briefly below, these models can serve a variety of functions. Very few studies have utilized laser scanning to study and model dinosaurs, but this type of work is becoming more common. Laser scanners are small enough to be carried to museum collections and exceedingly user-friendly, as specimens can be scanned quickly and easily, usually by passing a wand-like instrument around the surface of the fossil (see review in Wilhite, 2003). These scanners have recently been used to construct complete and accurate digital models of the "prosauropod" *Plateosaurus* (Gunga et al., 2007), the hadrosaurid *Edmontosaurus* (Sellers et al., 2009), the stegosaurid *Kentrosaurus* (Mallison, 2010c), and several species of theropods (Bates et al., 2009a) (Figs 5.2–5.4). Particularly powerful scanners, which implement light detection and range (LiDAR) imaging technology, have been brought into the field and used to map footprint sites and other in situ fossils that are too large or fragile to bring back to the lab (Bates et al., 2008a, 2008b) (Fig. 5.2). As with CT scanners,

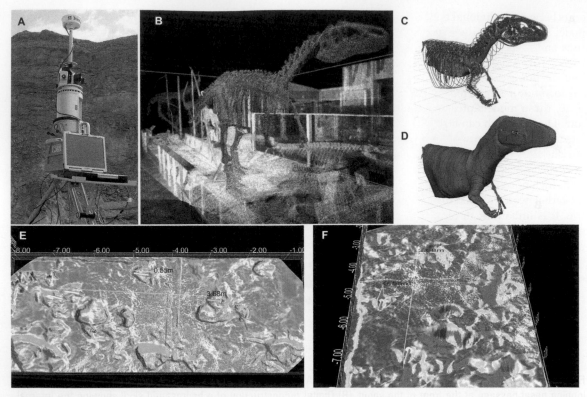

Figure 5.2 The use of LIDAR laser scanning in dinosaur research: (A) LIDAR apparatus in the field; (B–D) step-by-step process for digitizing an *Allosaurus* skeletal mount using LIDAR and constructing a digital model; (E, F) LIDAR digital maps of a dinosaur footprint site. All images courtesy of Dr Karl Bates.

these instruments provide digital three-dimensional images that can be manipulated in several different software programs and easily shared with colleagues.

Digital modeling

Internal and external CT scans and external laser surface scans provide high-resolution images that can be assembled into a digital model (Figs 5.2–5.4; see Plate 13). Oftentimes researchers construct models of an entire skeleton, based on a single well-preserved fossil that is scanned, or of a certain region of the anatomy, such as the skull or foot. These models are a stunning improvement on traditional two-dimensional drawings or photos of a fossil, as they can be rotated, examined from all angles, and digitally manipulated into a variety of poses and postures. Digital models are sometimes presented simply as an illustration to enliven an

anatomical description or as a web animation for public outreach and education. Often, however, these models are used as a springboard for studying aspects of dinosaur biology such as locomotion and speed, body mass, posture, and feeding strategies.

Some of the earliest attempts at explicit computer modeling of dinosaurs were presented by Stevens and Parrish (1999), whose "DinoMorph" computer package was used to digitally study the posture and range of neck motion in large sauropods. This modeling technique, which was reviewed in detail by Stevens (2002) and has since been used in additional analyses of sauropod and theropod biomechanics (Stevens and Parrish, 2005a, 2005b; Stevens et al., 2008), incorporates three-dimensional images of dinosaur specimens, usually generated through laser scanning or other image capturing technologies, into a computer animation program. The ranges of motion of the joints between individual bones are defined and the model is then virtually subjected to different postures and styles of locomotion (see Sellers et al., 2009, for a step-by-step description of this procedure). By doing this, realistic and unlikely postures and locomotor strategies can be identified. In essence, these models present an idealized abstraction of form, which can be used to assess function. For example, in their original paper Stevens and Parrish (1999) argued that sauropods such as *Diplodocus* and *Apatosaurus* could not extend their necks high into treetops, as had long been assumed, but rather were restricted to a low-browsing posture. It is important to remember that this result may or may not be correct – interpretations of function depend on the parameters assumed by the model, in this case the degree of motion between the joints – but the technique is valuable because it requires functional inference to be firmly rooted in a detailed understanding of form.

Computer modeling, both to visually reconstruct the skeletons of dinosaurs and to explicitly test functional hypotheses, is becoming more prevalent in dinosaur research. Gunga et al. (2007, 2008) constructed digital models of the "prosauropod" *Plateosaurus* and the enormous sauropod *Brachiosaurus*, used these to measure the precise surface area and volume of the skeletons, and then used this information to predict body mass (Fig. 5.3A). A similar exercise was performed by Bates et al. (2009a), who focused on theropod and ornithopod dinosaurs (Fig. 5.3B–D). They created digital models of five dinosaur skeletons and used these as scaffoldings around which internal organs, muscles, and skin were reconstructed. These more lifelike reconstructions, which included bones and associated soft tissues, were used to calculate body mass, identify the center of mass of the animal, and study the locomotory capabilities of each species by looking at plausible ranges of motion in the limbs. Recent work by Mallison (2010a, 2010b, 2010c) employed digital models of *Plateosaurus* and *Kentrosaurus* to test many aspects of functional biology, including whether the former species was primarily bipedal or quadrupedal and whether the latter could walk erect (Fig. 5.4; see Plate 13). As is evident, digital models such as these are becoming increasingly necessary when conducting rigorous explicit analyses of function.

Morphometrics

CT scanning, laser scanning, and digital modeling are techniques for imaging morphological form and studying aspects of anatomy, such as internal morphology or the articulations between bones, which are otherwise difficult to assess. But how can morphology, especially the overall shape of a structure or the body plan of an entire organism, be quantified? Anatomical descriptions of dinosaurs are full of generalized statements about anatomy. It is often said that a certain bone, such as the humerus, is long or robust, or that a certain flange on a bone is broad or narrow. Similarly, certain dinosaur species are often described as massive or small, and subgroups of dinosaurs, such as theropods, are said to be morphologically diverse or anatomically conservative. Taken at face value, these statements are not very informative. "Long" and "short" only make sense in a comparative framework (a human would look huge to a mouse but tiny to an elephant), and words such as "pronounced" and "narrow" may have different meanings to different researchers. It goes without saying that dinosaur researchers should strive for explicit descriptions of morphology whenever possible, and should aim to quantify features of the anatomy and make comparisons with other species. In some cases this is straightforward: it is easy to measure the length of a homologous bone in many species and demonstrate that, for instance, the humerus in one species is longer than that in another. But such quantification becomes trickier as more complex features, such as complete skulls or entire skeletons, are considered.

In cases such as these, morphometrics is an invaluable tool for quantifying morphological form and encapsulating anatomical variation between different species or structures (Fig. 5.5). Morphometric analysis is a powerful technique because it takes into account the morphological complexity of a specimen (such as a three-dimensional skull composed of 20 or more bones) and distills what seems to be an intractable quagmire of information into a simplified manageable model that can be described statistically. In essence, morphometrics takes a complex structure and reduces it to an abstraction that can easily be compared with other such abstractions. The exact procedures involved in this process vary, but in general three steps are

Figure 5.3 The use of digital models in dinosaur research. Digital models of the sauropod *Brachiosaurus* (A), the hadrosaurid *Edmontosaurus* (B), the ornithomimosaurian theropod *Struthiomimus* (C), and the tyrannosaurid theropod *Tyrannosaurus* (D). Image (A) reproduced with permission from Gunga et al. (2008); images (B–D) courtesy of Dr Karl Bates and modified from images in Bates et al. (2009b).

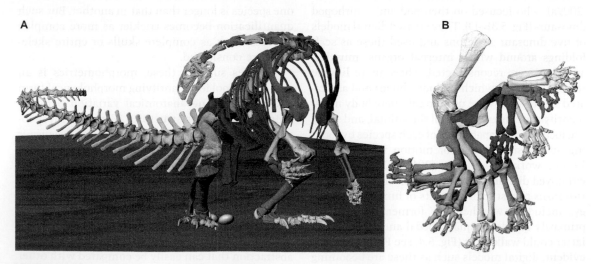

Figure 5.4 Dinosaur digital models used to study range of motion and body posture: (A) model of *Plateosaurus* in an inferred egg-laying posture; (B) range of motion in the forelimb of *Plateosaurus*. Images courtesy of Dr Heinrich Mallison.

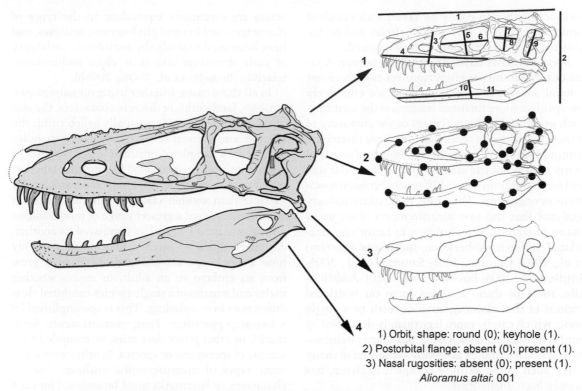

1) Orbit, shape: round (0); keyhole (1).
2) Postorbital flange: absent (0); present (1).
3) Nasal rugosities: absent (0); present (1).
Alioramus altai: 001

Figure 5.5 The use of morphometrics in dinosaur research. This figure illustrates four of the main methods for simplifying a complex structure (in this case, the skull of the tyrannosaurid *Alioramus*) into a model that can be compared statistically with other such models. These methods include representing the skull by several measurements (1), by plotted landmarks (2), by an outline (3), and by a table of discrete character scores (4). *Alioramus* skull courtesy of Frank Ippolito, American Museum of Natural History.

necessary (Fig. 5.5). First, morphological data must be acquired, usually in the form of two- or three-dimensional images generated from traditional photography or CT or laser scanning. Second, complex morphology must be simplified, most commonly by reducing a specimen to a set of basic measurements or an assemblage of plotted landmarks. Third, the simplified abstractions – perhaps a set of measurements or landmarks summarizing the skull shapes of an array of dinosaurs – are subject to statistical analyses that quantify the variability in shape and size between the specimens. These procedures are described in more detail in the general reviews of Rohlf and Marcus (1993), O'Higgins (2000), and Adams et al. (2004) and the books of Bookstein (1991), Elewa (2004), and Zelditch et al. (2004).

The second step in morphometric analysis – summarizing complex morphology with a simplified set of proxies – is the single most difficult, but important, part of the procedure. The quality of the analysis stands or falls on how well complicated morphology, with all its qualitative nuances, is distilled into a quantitative dataset. There are many techniques that researchers use, and these are constantly evolving with improvements in image capturing technology and statistical software (Fig. 5.5). Many studies in the older literature represented specimens with a series of measurements, such as the length, width, and thickness of a structure or the straight-line distances between various bones (Dodson, 1976; Chapman et al., 1981; Weishampel and Chapman, 1990). These techniques have the advantage of being simple and straightforward:

measurements can easily be taken with standard equipment such as rulers and calipers and no sophisticated imaging technology is required.

Measurements have drawbacks, however. A table of measurements often captures only the most general aspects of shape and may not effectively encapsulate more nuanced features of the anatomy, such as individual bony flanges or the curvature of structures. More problematic, raw measurements summarize information on both shape and size, and many morphometric studies wish to factor out size and focus solely on shape differences between specimens or organisms. Therefore, more contemporary analyses that use raw measurements often use a litany of statistical techniques to factor out size-related differences between specimens (Carrano et al., 1999; Carrano, 2001; Samman et al., 2005; Smith et al., 2005; Buckley et al., 2010). Additionally, most of these analyses focus on restricted regions of the anatomy, such as teeth or a single bone, which can be more legitimately described by simple measurements. Using only measurements to describe a complex structure composed of many parts, such as the skull or complete skeleton, has largely been abandoned.

Three other techniques are now commonly used to summarize the morphology of complex structures (Fig. 5.5). First, the outline of a specimen can be traced. This technique works well for simple structures such as leaves or trilobite carapaces, and a version of this method was used by Smith et al. (2005) to quantify variation in the shape of dinosaur teeth, but it is not ideal for studying extremely complicated structures such as skulls. Second, a structure can be reduced to a set of plotted data points, or landmarks. These landmarks, which are equivalent to plotted coordinates in two- or three-dimensional space, are usually placed at major sutures between bones or other important positions on a specimen (such as the most anterior point, the point of greatest curvature, etc.). This method is widely employed in the dinosaur literature and has been utilized in a number of studies (Chapman, 1990; Chapman and Brett-Surman, 1990; Dodson, 1993; Egi and Weishampel, 2002; Bonnan, 2004, 2007; Chinnery, 2004; Snively et al., 2004; Young and Larvan, 2010). Third, discrete characters, such as the presence or absence of structures, can be compiled and scored for an organism. These char-acters are essentially equivalent to the types of characters used in most phylogenetic analyses, and have been used to study the anatomical variability of early dinosaurs and their close archosaurian relatives (Brusatte et al., 2008a, 2008b).

In all these cases, whether using measurements, outlines, landmarks, or discrete characters, the aim of a morphometric study is usually to determine the amount and pattern of morphological variability present in a sample of organisms. For instance, a researcher may be interested in knowing whether one dinosaur subclade was more morphologically variable than another clade, or whether a certain subclade exhibited a greater range of morphologies during one time interval as compared to another. Alternatively, a researcher may want to study how a single species changed shape as it grew from an embryo to an adult, or assess whether males and females of a single species exhibited clear differences in morphology. This is accomplished by a two-step procedure. First, measurements, landmarks, or other proxy data must be compiled for a sample of specimens or species. In other words, the same types of measurements, outlines, discrete characters, or landmarks must be assessed for each species. When using the latter, consistent landmarks must be marked at homologous positions on a number of specimens. For instance, the ventralmost point on the premaxilla–maxilla suture and the most anterior tip of the snout, among other landmarks, may be plotted on each image of a theropod skull in a morphometric analysis of cranial shape in carnivorous dinosaurs.

Once available, the dataset of morphological proxies, scored across a range of specimens or species, is subjected to multivariate statistical analysis (Fig. 5.6). The most commonly used multivariate technique is principal components analysis (PCA), but other methods include principal coordinates analysis (PCO, also known as multidimensional scaling) and canonical variates analysis. All these techniques are similar in that they summarize information from the entire gluttony of measurements or landmarks and present a smaller, more manageable set of variables to describe each specimen. These variables are the equivalent to graphical coordinates, such as the x and y coordinates on a bivariate plot. Therefore, using these variables, specimens can be plotted in a morphospace: a mul-

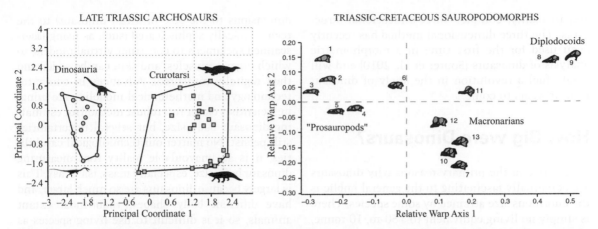

Figure 5.6 The use of morphometrics in dinosaur research. Two examples of morphospaces – visual "maps" in which specimens are plotted according to how similar/different their morphologies are from each other: (A) morphospace for Late Triassic archosaurs, including early dinosaurs and their contemporaries, derived from a discrete character dataset; (B) morphospace for Mesozoic sauropodomorphs, derived from a landmark-based morphometrics analysis. Image (A) compiled by the author based on figure in Brusatte et al. (2008b); image (B) based on figure in Young and Larvan (2010) and reproduced with kind permission from Springer Science + Business Media.

tidimensional graph composed of numerous axes that represent different aspects of morphological shape (Fig. 5.6). Based on the positions of specimens or species in morphospace, differences in morphology can be quantified, the spread of morphological variability can be assessed (usually by calculating range and variance statistics), and unpopulated regions of morphospace (which represent unrealized morphologies) can be identified.

Although dinosaurs were some of the first fossil vertebrates subjected to morphometric and morphospace analysis in the 1970s, this type of research has taken a back seat to phylogenetic and functional morphology studies in contemporary dinosaur paleontology. Nevertheless, morphometrics has been valuable in quantifying major morphological differences among certain dinosaur subclades and has been used as a framework for biomechanical and other functional studies. For instance, Chapman and Brett-Surman (1990) and Dodson (1993) used landmark-based morphometrics to identify significant differences in skull shape among different species of hadrosaurs and ceratopsians, respectively, whereas Young and Larvan (2010) used similar techniques to show that sauropodomorph cranial form became progressively more divergent in more derived species (Fig. 5.6).

Concerning functional morphology, Chinnery (2004) employed a morphometric technique to illustrate that ceratopsian postcranial elements became larger and more robust during the evolution of larger, more derived forms, presumably to offer additional body support, and Bonnan (2004) quantified differences in sauropod limb bone shape that he suggested related to locomotor differences between species. Dodson (1976) and Chapman et al. (1981) used morphometrics in an attempt to distinguish males from females in ceratopsian and pachycephalosaurian dinosaurs, and Egi and Weishampel (2002) quantified shape differences between juvenile and adult hadrosaurids.

This is but a small sample of the types of morphometric analyses that are possible, and similar studies promise to become more prevalent as new techniques are developed and the relevant software becomes more accessible. In particular, it is now possible to plot landmarks on three-dimensional images using standard desktop computers and software programs, a stunning technique that pioneers of coarse measurement-based studies would have found unfathomable. This method permits unprecedented quantification of dinosaur morphology, and obviates problems with plotting landmarks or outlines on two-dimensional projections of what are, in reality, complex three-dimensional structures. The three-dimensional method has recently been used for the first time in a morphometric analysis of dinosaurs (Souter et al., 2010) and will surely fuel a revolution in the study of dinosaur form in years to come.

How Big were Dinosaurs?

Surely, one of the primary reasons why dinosaurs are perpetually fascinating to the general public is the enormous size attained by some species. There is simply no living equivalent to a 50-m, 10-tonne, plant-guzzling sauropod dinosaur. Scientists, too, have long been interested in studying the range of dinosaur body size, in terms of both length and mass. Many fundamental questions have been the subject of intense debate, but remain without a convincing answer today. How were some dinosaurs able to achieve such great size? Did colossal dinosaurs such as sauropods or *Tyrannosaurus* grow faster than other dinosaurs? How early in the evolution of dinosaurs did truly massive size evolve? Each one of these questions is underpinned by an even simpler and more problematic question: how can scientists accurately estimate the body size of dinosaurs, especially the mass of colossal species with no modern analogues?

It is common to read vague statements about body size in the dinosaur literature, even in contemporary publications. Species are sometimes said to be roughly the same size as a certain modern animal, such as a cow or an elephant, and often it is simply stated that a certain species was of a certain approximate length or tonnage. The situation is even more egregious in the popular press. Every few months it seems as if another discovery is trumpeted as the "world's biggest dinosaur," and even less sensational fossils are often breathlessly compared to the sizes of buildings or buses. Hyperbole may have its place in promoting new discoveries to the general public, but from a scientific standpoint it is important to estimate the length and mass of dinosaur species using a consistent rigorous methodology that, hopefully, yields an accurate answer. Many such techniques have been proposed, ranging from the somewhat crass (such as constructing miniature models and scaling up their

dimensions to the size of a full dinosaur) to the technologically sophisticated (such as using laser-scanned specimens to construct virtual models, to which organs, muscles, and skin can be added and then digitally measured). Advances in computer technology and mathematical modeling are bringing scientists closer to a robust method for estimating dinosaur body size, but several uncertainties still persist. No matter our technological capabilities, it is an unavoidable reality that measuring dinosaur body size, especially mass, isn't easy. This is largely because dinosaurs are so much larger, and have different body shapes, than most extant animals, so it is difficult to use living species as a comparison.

The first repeatable, explicit, quantitative method for estimating dinosaur size was presented by Gregory (1905), who in a prescient paper in *Science* used a scale model to estimate the mass of the enormous sauropod *Apatosaurus*. Gregory's (1905) technique was straightforward and could easily be accomplished by any researcher with basic tools at his or her disposal. First, an accurate model of the dinosaur in question is produced. During Gregory's time such models were hand sculpted from clay or plaster, but today they can be digitally assembled using high-resolution CT or laser scans in a computer animation program. Second, the scale of the model compared with the actual specimen is measured. Third, the volume of the model is determined. This can be assessed digitally with modern techniques, but in Gregory's time the most common methods included immersing the model in sand or water and noting how much material was displaced. Fourth, the volume of the model is multiplied by the assumed density (mass per volume) of a living dinosaur, usually held to be approximately equivalent to the density of water (1000 kg/m^3), and then multiplied by the scaling factor of the model to yield the estimated mass of the full-sized dinosaur.

This method was followed by Colbert (1962), who constructed and analyzed a representative set of scale models for several major dinosaur groups, and later by Alexander (1985, 1989) and many other authors. Indeed, before the advent of computers, this technique was the only available method for estimating dinosaur mass with any type of repeatable quantitative rigor. Colbert's (1962) study reported a mass of approximately 7 tonnes for

Tyrannosaurus, 8.5 tonnes for *Triceratops*, 28 tonnes for *Apatosaurus*, and a jaw-dropping 78 tonnes for *Brachiosaurus*. Not all dinosaurs, however, were found to be heavyweights: the primitive ceratopsian *Protoceratops*, for instance, was estimated at only 177 kg. Similar values were reported by Alexander (1985), although in some cases estimates for the same species were widely divergent in both studies, due to slight differences in the shapes of the models used. This underlines one of the fundamental flaws of this method: there is always some subjectivity in creating scale models, especially those made by hand, and even subtle differences can lead to enormous differences in estimated mass because the model itself must usually be scaled up by several orders of magnitude. This can even be true if two researchers agree on every detail of a reconstructed skeleton, because soft tissues such as muscle and skin must be overlain on the skeletal chassis. Fossils can very rarely guide these sorts of decisions, because soft tissues are rarely preserved, so this aspect of model construction is almost entirely conjectural.

Additionally, the assumption that all dinosaurs had the same density as water may not be a safe supposition. It is true that the majority of an animal's body is composed of water, but the skeletons of many dinosaurs were invaded by a maze of pneumatic air sacs that hollowed out the interior of many bones, especially the skulls and vertebrae. In short, these pneumatic sacs replaced heavy bone with air, which essentially has no mass. The presence of extensive air sacs, therefore, probably would have substantially lightened the skeleton, and for this reason alone it is likely that many mass estimates presented by Colbert (1962) and Alexander (1985), especially the astronomically high estimates for *Brachiosaurus* and other sauropods, are overambitious. Indeed, an interesting digital experiment by Witmer and Ridgely (2008) provides a specific example of this problem. They used CT scans to build accurate three-dimensional digital models of the skulls of *Tyrannosaurus* and *Majungasaurus*, which included the full array of cranial pneumatic sinuses. By measuring the volume of the skull with the air-filled sinuses intact, and then comparing with the volume calculated when the sinuses were filled with bone, they found that the actual skull had approximately 15–20% less bone

by volume than the theoretical solid skull. This would translate into a real skull mass 5–10% lower than that assumed using the density of water.

Although still discussed in the literature and sometimes utilized in rare cases, the modeling techniques of Gregory (1905), Colbert (1962), and Alexander (1985) are now known to be problematic. More recent work has utilized a combination of high-resolution imaging technologies (CT and laser scans), computer animation software, and robust statistical techniques to build and analyze more accurate digital models. Although some problematic assumptions remain, such as the amount and types of soft tissue to add to a skeleton, these thorny problems can be assessed by sensitivity analyses that vary parameters such as muscle mass, soft tissue density, and extent of internal air sacs (see Henderson and Snively, 2004; Gunga et al., 2007; Bates et al., 2009a, 2009b) (Fig. 5.7). This allows the researcher to get a quantitative handle on the amount of variability introduced by different assumptions. In some cases, estimated masses vary widely when different assumptions are employed, whereas in some models even a wide range of assumed soft tissue and air sac parameters make little difference in the predicted mass.

These new breeds of digital models are usually built in one of three ways. The most common approach is to three-dimensionally laser scan an entire skeleton, import the data into a computer animation program, and digitally build a model that can then be statistically analyzed. This procedure was followed by Gunga et al. (2007), who analyzed the body mass of the "prosauropod" *Plateosaurus*, and by Bates et al. (2009a, 2009b), who studied theropod body mass, and its implications for locomotion, by building models of several genera, including *Acrocanthosaurus* and *Tyrannosaurus* (see Fig. 5.3). Surprisingly, their estimated masses for *Tyrannosaurus* were about the same magnitude – 7 tonnes – as those estimated by the handmade scale modeling technique of Colbert (1962) and Alexander (1985). *Allosaurus*, however, was estimated at approximately 1500 kg, approximately 30% less than estimated by Colbert (1962). A second method, the so-called "photogrammetrical" approach, uses a computer-aided design program to combine several high-resolution photographs, taken from various angles, into a composite

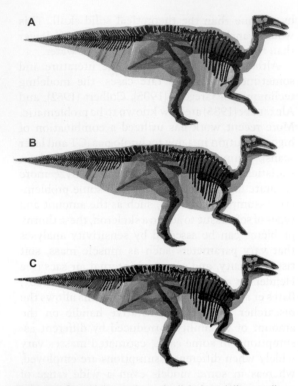

Figure 5.7 Sensitivity analyses in digital modeling studies. The images depict three digital models of the hadrosaurid *Edmontosaurus* with varying degrees of muscle and soft tissue: (A) rotund, (B) moderate, and (C) gracile. Images courtesy of Dr Karl Bates and modified from Bates et al. (2009b).

three-dimensional restoration of the skeleton in question (Gunga et al., 1995, 1999, 2008). When Gunga et al. (2008) utilized this technique to estimate the mass of *Brachiosaurus*, they generated a prediction of 38 tonnes, less than half of what was estimated by Colbert's (1962) procedure (Fig. 5.3A).

Finally, a third method was outlined by Henderson (1999), which uses two-dimensional reconstructions of an animal in lateral and dorsal views to create a three-dimensional computer model (Fig. 5.8). A similar technique presented by Seebacher (2001) utilizes body outlines that are fitted to polynomial equations, which can then be integrated using calculus to predict body mass. Both methods were validated using extant animals, whose actual masses were estimated extremely well by the various mathematical techniques. Henderson

(1999) also recovered an estimated mass for *Tyrannosaurus* similar to that reported by Colbert (1962) and Bates et al. (2009a), but reported a mass of 13 tonnes for *Diplodocus*, which is larger than the 10 tonnes predicted by Colbert (1962). Seebacher (2001) reported a mass of only 952 kg for *Allosaurus*, less than half of that estimated by Colbert (1962). Similarly, his estimate for *Brachiosaurus* (28 tonnes) is substantially less than Colbert's, whereas his prediction for *Diplodocus* (20 tonnes) is much greater than that of Colbert's. *Tyrannosaurus* was estimated at approximately 6.7 tonnes, of the same general magnitude as the mass estimates recovered by all methods. Therefore, despite all the uncertainties in estimating dinosaur mass, we can be quite confident that *Tyrannosaurus* had a mass in the range of 5–7 tonnes.

But what about the body mass of dinosaurs that are not known from complete, or even near complete, skeletons? This is the reality for the vast majority of dinosaur species. All the above modeling techniques, both the coarse methods of Gregory (1905) and the more sophisticated computer and mathematical approaches that have gained traction in recent years, require a nearly complete skeleton. When only a few fragmentary fossils are available it may be possible to roughly compare the fragmentary material to homologous bones in a more complete specimen whose mass has been estimated by modeling. For instance, if a radius the size of an average *Allosaurus* radius is all that is known from a new species, it is a good ballpark guess that this new species was approximately the same mass as *Allosaurus*. These crass approximations, however, are only sufficient for general characterization of a specimen. If one wishes to study body size in more detail – for instance, if one aims to assess trends in body size evolution through the history of a subclade or how body mass changed during growth – more accurate estimations are needed.

In these cases, regression equations are useful. Body mass estimates derived from complete specimens can be plotted against various measurements of individual bones from those specimens, creating a bivariate plot and, if the data are clean enough, statistically significant regression equations (Fig. 5.9). Measurements from fragmentary skeletal material can be inserted into these equations and body mass predicted. Some of the most useful

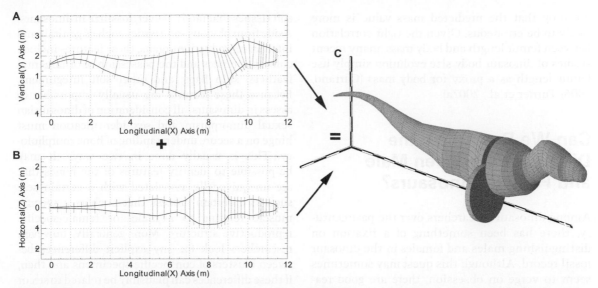

Figure 5.8 A mathematical method for estimating the body masses of dinosaurs. A theropod skeleton is shown in lateral (A) and dorsal (B) views, and these images can be combined mathematically into a three-dimensional digital model used to estimate body mass. Modified from Henderson (1999). Used with permission from the Paleontological Society.

regression equations were presented by Anderson et al. (1985) and Christiansen and Fariña (2004), and many have been verified using living vertebrates. These studies found that the most confident predictor of body mass is either femur length or femur circumference (Fig. 5.9). In both living animals and extinct dinosaurs for which good mass estimates are available, there is a tight correlation between both of these femoral measurements and body mass, meaning that if one is able to measure the femur in a fragmentary specimen then body mass can be predicted with confidence and rigor. Other measurements, such as the length of the tibia or fibula, can also be used to predict body mass, but there is not such a tight correlation between these measurements and mass in complete specimens,

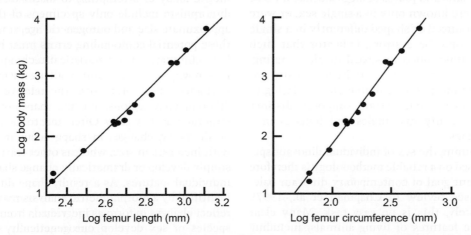

Figure 5.9 Plots showing how body mass of theropod dinosaurs can be estimated quite confidently based on either femur length or circumference. Data from Christiansen and Fariña (2004).

meaning that the predicted mass value is more likely to be erroneous. Given the tight correlation between femur length and body mass, many recent studies of dinosaur body size evolution simply use femur length as a proxy for body mass (Carrano, 2006; Turner et al., 2007a).

Can We Recognize the Difference Between Male and Female Dinosaurs?

Among dinosaur researchers over the past century, there has been something of a fixation on distinguishing males and females in the dinosaur fossil record. Although this quest may sometimes seem to verge on obsession, there are good reasons for being able to dependably identify the sex of dinosaur specimens. Most generally, understanding dinosaur sex offers tantalizing insight into dinosaur biology and ecology. Information on the sex and ontogenetic age of individual specimens would allow researchers to study the demographics of dinosaur populations, and being able to distinguish whether specimens are males or females would enable recognition of anatomical and biological differences between sexes (referred to as "sexual dimorphism"). A firm recognition of sex may help researchers understand the function of so-called "bizarre structures" such as the horns of ceratopsians and plates of stegosaurids. If these features were known only in a single sex, or were more pronounced or shaped differently in a single sex, this would be a strong indicator that such structures functioned in sexual display, mating, or sex-specific biological activities. Finally, understanding morphological differences relating to sex may reveal that certain supposed distinct species are really just males and females of a single species.

Determining the sex of individual dinosaur specimens based on a reliable methodology is therefore an important goal of contemporary dinosaur paleontology (see review in Chapman et al., 1997). Unfortunately, this is no easy task. Many clear sex-specific features of living animals, including the birds and crocodiles that form the extant phylogenetic bracket for dinosaurs, are composed of soft tissues that rarely, if ever, fossilize in dinosaurs. Aside from obvious soft tissues such as genitals and internal reproductive organs, these features include color differences and the possession of ornamental feathers, skin flaps, and other soft integument. Because these features are usually impossible to assess in dinosaurs, all consideration of dinosaurian sexual dimorphism and sex identification must hinge on a secure understanding of bone morphology. This is usually done in two ways. It may be possible to identify features of the bones that are unequivocally associated with a certain sex, such as a certain process that anchors a specific muscle relating to a male- or female-specific reproductive structure. More generally, however, researchers look for any explicit differences between clusters of conspecific specimens and then, if these differences can plausibly be related to sex or reproduction, interpret them as possible instances of dimorphism.

In order to plausibly identify variability that may be caused by sexual dimorphism, researchers must analyze large samples of specimens and find clear-cut differences in size, shape, or possession of discrete features such as crests or horns. There are many methodological limitations that make this exercise difficult. Morphometric techniques, such as those outlined above, are invaluable in objectively recognizing size and shape differences, but the utility of these analyses depends on the quality of data that are analyzed. It is critical that morphometric analyses attempting to distinguish sexual dimorphism include only specimens of the same approximate size and ontogenetic age; if not, then these potential confounding errors must be somehow eliminated using statistical techniques. The reason why is obvious. Almost all organisms exhibit changes in absolute size, the relative sizes of different body regions, and the shapes of various structures as they grow. Often, due to the principle of allometry, changes in shape occur in tandem with increases in size, whereas other features may simply develop or dramatically change shape as an individual matures. As a result, shape differences identified by a morphometric analysis may simply reflect changes in form as individuals from a single species or sex develop ontogenetically, and not discrete differences between the sexes (Chapman et al., 1997).

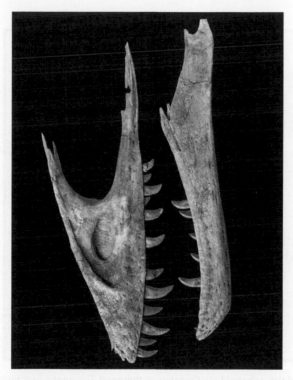

Plate 1 The maxilla and dentary of the tyrannosaurid theropod *Alioramus altai* from the Late Cretaceous of Mongolia. The maxilla is 43 cm long anteroposteriorly. Photograph by Mick Ellison.

Plate 2 A life reconstruction of the colossal Late Cretaceous theropod *Tyrannosaurus*. Illustration by Jason Brougham.

Plate 3 An ontogenetic growth sequence of the Late Cretaceous theropod *Tyrannosaurus,* showing how an individual changed in size and body proportions as it grew from a hatchling into a 6-tonne, 13-m long adult. Illustration by Jason Brougham.

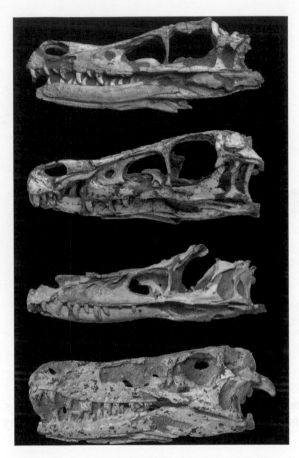

Plate 4 The skulls of the dromaeosaurids *Velociraptor* (top three images) and *Tsaagan* (bottom image). Photograph by Mick Ellison.

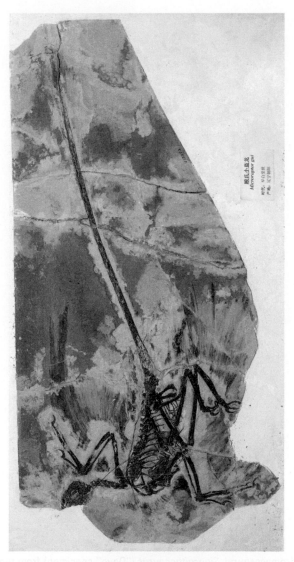

Plate 5 The small, four-winged, feathered dromaeosaurid *Microraptor* from the Early Cretaceous of China. Photograph by Mick Ellison.

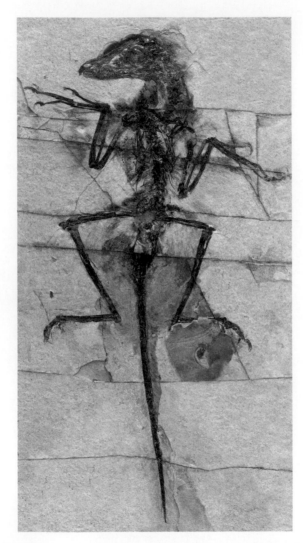

Plate 6 The small feathered dromaeosaurid *Sinornithosaurus* ("Dave" specimen) from the Early Cretaceous of China. Photograph by Mick Ellison.

Plate 7 Close-ups of the filamentous feathers along the head (top) and more complex feathers along the forearm (bottom) of the small feathered dromaeosaurid *Sinornithosaurus* ("Dave" specimen) from the Early Cretaceous of China. Photograph by Mick Ellison.

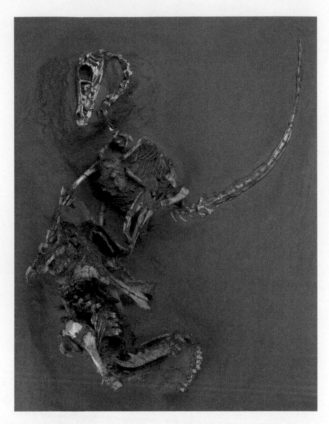

Plate 8 The "fighting dinosaurs" specimen from the Late Cretaceous of Mongolia, in which a dromaeosaurid (*Velociraptor*) is preserved in combat with a ceratopsian (*Protoceratops*). Photograph by Mick Ellison, with assistance from Denis Finnin.

Plate 9 A life reconstruction of the dromaeosaurid theropod *Velociraptor*, one of the closest dinosaurian relatives to birds. Quill knobs on the ulna (forearm bone) of one *Velociraptor* specimen, as well as the possession of feathers in close relatives, strongly indicates that this familiar theropod possessed a coat of feathers. Illustration by Jason Brougham.

Plate 10 A life reconstruction of a parent (the oviraptorosaur theropod *Citipati*) protecting several of its offspring. Illustration by Jason Brougham.

Plate 11 A life reconstruction of a parent (the troodontid theropod *Troodon*) brooding a nest of its eggs. An adult *Troodon* found fossilized in association with a nest indicates that *Troodon* actively brooded its eggs, and various lines of evidence (especially comparison of the ratio of body size of the adults and the size of the egg clutch with those ratios in living birds) suggest that males may have been the primary caregivers. The foreground plant is *Sapindopsis* sp. from the Late Cretaceous Two Medicine Formation of Montana, USA. Illustration by Jason Brougham.

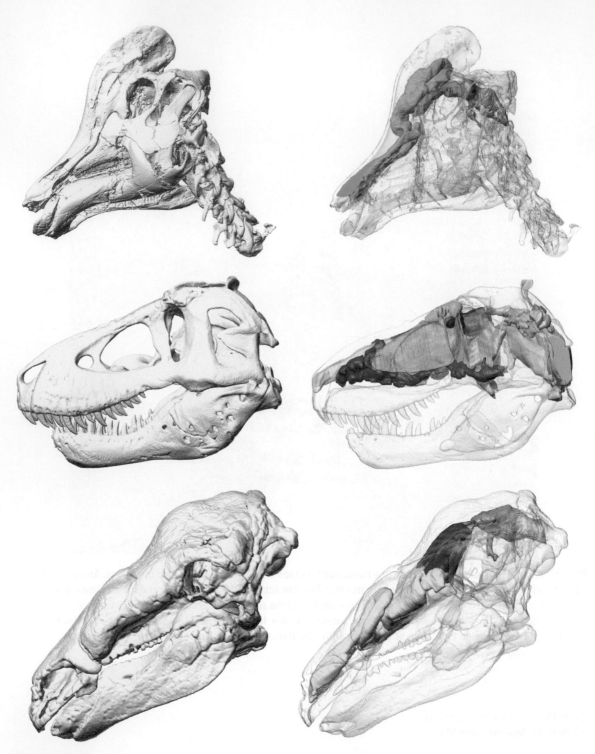

Plate 12 Digital models of the external and internal skull morphologies of a lambeosaurine hadrosaurid (top), tyrannosaurid theropod (middle), and ankylosaurid ornithischian (bottom), as revealed by CT scanning. The brain of the lambeosurine and ankylosaurid are shown in blue and the other colors (in all images) denote internal spaces such as sinuses and the nasal cavity. Images courtesy of WitmerLab at Ohio University, and modified with permission from Evans et al. (2009) and Witmer and Ridgely (2008, 2009).

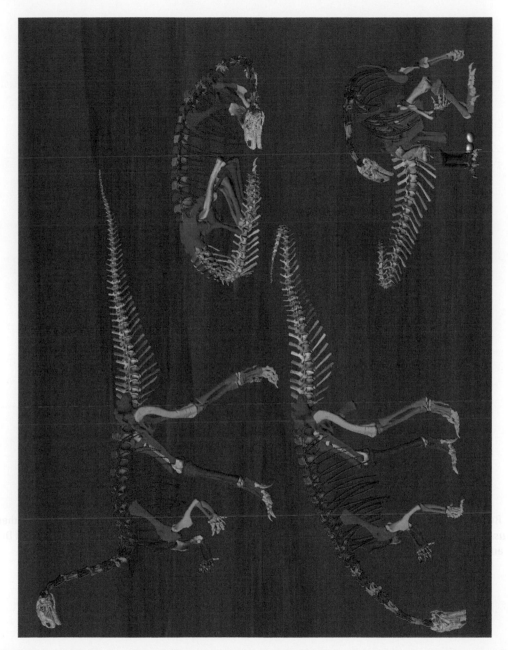

Plate 13 Digital models of the Late Triassic basal sauropodomorph *Plateosaurus* in a variety of poses. Images courtesy of Dr Heinrich Mallison. See Mallison (2010a, 2010b) for more details.

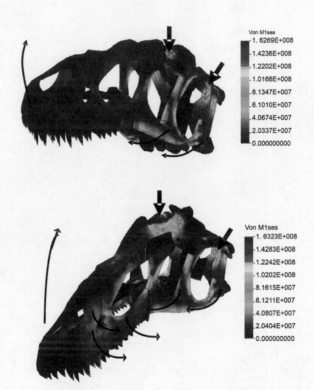

Plate 14 Results of a finite element analysis used to study the skull strength and biting behavior of the predatory theropod *Allosaurus*. Hotter colors (more red on the scale on the right) indicate regions of higher stress. Image courtesy of Dr Emily Rayfield and modified from Rayfield (2005b), with permission.

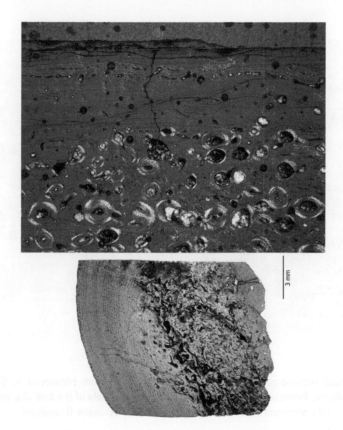

3 mm

Plate 15 Histological thin sections of the dwarf sauropod *Europasaurus*. Images courtesy of Dr Martin Sander.

Plate 16 A paleoecological reconstruction of the dinosaur-rich ecosystem preserved in the middle Cretaceous (c.112 million years ago) Antles Formation of Oklahoma, USA. Four individuals of the bird-like dromaeosaurid theropod *Deinonychus* pursue a juvenile ornithopod *Tenontosaurus*. Illustration by Jason Brougham.

Furthermore, two other problems are endemic to many studies of sexual dimorphism in the dinosaur fossil record. First, in order to identify differences in form that may plausibly be due to sexual dimorphism, it must be certain that those features are authentic and not simply a figment of crushing or poor preservation. For instance, it is often said that a certain species has a "robust" and "gracile" morphotype, but in some cases these differences may be solely preservational and not reflect genuine biological variability. Second, robust inference of sex-related differences must be based on a large sample of individuals that, ideally, comes from a single population. Otherwise, geographical variation, or subtle changes in the morphology of a single species that evolve over time, could be mistaken for sex-specific differences. In some cases paleontologists can study rapidly accumulated bonebeds, which plausibly represent a single population that lived and died together, but often specimens from across a large geographical area or a temporally extensive formation must be pooled to achieve the large sample sizes required for rigorous statistical analysis. This is not ideal. For instance, a study of shape variation in *Tyrannosaurus*, using specimens from Upper Cretaceous deposits across western North America, may identify statistically supported clusters of specimens. However, do these clusters represent males and females, or could they be geographic variants or even different species or subspecies that were separated from each other in time?

Studying sexual dimorphism in dinosaurs, therefore, is not straightforward. However, all the difficulties inherent in identifying dimorphism have not deterred researchers. Early workers, such as Nopcsa (1929), observed a sample of fossils, noted differences, and simply guessed that a certain feature or general morphotype may be due to sexual dimorphism. These studies were groundbreaking, and set an importance precedent for future work, but the approach was crude and sometimes resulted in embarrassing errors. Nopcsa, for instance, suggested that all crested lambeosaurine hadrosaurids were males and the crestless, or more modestly crested, saurolophines were females. Even at that time, however, most other researchers scoffed at this suggestion and favored the explanation that is obvious today: lambeosaurines and saurolophines

are different subclades within Hadrosauridae, and their morphological differences are due to the fact that they are only distantly related to each other. Although superficial comparisons such as Nopcsa's have largely been superseded by more objective statistical techniques, modern researchers often make similar general comparisons when faced with small sample sizes. For instance, Colbert (1989, 1990) and Carpenter (1990) argued that the theropods *Coelophysis* and *Tyrannosaurus*, respectively, could be separated into robust and gracile morphs. These conclusions were based on naked-eye observation of a handful of fossils and were not corroborated by detailed statistical analyses of numerous specimens, which would have verified whether the supposed robusticity differences between specimens were quantitatively significant. Furthermore, the specimens analyzed were not all the same general size and it was unknown whether they were all approximately the same age.

Over the past several decades researchers have become more explicit and quantitative when attempting to study dimorphism. Most current analyses assess large samples of specimens using morphometric techniques, meaning that specimens are objectively compared with each other using either measurements or landmarks (see above). The first meticulous studies of this nature were presented by Dodson (1975, 1976), who compiled numerous measurements on skulls of Late Cretaceous hadrosaurids and the basal ceratopsian *Protoceratops*. These measurements were assessed with a battery of statistical techniques, including multivariate analysis, which identified statistically significant clusters of specimens that were interpreted as likely males and females. Identification of plausible sexual dimorphs corroborated earlier ideas of Brown and Schlaikjer (1940), Kurzanov (1972), and Hopson (1975) that the fantastic crests of hadrosaurids and frills and horns of ceratopsians were probably subject to sexual variation.

Additional studies soon followed on the heels of Dodson's (1975, 1976) pioneering analyses. Dodson's (1976) study of *Protoceratops* was followed by a similar morphometric analysis, this time using landmarks instead of raw measurements, and similar clusters of specimens were recovered (Chapman, 1990). Both studies indicated that one sex, probably males, had a longer and wider frill and more prominent nasal horns. Lehman's (1990) morphometric analysis of *Triceratops* also found significant differences in horn and frill morphologies among what were once considered different species, but what he argued to represent males and females of one or two species. In particular, one morphotype had more erect horns, whereas another cluster of specimens had horns that inclined forwards. Recent work, however, has shown that the horn orientation of *Triceratops* changed from forward-sloping to erect as an individual became mature (Horner and Goodwin, 2006). This underscores a major, but understandable, flaw in the morphometric studies of Dodson (1976), Chapman (1990), and Lehman (1990): there was no attempt to analyze only those specimens of a certain age class. Determining the age of a specimen was exceedingly difficult at the time of these studies, as histological techniques that are widely used today were not yet available. Reanalyzing these trailblazing morphometric studies with reference to new ontogenetic data, and assessing new case studies in a combined morphometric and histological framework, promise to be rich avenues for future research on sexual dimorphism.

Morphometric techniques have been used to assess possible dimorphism in a number of additional studies. In another pioneering study, Chapman et al. (1981) collected several measurements on a number of specimens of the pachycephalosaurid *Stegoceras*, subjected them to multivariate statistical analysis, and found evidence for significant clusters of highly domed and more flat-headed specimens, which they considered as plausible males and females, respectively. Weishampel and Chapman (1990) used multivariate analysis of femoral measurements in the "prosauropod" *Plateosaurus* to argue that subtle sexual variation was present, whereas Raath (1990) used more simple statistical techniques to identify potential dimorphic variation in femur robusticity in a population of the primitive theropod *Syntarsus*. Recently, Barden and Maidment (2011) used geometric morphometrics to identify two major classes of femur shape in specimens of the stegosaur *Kentrosaurus*, which they showed were independent of size differences between the specimens and interpreted as sexual dimorphism. By and large, however, these studies suffer from the same drawbacks as those discussed

above: the simultaneous analysis of juvenile and adult specimens, the pooling of specimens from a wide time and geographic interval instead of a single population, and little consideration of the deformation of specimens. Because of these issues, there is not a single clear-cut case of sexual dimorphism that has been unequivocally demonstrated by a morphometric analysis.

Morphometrics, however, may not be the ideal method for identifying dinosaur sex. All these analyses aim to identify statistically significant clusters of specimens that are characterized by a combination of certain morphological features (such as erect horns, broad frills, etc.). Even if these analyses are able to pinpoint such clusters, it is always possible that they reflect differences other than sex (such as different species or ontogenetic stages). In other words, it is not always straightforward to interpret clustering in a morphometric plot and, with this in mind, it is important to search for explicit features of the dinosaur skeleton that may be unequivocally linked to sex. For instance, perhaps males possess a certain set of muscles to control their reproductive organs that leaves a clear osteological correlate on a bone, or the architecture of the female pelvis may be different from that of males in order to facilitate egg-laying. These features, if correctly identified, would be a one-to-one correlate of sex, and would not require large sample sizes or statistical techniques to validate their reality.

Searching for explicit indicators of sex has become a major theme in dinosaur research and several suggestions have been offered. In order for a hypothesized sexual indicator to be valid, however, there must be a convincing reason why that feature should only be present in a male or female. Usually this is supported by identifying the same feature in birds and crocodiles, the closest living relatives of dinosaurs. Applying this strict test has helped validate some potential sexual indicators in dinosaurs, whereas it has exposed others as invalid. For instance, in perhaps the most familiar case in the literature, Larson (1994) suggested that the first tail chevron of male dinosaurs was larger than that of females, in order to anchor the "penile retractor muscles," and occupied a more anterior position on the tail than in females, lending females more space for storing and laying eggs. Therefore, a dinosaur specimen with a tiny, posteriorly positioned first chevron could be identified as female. However, an explicit study of chevron size and position in living crocodiles demonstrated that the first chevron attached in the same position along the tail regardless of sex, and that there was no clear difference in first chevron size or shape between males and females (Erickson et al., 2005). Because Larson's (1994) hypothesis cannot be demonstrated in living animals, there is little reason to believe that it can be used to identify the sex of a dinosaur specimen. Similarly, Carpenter (1990) suggested that female *Tyrannosaurus* had a wider pelvis, especially more divergent ischia, to accommodate a larger passage for eggs. However, a recent neontological study by Prieto-Márquez et al. (2007) showed that in living alligators it is actually males that have a wider pelvis, and females a deeper pelvis, but that these differences only become apparent in sample sizes much larger than those available for most dinosaur species.

Seeing as how neontological comparisons have wounded the hypotheses of Carpenter (1990) and Larson (1994), both of which gained considerable currency in the dinosaur literature throughout much of the 1990s and 2000s, the pressing question remains: are there any clear osteological correlates of sex in dinosaurs? Only two features remain as unequivocal indicators of sex, although each of these is difficult to identify in practice. First, Sato et al. (2005) described a theropod dinosaur specimen with fossilized eggs in its oviduct, which is an obvious sign that this specimen is a female. However, very few dinosaur fossils are found this way, rendering this feature essentially useless for determining the sex of the vast majority of dinosaur individuals. Second, Schweitzer et al. (2005b) described a bizarre, densely mineralized, highly vascularized tissue lining the internal cavity of a *Tyrannosaurus* femur. A similar bony tissue was later reported in a specimen of the ornithopod *Tenontosaurus* (Lee and Werning, 2008) (Fig. 5.10). Based on extensive comparisons with living animals, Schweitzer et al. (2005b) identified this tissue as medullary bone, a unique and ephemeral tissue that extant birds use as an easily accessible calcium source for shelling their eggs. As only females lay eggs, the possession of this tissue is a surefire indicator that a specimen is female. However, using medullary bone to identify a female specimen also has its difficulties. It is only present when a female is ovulating, and therefore is not likely to be preserved very often in the fossil record, and its iden-

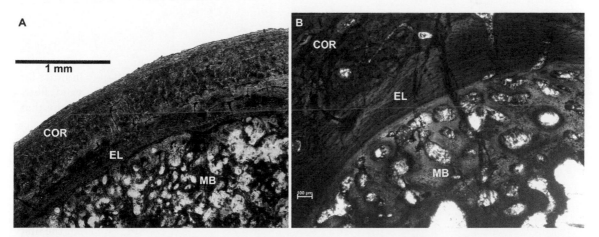

Figure 5.10 Medullary bone, a unique tissue only present in ovulating females and used as a ready source of calcium for shelling eggs, in a living pelican (A) and the ornithischian dinosaur *Tenontosaurus* (B). COR, outer bone cortex, composed of compact bone, which provides most of the structural support for the bone; EL, endosteal lamina, a thin layer of bone that usually lines the marrow cavity (when the marrow cavity is not actively being expanded or modified); MB, medullary bone, which in these specimens lines the marrow cavity. Note the striking similarity between the medullary bone of the pelican and *Tenontosaurus*. Images copyright of Sarah Werning.

tification requires destructive thin sectioning of bone. More worrisome, Chinsamy and Tumarkin-Deratzian (2009) argued that some diseases can create pathological bony tissue that is superficially similar to medullary bone, and the two can easily be confused in fossil specimens. Therefore, paleontologists are still lacking unequivocal, accessible, consistent features for identifying sex in dinosaurs, and the search for these indicators will likely continue for quite some time.

Conclusions

Understanding and quantifying the form of dinosaurs is critical for studies of behavior, locomotion, feeding, and growth. New technologies such as CT scanning and laser scanning allow paleontologists to study the anatomy of dinosaurs, especially internal spaces such as the brain and cranial sinuses, in unprecedented detail. Sophisticated computer software permits the construction of digital models,

which are exceedingly useful in studies of locomotion, and morphometric analysis and other statistical techniques allow the morphology of dinosaurs to be quantified, simplified into a manageable model, and compared across large samples. This toolkit of techniques, which is rapidly expanding and becoming more accessible with the advent of new technologies, is useful for addressing two perpetual questions in dinosaur paleontology. How big were dinosaurs? Can we tell male and female dinosaurs apart? Dinosaur body mass can now be estimated with quantitative rigor using computer models and meticulous mathematical techniques. Recognizing differences between males and females, and identifying specific specimens as belonging to a certain sex, still remain difficult. The search for explicit indicators of sex in the dinosaur fossil record will likely remain a major focus of future research, and the integration of morphometric techniques with a greater understanding of the ontogenetic age of individual specimens is a promising area of study.

6 Locomotion and Posture

How did dinosaurs move? This simple question captures the imagination of scientists and the general public, and has spawned limitless speculation from Hollywood animators and artists. Understanding how dinosaurs stood and moved is integral to understanding how they functioned as living animals, and it is no surprise that generations of scientists have been infatuated with studying dinosaur locomotion. Having some knowledge about posture and gait is also essential for bringing dinosaurs to life on the big screen or in the pages of a children's book, and for this reason high-quality artists and animation studios are particularly interested in the latest scientific research on dinosaur anatomy and locomotor habits (although oftentimes Hollywood portrayals of dinosaurs are based on a heavy dose of speculation with only a pinch of science).

Unlocking the mysteries of dinosaur locomotion is therefore one of the most sought-after goals in contemporary dinosaur research. Several major questions have generated significant debate. How fast and agile were dinosaurs, and which species could run at rapid speeds? Which types of dinosaurs were bipedal or quadrupedal, which postural condition was primitive for dinosaurs, and how many times did each postural condition evolve independently within the dinosaur clade? Were the long necks of sauropods held aloft, and able to reach high into the treetops, or limited to a more horizontal posture? Could some non-avian dinosaurs fly or glide and, if so, did they navigate the skies in the same style as modern birds? Unfortunately, answering these questions is rarely easy. Aside from the fact that living dinosaurs (other than birds) cannot be observed directly, several problems hinder studies of dinosaur locomotion. Dinosaurs came in a wide variety of shapes and sizes, so their locomotor habits cannot be studied uniformly. More problematic, the soft tissues that power a moving animal, most importantly muscles and ligaments, are rarely fossilized and must be hypothetically reconstructed for dinosaurs. Finally, there is still great uncertainty about body mass, center of mass, and structure of limb articulations in most dinosaurs, all of which are essential information when studying posture, speed, and agility.

Despite these problems and uncertainties, questions about dinosaur locomotion and posture remain interesting, relevant, and important to address. Simply put, it is necessary to understand at least basic facts about dinosaur movement and posture in order to study the biology of dinosaurs as living creatures. Although many of the above uncertainties will never be completely eliminated, dinosaur paleontologists can be honest about the limitations of the fossil record and focus on specific tractable questions that are testable. This chapter will outline the toolkit of methods that scientists use to study dinosaur locomotion and test explicit hypotheses about how dinosaurs moved and stood. These methods, some of which are extremely sophisticated and utilize advanced mathematics and the latest in computer technology, are useful in outlining the major body postures and plausible locomotor styles employed by different dinosaurs. They also help address a few major questions about dinosaur locomotion and posture that have been the subject of decades of speculation and debate. The chapter closes with a look at four issues in particular: the speed of large carnivorous dinosaurs, the putative flying and burrowing lifestyles of some dinosaurs, and the neck posture of sauropods.

Methods for Studying Locomotion and Posture

Gross skeletal anatomy

The most important source of information on dinosaur posture and locomotion comes from the most abundant component of the dinosaur fossil record: the bones themselves. If a skeleton is complete enough, it is often straightforward to determine whether a dinosaur was bipedal or quadrupedal, whether it was bulky or svelte, and whether it had the long slender limbs of an agile runner or the robust, stocky, weight-bearing limbs of a plodder. These are only very basic observations, however, and in reality most dinosaurs do not fall into obvious categories of "gracile" or "bulky," or their limbs are not clearly "robust" or "long." It can even be difficult to assess whether a species always, or even occasionally, walked on either two or four limbs. More sophisticated approaches are clearly needed to study dinosaur locomotion in detail, but it is always important to remember that simple obser-

vations should be considered first. For instance, it is obvious that lithe long-legged but short-armed theropods were bipeds and bulky columnar-limbed sauropods were quadrupeds, and it is no great leap to conclude that theropods could probably move faster and behave more athletically than sauropods of similar or greater size.

Other observations of gross anatomy are more informative. For instance, there are several explicit features commonly seen in living animals that run fast and/or possess great endurance which enable a far-roaming lifestyle (so-called "cursorial" species). These characteristics include long limbs, longer distal limb segments than proximal segments (e.g., longer tibia than femur), well-developed attachment sites for the protractor and retractor musculature that powers the limb (moving it forward and backward, respectively), feet (and hands in quadrupeds) with reduced outer digits and a long central digit, and hinge-like joints between different segments of the limb (Coombs, 1978b; Hildebrand and Goslow, 1998). Other features are hallmarks of slower, more lumbering, less far-ranging species (so-called "graviportal" animals), such as robust columnar limbs, short limbs with especially short distal segments, and feet and hands (in quadrupeds) that are short, stout, and arranged in a colonnade-like semicircle to more effectively support the weight of the animal rather than provide thrust for rapid locomotion. If some or all of the first set of features are present in a dinosaur skeleton, it is a good indicator that the species in question was, in the most general terms, athletic and capable of fast speeds. If some or all of the more graviportal features are observed, it is likely that the species was slower and handicapped by less efficient locomotion. However, the mere possession of these features says nothing about absolute speed.

Furthermore, the manner in which bones articulate give information on the range of motion of certain parts of the skeleton. This is critical in assessing many important functional questions, such as whether the forelimbs of some putatively quadrupedal species could rotate to the degree necessary to reach the ground, whether sauropods could raise their necks high, whether the hands of some species were used for grasping or as a weight-supporting hoof, and whether some species were able to extend their limbs outward to glide or use their limbs to dig a burrow. Range of motion studies are common, and frequently implemented using either manual manipulation of casts or computer modeling programs (Stevens and Parrish 1999; Carpenter, 2002: Stevens, 2002; Senter, 2006a, 2006b). However, it must always be remembered that limb joints in a living animal are not composed solely of bone, but also of soft tissues such as cartilage, tendons, ligaments, and muscles, none of which are commonly preserved in the fossil record and must be reconstructed by researchers. Recent work indicates that thick pads of cartilage capped the long bones of many dinosaurs (Holliday et al., 2010); these would have played at least some role in joint movement, but the effect of cartilage and other unobservable soft tissues on range and degree of joint motion is very difficult to quantify.

Bone measurements and comparisons with living analogues

General observations of bones and skeletons can be informative when studying dinosaur locomotion, but it is often more illuminating to analyze the size, shape, and strength of the bones in quantitative detail. Meticulous information on the morphology of bones, especially those of the limbs (which bear the weight of the animal and propel the animal during locomotion), can give keen insight into the posture, relative speeds, and agility of dinosaur species. These approaches are especially powerful when similar information is available for living terrestrial vertebrates whose speed and athletic performance can be directly observed. This enables validation of the relationship between certain explicit speeds and behaviors in living animals and quantitative features of their limbs (such as ratios of different limb components and measures of bone strength). For instance, if the strength of the femur in mediolateral bending is always above a certain threshold in living animals that can run, it would be reasonable to conclude that a dinosaur with a measured bone strength above this cutoff could probably also run.

Of course, the utility of these approaches depends on the choice of living analogues. Extant mammals are usually chosen as an analogue, and for good reason. Carrano (1998) demonstrated that mammals, especially those of mid to large size, share with dinosaurs an upright limb posture, slender hindlimb proportions (especially the gracility of the femur), and extremely similar ratios between different bones of the hindlimb (Fig. 6.1). Living birds, on the other hand, have hindlimb ratios and dimensions quite distinct from those of living mammals and non-avian dinosaurs, which Carrano (1998) persuasively argued was due to basic postural differences in the two groups. Living mammals hold their femur more vertically during locomotion and limb motion is primarily driven by the hip muscles. On the contrary, birds exhibit a unique posture among extant vertebrates in which the femur is held nearly horizontally and moves little during walking, as most limb movement occurs at the knee joint instead of the hip. Therefore, similarities in limb dimensions and robusticity between living mammals and Mesozoic dinosaurs suggests that they had similar postures and styles of locomotion, whereas birds, despite having evolved from theropod dinosaurs, had a distinct style, which may have originated after the earliest birds diverged from their theropod ancestors (Carrano, 1998). Mammals are therefore the better living analogue

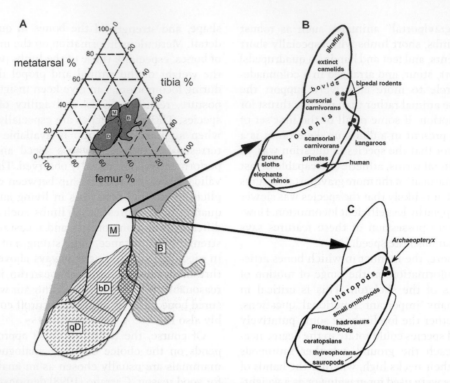

Figure 6.1 Mammals as modern analogues for dinosaurs. (A) Ternary plot, based on the proportions of different parts of the hindlimb, showing that dinosaurs (qD, quadrupedal dinosaurs; bD, bipedal dinosaurs) have similar limb proportions to living mammals (M) but quite different proportions from living birds (B). The close-ups show where specific groups of mammals (B) and dinosaurs (C) fall out in the ternary plot. Modified from Carrano (1998). Used with permission from the Paleontological Society.

for studying dinosaur locomotion (see also Carrano and Biewener, 1999; Christiansen, 1999; Carrano, 2001). That being said, it is important to remember that living mammals do not encapsulate the full range of body sizes and shapes of Mesozoic dinosaurs. Comparisons with living mammals, or any other extant animals, are by no means a foolproof tactic, but are surely more explicit and testable than speculation based on subjective observation of dinosaur skeletons.

Several examples of quantitative studies of dinosaur bones, interpreted in a comparative context with modern analogues, have been presented in the literature. Many authors have examined bone strength, or the ability of a limb bone to withstand bending along its long axis (Alexander, 1985; Farlow et al., 1995a; Christiansen, 2000a). A strength indicator can be calculated by taking into account body mass and the length and cross-sectional area of the

bone in question. Generally, animals with relatively larger strength indicators are able to endure progressively stronger forces on the bones, suggesting progressively greater agility and speed. Comparisons of strength indicators in many species of dinosaurs can help indicate which dinosaurs were likely faster than others. Furthermore, comparisons with living mammals can give a ballpark, qualitative estimate for the actual speed of a dinosaur, because two animals with a similar posture and a similar strength indicator probably had similar locomotor abilities. Alexander (1985) found that the four dinosaurs he studied – *Apatosaurus, Diplodocus, Triceratops,* and *Tyrannosaurus* – had much lower bone strengths than living buffaloes and ostriches, and therefore none of them were likely to have been able to run fast. Furthermore, *Apatosaurus* had a bone strength value similar to that of a living elephant, suggesting that the two animals had a similar athletic ability. Alexander's (1985) calculations were based on questionable mass estimates derived from scale models of often incomplete specimens, but a more recent analysis utilizing better fossil material agreed that *Tyrannosaurus* could not run rapidly (Farlow et al., 1995a). A third study analyzed a much greater number of dinosaur species, and found that strength indicators were larger for small theropods, such as *Coelophysis*, than for large carnivores, such as *Tyrannosaurus* (Christiansen, 2000a). Therefore, small theropods were probably relatively faster than their larger cousins.

Studies of bone strength may indicate the relative speeds of dinosaurs, and perhaps a coarse estimate of absolute speed by comparison to living mammals, but the sources of potential error are enormous. Most problematic, there it not always a clear link between bone strength and speed in living animals, rendering comparisons between dinosaurs and living analogues tenuous at best (Hutchinson and Allen, 2009). With this in mind, a second type of quantitative analysis has aimed to understand dinosaur locomotion more generally. Is it possible to group dinosaurs into a series of locomotor categories akin to those often used to describe living mammals? Extant mammals are commonly grouped into one of four classes – cursorial, sub-cursorial, mediportal, and graviportal – ranging from the fastest, most athletic, and most far-ranging species (cursorial) to the slowest, most plodding, and least far-ranging (graviportal). Although these terms are difficult to actually define, it is generally held that more cursorial animals have a faster top speed, longer stride, and more stamina than more graviportal animals. These differences in locomotor performance are thought to be reflected in the skeleton: more cursorial animals generally have longer distal limb segments, gracile limb bones, and hinge-like joints, whereas more graviportal species have shorter and stockier limbs (see Farlow et al., 2000 for a summary).

Coombs (1978b) attempted to assign dinosaurs to these same four categories, by measuring the ratios of various limb bones and using these values to construct a morphospace into which dinosaurs and mammals could both be plotted. He found that sauropods and stegosaurs were graviportal and hence among the slowest of the dinosaurs, whereas small theropods and small bipedal ornithischians

were swift and/or far-roaming cursors. Other dinosaurs, including large theropods, ornithopods, and "prosauropods," were somewhere in between these two ends of the spectrum. However, Carrano (1999) pointed out a vexing problem with this type of analysis: limb ratios and other anatomical features used to bin taxa into a locomotor category are not discrete characters, but vary continuously. Speed and agility also comprise a spectrum: some animals are very fast, some slow, and there are many in between. Therefore, it is somewhat arbitrary to break a continuum of either limb measurements or locomotor abilities into discrete categories.

Because of these qualms, Carrano (1999) provided an improved and updated version of Coombs' (1978b) analysis. He used a much larger sample of dinosaur species, a greater array of measurements, and sophisticated multivariate techniques to take into account information from all measurements and condense them into a single representative value for each taxon. These values were plotted along a single morphospace axis, with end points representing the fastest, most agile, far-roaming taxa and the slowest, most lumbering, narrowest-roaming taxa, respectively (Fig. 6.2). All dinosaurs were plotted along this axis, and in this way their relative locomotor abilities could be assessed in a comparative fashion, without the need to bin species into discrete categories. This analysis found sauropods, ankylosaurs, stegosaurs, and ceratopsians to be the slowest and least agile (most graviportal in traditional terminology) and many small coelurosaurian theropods to be the fastest and most agile (most cursorial). Carrano (1999) also provided a similar analysis of relative speed and agility in living mammals, which once again illustrated that dinosaurs and extant mammals likely had a similar range of locomotor abilities.

Footprints and trackways
Consideration of skeletal anatomy usually gives only a relative indication of the locomotor habits of dinosaurs and extremely coarse estimates for absolute speed based on comparisons to imperfect living analogues. Fossil footprints provide a powerful source of data for calculating the absolute speed of a specific animal moving at a specific moment in time (Alexander, 1976, 1989). The stride length between individual footprints in a trackway depends

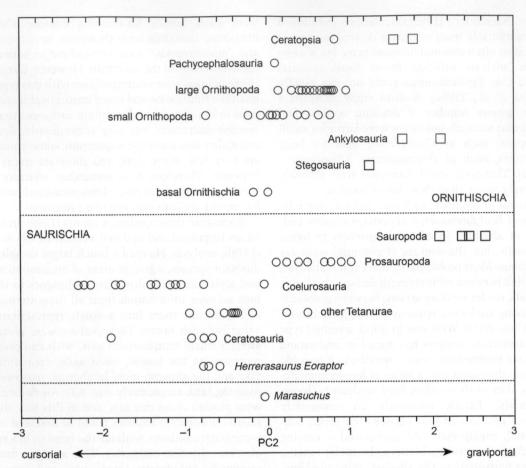

Figure 6.2 The cursorial–graviportal spectrum in dinosaurs. The single axis on the bottom denotes the relative cursoriality of dinosaur species, based on limb proportions. Those species farther to the left are relatively more cursorial than those farther to the right. Modified from Carrano (1999), and reproduced with permission from Cambridge University Press.

on both the size and speed of the trackmaker. Animals that are moving faster, or those with longer legs, will produce greater strides (Fig. 6.3). If a researcher has information on both the spacing between tracks and the limb length of the trackmaker, these values can be entered into the following equation to determine an estimate of speed:

$$\text{Speed} = 0.25(\text{acceleration due to gravity})^{0.5}$$
$$\times (\text{stride length})^{1.67} \times (\text{hip height})^{-1.17}$$

It is usually difficult to know the hip height of a specific trackmaker, because skeletons are rarely preserved in association with tracks. In this case, it is common to estimate the hip height as four times the footprint length, based on a similar ratio in living mammals and dinosaurs known from complete skeletons (Alexander, 1989). Acceleration due to gravity is reasonably assumed to be the same as in the modern world (9.81 m/s²). Inserting the necessary measurements into the equation will produce a speed estimate (in meters per second) for the trackmaker, at the time when it was making the tracks in question.

Speed calculations based on trackways have been compiled for several groups of dinosaurs, and these are reviewed by Alexander (1989, 2006), Thulborn (1990), Irby (1996), Farlow et al. (2000), and Hutchinson (2005). As would be expected, different trackways show that the same general

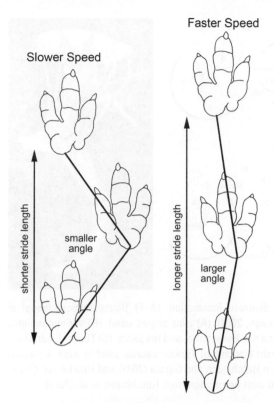

Figure 6.3 Schematic illustration showing how speed can be estimated from trackways. The footprint outline is of the characteristic Early Jurassic theropod ichnogenus *Eubrontes*. Footprint image modified from Milner et al. (2009).

type of dinosaur could move at different speeds. Most measured trackways, including those of theropods, sauropods, and ornithopods, were made by slow-moving animals walking at speeds of 1–4 m/s (2.2–9 mph). Some exceptional trackways indicate that ornithopods and large theropods could move faster, perhaps up to speeds of 11–12 m/s (approximately 25 mph; Farlow, 1981; Day et al., 2002). There is no clear trackway evidence, however, that sauropods could move at this speed, or that large theropods had a suspended phase in their locomotory cycle in which both feet were off the ground at the same time (the technical definition of running).

The use of trackways in estimating trackmaker speed and maneuverability, although an insightful tool, is hampered by several biases and uncertainties. First, the relationship between stride length, hip height, and speed is valid for living mammals,

but this correlation may not hold true for dinosaurs. Second, even among living mammals, there is quite a lot of variability in the relationship between stride length and speed. When living mammals are observed and their stride lengths and actual speeds measured, it has been shown that when the observed stride length is entered into the above equation it may predict a speed estimate that differs from the true speed by up to a factor of two (Alexander, 2006). Third, the assumption that dinosaur hip height is approximately four times the length of the footprint is a coarse generalization. This relationship does not hold for all dinosaur groups (Thulborn and Wade, 1984; Henderson, 2003), and the size and shape of the preserved footprint may not always reflect the true morphology of the foot due to vagaries of preservation and the cohesiveness (or lack thereof) of the sediment (Gatesy et al., 1999; Henderson, 2006a; Falkingham et al., 2010).

With these caveats in mind, speed estimates derived from trackways should not be treated as accurate. However, at a general level, trackways can clearly distinguish whether an animal was moving relatively slow or fast. For instance, it is a sound conclusion that, based on known trackways, sauropods could not move rapidly and large theropods could not attain the blazing speeds portrayed in *Jurassic Park* and other popular depictions.

Biomechanical modeling

The most powerful and promising method for studying dinosaur locomotion is biomechanical modeling (Hutchinson and Gatesy, 2006; Hutchinson and Allen, 2009; Sellers et al., 2009; Hutchinson, 2011) (Fig. 6.4). The advent of new computer technology and statistical methodologies now allows dinosaur speed, posture, and gait to be studied in great detail. Digital models of dinosaurs can be built, using information from fossilized bones and footprints, soft tissues can be added to the skeletal chassis, and the fleshed-out model can be put through a range of gymnastic routines that can help assess which types of behaviors, postures, and speeds were realistic. A great benefit of these approaches is that many of the uncertainties of the fossil record can be taken into account via sensitivity analyses, which assess the degree to which certain inferences are robust given the many realistic biases faced by paleobiologists. Instead of

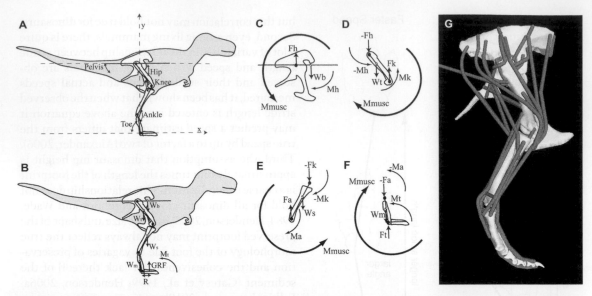

Figure 6.4 The use of biomechanical modeling in studying dinosaur locomotion. (A–F) Biomechanical model of *Tyrannosaurus* (from Hutchinson and Garcia, 2002 and Hutchinson, 2004): (A) joint angles used, (B) various weights and forces on the model, and (C–F) muscle moment arms about the hip, knee, ankle, and toe joints. (G) Three-dimensional biomechanical model based on muscle reconstruction of the right hindlimb of *Tyrannosaurus*, used to assess running mechanics (from Hutchinson et al., 2005). (A–F) Reproduced from Hutchinson and Garcia (2002) and Hutchinson (2004), with permission from Nature Publishing Group. (G) Reproduced with permission from Hutchinson et al. (2005).

building a single model or making a single calculation, researchers can perform a set of analyses that take into account a wide range of possible values for many parameters that must be estimated from incomplete fossil material, such as body mass, limb articular geometry, and soft tissue size and morphology. If a certain result emerges from all or most of these sensitivity studies – for example, the conclusion that *Tyrannosaurus* was not a fast runner or that hadrosaurs ran more efficiently on two legs instead of four – it can be considered strongly supported, even more so if it dovetails with results from the methods described above.

Many types of biomechanical models, ranging from the simple to the mind-numbingly complex, have been used to study dinosaur locomotion in recent years. Some of these are two-dimensional renderings of limited regions of the anatomy, such as the hindlimb (Hutchinson and Garcia, 2002; Hutchinson, 2004; Sellers and Manning, 2007). Others utilize three-dimensional renderings of the hindlimb (Snively and Russell, 2002; Hutchinson et al., 2005, 2007; Gatesy et al., 2009) or the entire

skeleton (Henderson and Snively, 2004; Sellers et al., 2009). Oftentimes these models are used to predict the maximum speed that dinosaurs could attain, and this has been the main focus of important contributions by Hutchinson and Garcia (2002), Sellers and Manning (2007), and Sellers et al. (2009) (Fig. 6.4). Other models have been marshaled to address specific questions, such as the range of potential limb postures (Gatesy et al., 2009), the size and lengths of limb muscles (Hutchinson and Garcia, 2002; Hutchinson, 2004; Hutchinson et al., 2005), the ability of a dinosaur to turn at various speeds (Henderson and Snively, 2004; Hutchinson et al., 2007), and the strength of certain bones under different locomotion-induced stresses (Snively and Russell, 2002). Many of these analyses build on each other and provide reciprocally illuminating information. Gatesy et al.'s (2009) study of realistic limb postures, for instance, helps provide plausible postural configurations, which must be input as a parameter into models that attempt to calculate speed (Hutchinson and Garcia, 2002; Sellers et al., 2007).

Clearly, biomechanical modeling is an active area of research and continued advances in technology will allow more powerful, and hopefully accurate, studies of dinosaur posture and speed. The study of dinosaur locomotion is becoming increasingly dominated by explicit, quantitative, biomechanical modeling studies, and this is a positive development. As persuasively argued by Hutchinson and Allen (2009) in their excellent review of theropod biomechanics, the study of dinosaur locomotion has "matured" to the point where imprecise arguments based solely on anatomy are getting left behind in favor of more sophisticated modeling approaches. Locomotion in living animals is complicated: it is controlled by body posture, muscle and tendon dynamics, neurological functions, and numerous other constraints and parameters, only some of which are related to hard-tissue anatomy. Simply observing, measuring, and describing a dinosaur skeleton is not sufficient for understanding locomotor function, except in the most general terms. In light of the complexities of locomotor behavior and the uncertainties of the fossil record, quantitative biomechanical modeling is now a requirement for refined study of dinosaur locomotion.

The Evolution of Posture and Locomotion in Dinosaurs

What are the major patterns in locomotor evolution in dinosaurs, how many times did certain postures evolve, and how did the major groups of dinosaurs stand and move? Questions like these can be addressed with the toolkit of approaches described above, including consideration of skeletal anatomy and reconstructed muscles, comparisons with living analogues, examination of trackways, and the use of explicit computer modeling techniques. Information on posture and locomotion derived from these sources can then be interpreted in a phylogenetic context, fleshing out major trends and events in locomotor evolution during dinosaur history. This is an important exercise for understanding the origin of the postural and locomotor features that are unique to birds among living vertebrates, such as a bipedal stance in which most hindlimb motion is centered at the

knee joint and arms that are modified into wings. As with many "stereotypical" avian features, many of these characters originated in a stepwise fashion during the long evolutionary history of dinosaurs, not in a profound burst of morphological change associated with the origin of flight (Hutchinson and Gatesy, 2000; Gatesy, 2002; Hutchinson, 2006; Dececchi and Larsson, 2009; Hutchinson and Allen, 2009).

Primitive dinosaurs and their closest relatives

Dinosaurs are but one member of the more inclusive archosaur clade, which also includes living crocodiles and an array of extinct species restricted to the Mesozoic. The ancestral archosaur was a quadrupedal sprawling animal that was probably lightly built (Ewer, 1965; Parrish, 1987; Sereno, 1991a). Because its limbs sprawled to the side of its body, the force generated when the feet (and hands) contacted the ground was located lateral to the hip joint, which would tend to deflect the femur away from the body midline during locomotion (abduction) (see Fig. 1.15). As a result, large adductor muscles were needed to both pull the femur inward during the step cycle and support the body weight of the animal (Hutchinson and Gatesy, 2000). This postural and muscular configuration changed in the earliest dinosauromorph cousins of dinosaurs, whose more upright stance is reflected in their gross skeletal anatomy and their footprint record (Sereno, 1991a; Sereno and Arcucci, 1993, 1994; Ptaszyński, 2000; Carrano and Wilson, 2001; Brusatte et al., 2011a). Associated with the shift to a more erect stance was the development of a simple hinge-like ankle that restricted hindlimb motion to a parasagittal plane, because the ancestral rotary ankle that permits a wider range of motion necessary for sprawlers was no longer required (Parrish, 1987; Sereno and Arcucci, 1990; Novas, 1996) (see Figs 1.17 and 1.18). Because the limbs were now positioned directly underneath the body, the ground force was located medial to the hip joint, which would tend to bring the femur close to the midline during the step cycle (see Fig. 1.15). Therefore, larger abductor muscles were needed to pull the limbs outward, to ensure that they did not cave inward during locomotion (Hutchinson and Gatesy, 2000).

All dinosaurs therefore evolved from a common ancestor that walked upright and had large femoral abductor muscles to support the body. Furthermore, this ancestor would have had extensive hindlimb retractor muscles, which stretched from the long tail through the deep brevis fossa on the ventral surface of the ilium (a diagnostic feature of dinosaurs) and the fourth trochanter of the femur (Gatesy, 1990). The protractor muscles, which bring the limb forward, and the extensor muscles, which support body weight and straighten the limb, were also quite large, as indicated by the large and rugose femoral lesser trochanter to which they attached. As is evident by the large amount of muscle attaching to the pelvis, most of the motion between individual components of the hindlimb was centered at the hip joint. This basic bauplan – large abductor muscles, large protractor and retractor muscles, and hip-driven locomotion – was probably the ancestral body type for each of the three major dinosaur clades (theropods, sauropodomorphs, and ornithischians). However, several postural and myological modifications occurred within these groups, as will be reviewed below. One of the most common series of changes, which evolved in parallel many times in different groups, involved enlargement of the lesser trochanter and elongation of the ilium, both of which served to further expand the protractor muscles (Carrano, 2000).

However, one very important feature of the common dinosaurian ancestor remains unknown. It is unclear whether this ancestor was bipedal (habitually or occasionally) or quadrupedal, as some of the closest relatives to dinosaurs walked on two legs and others on four (Sereno and Arcucci, 1993, 1994; Dzik, 2003; Nesbitt et al., 2010; Brusatte et al., 2011a). It is also unclear exactly how fast this common ancestor was, and what modern analogue it may compare to. Some trackways of close dinosaurian relatives exhibit large stride lengths, suggesting that at least some dinosaur cousins, and perhaps the common ancestor itself, were capable of high speed and agility (Ptaszyński, 2000; Brusatte et al., 2011a). Whether these early dinosaurs and close relatives were any faster or more agile than contemporary bipedal and upright-walking crocodile-line archosaurs, such as *Effigia* and *Poposaurus* (Nesbitt and Norell, 2006; Schachner et al., 2011a), is difficult to assess, but demands further study.

Ornithischians

The earliest ornithischians known from the fossil record were small, slender, bipedal animals (Thulborn, 1972; Norman et al., 2004c; Butler et al., 2007). More derived ornithischians, however, diversified into a dizzying array of shapes and sizes. Locomotor and postural evolution in ornithischians has been the subject of much less research than that devoted to saurischians, but several general trends are apparent.

Obligate quadrupedalism evolved at least twice in ornithischians: once in thyreophorans and once in ceratopsians (Coombs, 1978b; Sereno, 1997; Carrano, 1998). Basal members of these clades were bipedal, or perhaps facultatively quadrupedal (i.e., they could walk on two or four legs, depending on their speed and conditions of the substrate). More derived members of each group evolved both a quadrupedal body posture and large body size, although it is unclear if the two are correlated. However, their postures differ in detail. The hands and feet of stegosaurids, and probably ankylosaurids as well, are modified into stout, compact, weight-bearing structures (Galton and Upchurch, 2004b; Vickaryous et al., 2004; Senter, 2010). In particular, the metacarpals of stegosaurs are arranged in a vertical, tube-like colonnade, as is also the case in colossal sauropods and many living mammals that cannot run fast and which use their limbs primarily for weight support (Senter, 2010). It is no surprise, therefore, that Carrano's (1999) multivariate study of limb measurements found stegosaurs and ankylosaurs to be among the most graviportal dinosaurs (Fig. 6.2).

Ceratopsians, on the other hand, have more elongate and gracile hands and feet that are not arranged in a compact colonnade, and these seem to have been retained from smaller, more cursorial, and probably bipedal ancestors (Fujiwara, 2009) (Fig. 6.5). Evidence from narrow-gauge trackways and articulations between well-preserved limb bones indicate that ceratopsians probably held themselves with an upright posture and did not sprawl as was long thought (Paul and Christiansen, 2000; Fujiwara, 2009) (Fig. 6.5). Whether this means that ceratopsians could run rapidly, using a "rhino-like" style as hypothesized by Paul and Christiansen (2000), is open to debate. However, Carrano (1999) found ceratopsians to be somewhat interme-

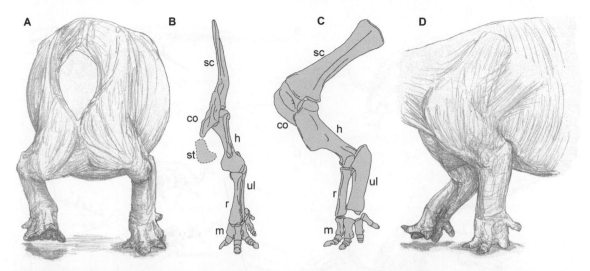

Figure 6.5 The body posture and locomotion of *Triceratops*. The upright posture of this familiar quadrupedal ornithischian, revealed by new fossil discoveries, is shown in anterior (A, B) and left lateral (C, D) views. Modified from Fujiwara (2009), and reproduced with permission.

diate on the cursorial–graviportal spectrum, and more cursorial than stegosaurs, ankylosaurs, and sauropods (Fig. 6.2). Therefore, it is likely that ceratopsians were blessed with at least moderate speed and maneuvcrability.

Several other clades and grades of ornithischians were probably facultatively quadrupedal, including derived ornithopods such as hadrosaurids and iguanodontids (Fig. 6.6). Basal ornithopods such as *Hypsilophodon* were generally small, slender, bipedal, and able to achieve high speeds, judging by their long distal hindlimbs and elongate feet with prominent central digits (Galton, 1974; Norman et al., 2004a). In contrast, the overall skeletal morphology of more derived ornithopods, such as iguanodontids and hadrosaurids, is indicative of a less cursorial animal that could walk quadrupedally, at least at times (Fig. 6.6). The forelimb is proportionally longer relative to the hindlimb than in basal ornithopods and anchored robust muscles (Dilkes, 2000). The hands have three prominent central digits which end in blunt hoof-like unguals, and exceptionally preserved soft tissue impressions indicate that these digits were bound together into a single functional unit. The outer digits, however, were smaller and not bound together, and likely retained some grasping function. The feet of these derived forms were heavily modified relative to the slender

cursorial morphology in basal taxa. The individual metatarsals and phalanges are stouter and more robust and the unguals are modified into hooves (Moreno et al., 2007) (Fig. 6.6). Finally, a litany of footprint evidence indicates that some derived ornithopods walked bipedally and some quadrupedally, and that they could even switch postures during a stroll (Lockley, 1992; Lockley and Hunt, 1995; Wright, 1999).

This battery of evidence strongly suggests that derived ornithopods could walk bipedally and quadrupedally, had hands that were used for both weight support and grasping, and walked on the tips of their toes (unguligrade posture) instead of on their entire pedal digits (digitigrade posture). Regardless of whether an ornithopod was walking bipedally or quadrupedally, it is likely that these animals were capable of attaining high speeds. Carrano (1999) found large ornithopods to occupy an intermediate position on the cursorial–graviportal spectrum, in a similar position as large-bodied theropods and some "prosauropods" (Fig. 6.2). Measured trackways indicate that some ornithopods moved at speeds of approximately 8 m/s (about 18 mph; see review in Hutchinson, 2005), whereas computer biomechanical models suggest that maximum speeds may have been as high as 17 m/s (38 mph), which is much greater than the predicted maximum speed

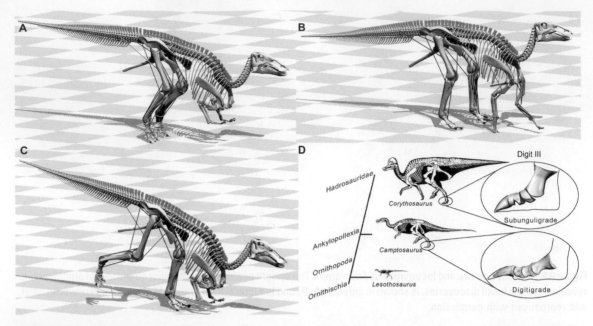

Figure 6.6 The body posture and locomotion of ornithopods: (A–C) still shots of a digital model of the hadrosaurid *Edmontosaurus* in different poses (hop, gallop, and run, respectively); (D) schematic illustrating a simplified phylogeny of ornithischians with digitigrade and subunguligrade forms indicated. Images (A–C) courtesy of Dr William Sellers and modified from the model presented by Sellers et al. (2009); image (D) modified from Moreno et al. (2007), and reproduced with permission.

of large theropods (Sellers et al., 2009). It is unclear whether these animals could move faster as bipeds or quadrupeds, although the computer models find equivocal evidence for faster bipedal speeds. Footprints, however, suggest that bipedalism may have been more common at lower speeds (Lockley, 1992). Clearly, this question remains open and deserves further study.

Sauropodomorphs

There is no doubt that enormous sauropods such as *Brachiosaurus* and *Diplodocus* were slow lumbering animals incapable of the speeds achieved by bipedal theropods and ornithischians (Fig. 6.7). The limbs of sauropods are columnar, the distal limb segments are short and stocky, and the hands and feet are arranged in compact colonnades, all of which demonstrate that sauropods prioritized body support over cursorial locomotion (Upchurch et al., 2004; Henderson, 2006b). This is no surprise, given that some species may have reached absurd masses of 75 tonnes or greater (Alexander, 1985, but see

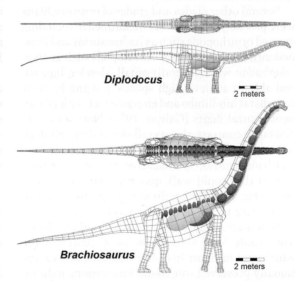

Figure 6.7 The body posture and locomotion of sauropods. Computer models of *Diplodocus* and *Brachiosaurus* in dorsal and lateral view, used by Henderson (2006b) to study aspects of sauropod locomotion. Images courtesy of Dr Don Henderson.

above). Carrano (1999) found sauropods to be the most graviportal of dinosaurs (Fig. 6.2), and footprint evidence indicates that sauropods commonly walked at speeds of 1–2 m/s (2–4 mph), which is an order of magnitude lower than the speeds implied by some trackways of theropods and ornithischians (Alexander, 1989; Hutchinson, 2005).

All sauropods were quadrupedal, and although they likely were able to rear up on their hindlegs to avoid predators or gain access to treetop vegetation, it would have been impossible for any species to walk bipedally for more than a few steps. So-called "manus-only" trackways raise the intriguing possibility that sauropods may have, at least occasionally, immersed themselves in water, whose buoyancy would have tipped the animal forward so that only its hands made contact with the lake or seabed (Lee and Huh, 2002). Perhaps sauropods ventured into water to feed on aquatic plants, or for some other reason. However, Henderson (2004) cautioned against an aquatic interpretation of manus-only trackways by mathematically showing how the large bulk of sauropods would have made them extremely unstable, and liable to tipping belly-up, in water. Furthermore, Falkingham et al. (2011) used computer simulations to demonstrate that quadrupedal sauropods walking on firm ground could produce manus-only trackways in certain situations, such as when more weight is being held by the manus than by the pes. Therefore, manus-only trackways are not clear evidence of aquatic habits.

Although the locomotor habits of sauropods are generally well understood, many questions remain about the posture and speed of "prosauropods," the basal grade of sauropodomorphs that were extremely common in the Late Triassic and Early Jurassic. The most basal sauropodomorphs, such as *Saturnalia* and *Thecodontosaurus*, were probably bipedal, due to their gracile hindlimbs with long distal segments and relatively short forelimbs (Upchurch et al., 2004). The locomotion and posture of larger and more derived "prosauropods," such as *Plateosaurus*, *Massospondylus*, and *Melanorosaurus*, are more difficult to reconstruct, and scientists have long disagreed about their locomotor habits. These species generally have proportionally longer forelimbs than basal sauropodomorphs, but these were not modified for weight support as in sauropods

and, as shown by recent computer modeling of *Plateosaurus*, had such a small range of motion that they may not have been able to reach the ground, at least in some taxa (Mallison, 2010a, 2010b) (see Fig. 5.4). Additionally, *Plateosaurus* apparently could not pronate its wrists, meaning that the hand could not rotate to contact the ground (Bonnan and Senter, 2007). However, other "prosauropods" were capable of pronation (Bonnan and Yates, 2007). These species, which also had longer forelimbs, were surely capable of quadrupedalism at least occasionally. Footprint evidence also suggests that some "prosauropods" could walk quadrupedally, although it is difficult to identify the tracks of basal sauropodomorphs with certainty (Baird, 1980).

In sum, current evidence indicates that some "prosauropods" could walk only bipedally, others could walk quadrupedally (either some or most of the time), and the most derived species, which are the closest relatives to true sauropods, were restricted to a weight-bearing, columnar, quadrupedal stance (Bonnan and Yates, 2007). Locomotor features among basal sauropodomorphs therefore evolved in a piecemeal fashion, culminating in the aberrant body plan of sauropods. With this in mind, it is foolish to attempt to group all, or even most, "prosauropods" into a few discrete locomotor and postural categories. Indeed, Carrano's (1999) multivariate study indicates that "prosauropods" possessed a wide range of locomotor abilities, as they are spread out across a large swath of the cursorial–graviportal spectrum (Fig. 6.2). Most species are somewhat intermediate on this plot and none fall on the cursorial side. Those "prosauropods" that were bipedal were likely among the slowest (least cursorial) of all bipedal dinosaurs. However, quadrupedal species were among the most cursorial of the dinosaur quadrupeds, and likely much swifter and more agile than thyreophorans, ceratopsians, some ornithopods, and all sauropods.

Theropods

The vast majority of published studies of dinosaur locomotion focus on theropods, undoubtedly due to perpetual interest in the origin of birds and the advent of unique avian functions such as flight, wrist folding, and knee-driven hindlimb mechanics. The evolution of avian-style locomotion was the subject of masterful reviews by Gatesy (2002) and Hutchinson and Allen (2009), whereas more general issues of non-avian theropod locomotion were summarized by Farlow et al. (2000). These papers should be consulted for detailed information, but their overarching themes are twofold. First, most non-avian theropods probably stood and moved in a similar manner. Second, individual components of the signature avian flight stroke and knee-driven hindlimb system evolved piecemeal across the theropod family tree, not in a sudden burst of anatomical transformation associated with the origin of birds. These conclusions have emerged from decades of careful study, based on the integration of anatomical data, computer modeling approaches, and a well-understood phylogenetic framework. Studies of theropod locomotion also benefit from one great asset not available for ornithischians and sauropodomorphs: living birds, the descendants of Mesozoic theropods, can be studied directly in the lab and the wild. This provides an empirical framework for interpreting postures, speeds, joint articulations, bone strengths, muscle masses, and many other important aspects of locomotion in dinosaurs, which otherwise must be studied in a more hypothetical manner.

All non-avian theropod dinosaurs possessed the same general body plan: they were bipedal, stood upright, walked on their digits, and but for a few exceptions had gracile hindlimbs and three functional toes that impacted the substrate during locomotion. Interestingly, despite a great range in body size, Mesozoic theropods had very little variation in their hindlimb proportions, at least compared to the great variety of limb proportions present in living birds (Gatesy and Middleton, 1997). In other words, the non-avian theropod body plan was quite conservative, and all (or most) species were well adapted for agile bipedal locomotion. Indeed, theropods possess a full complement of cursorial skeletal features, such as long and gracile hindlimbs with especially elongate distal elements, and it is no surprise that they are found to be the most cursorial dinosaurs in Carrano's (1999) multivariate study (Fig. 6.2). That being said, Carrano (1999) showed that theropods occupied a wide spread on the cursorial–graviportal spectrum, and some larger species were approximately as cursorial as "prosauropods" and many ornithopods. Although all

theropods had cursorial features, it seems as if big theropods were the least cursorial, whereas small theropods were more cursorial.

Exactly how fast, in absolute terms, were non-avian theropods? Trackway and computer simulations suggest that theropods were probably, on average, the fastest of all Mesozoic dinosaurs. Some exceptional tracksites indicate that some species could run at speeds of more than 12 m/s (equivalent to 27 mph) (Farlow, 1981; Alexander, 1989; Thulborn, 1990; Irby, 1996; Day et al., 2002). The great majority of tracksites, however, record the motion of walking theropods that sauntered at speeds of 2–7 m/s (approximately 4–16 mph). Therefore, trackways are agnostic about the maximum attainable speeds of theropods (or any dinosaurs, for that matter), and are insufficient for determining whether theropods were usually faster than their contemporaries, including potential prey species.

These questions can be addressed using innovative computer simulations presented by Sellers and Manning (2007). They estimated the absolute speed of five theropods using two-dimensional biomechanical models of the hindlimb, whose joint motions and soft tissues were varied using sensitivity analyses. Their results suggested an inverse relationship between body size and speed: the tiny theropod *Compsognathus* had an estimated maximum speed of 18 m/s (40 mph), whereas *Tyrannosaurus* was estimated at 8 m/s (18 mph). A follow-up study found that the hadrosaurid *Edmontosaurus*, a contemporary of *Tyrannosaurus*, potentially reached speeds of 17 m/s (38 mph) which, if true, means that *Tyrannosaurus* could not outrun at least one potential prey species (Sellers et al., 2009). It is important to remember that these models have limitations – for instance, the feet and toes are represented by a single point without muscles and tendons, and all tendons are represented as the same length, regardless of pose – so their results should be treated as rough estimates instead of precise truths. However, these results are currently the best available and are based on the most sophisticated and powerful set of explicit methods. Furthermore, the sensitivity analyses of Sellers and Manning (2007) indicate that their margin of error for any estimate of maximum speed is approximately 50%, which although seemingly high, is much better than the 200% error margin for speed estimates based on trackways (Alexander, 1989, 2006).

The evolution of "avian" locomotor characters on the line to birds

The dense fossil record of theropod dinosaurs, coupled with their well-understood genealogy, presents paleontologists with a remarkable opportunity to understand the evolution of "stereotypical" avian locomotor habits. Birds are unique among living vertebrates in their forelimb flight stroke, powered by enormous arms that can fold against the body when not in use, and a knee-driven style of hindlimb locomotion in which the femur is held horizontally and most back-and-forth limb motion occurs at the knee (not the pelvis). How exactly did these novel locomotor strategies evolve? Did they originate in birds or their immediate ancestors, or were they also present, either in total or in part, in the theropod ancestors of birds? If the latter is true, how were the individual components of these locomotor systems – the numerous anatomical features working in concert to enable flight and the knee-driven stance – assembled into the coherent, highly derived flight stroke and knee-driven hindlimb posture of living birds? These major questions have been the focus of decades of research. As has been articulated by the expert reviews of Gatesy (2002) and Hutchinson and Allen (2009), it is clear that these signature avian locomotor strategies did not evolve quickly or in association with the origin of birds, but were rather pieced together, little by little, over millions of years of theropod evolution, eventually culminating in the stereotypical locomotor functions of living birds. Here, some major patterns in locomotor evolution on the line to birds are reviewed, but readers should consult the above papers for more details.

The powered flight of living birds is enabled by a suite of anatomical features, including elongate forelimbs that can move with a wide range of motion, a wrist that folds the wing against the body, and aerodynamic flight feathers (Middleton and Gatesy, 2000; Gatesy, 2002). Exactly how the flight stroke evolved – in other words, how non-flying theropods developed into animals capable of taking to the skies – has been the subject of endless debate and speculation. For decades the debate centered on a simple dichotomy: did flight evolve in cursorial

ground-dwelling theropods ("ground-up") or arborial gliding theropods ("trees-down") (Gauthier and Padian, 1985; Padian and Chiappe, 1998; Shipman, 1998; Gatesy, 2002; Dial, 2003a, 2003b; Hutchinson and Allen, 2009; Dececchi and Larsson, 2011). This interminable debate is now recognized as simplistic, outdated, and based on a false dichotomy between arboreal and cursorial origin theories.

It is likely that the origin of avian flight, or more properly the origin of the avian flight stroke, is more complicated. Primitive theropods were undoubtedly ground-dwelling animals that used their hands to catch and grasp prey. Some of this arm and wrist mobility was surely retained in birds, but further modifications were needed to enable a coherent powered flight stroke. The observation that juvenile extant birds can use even rudimentary wings to help ascend steep slopes, because wing-flapping forces literally push the animal against the substrate, provides an exciting new way of thinking about the origin of flight (Dial 2003a, 2003b; Dial et al., 2008). When behaving this way, birds do not simply ascend from the ground or descend from the trees, but exhibit something of a combination behavior: they are using flapping motions to run, but also maintain balance on a complex three-dimensional substrate (analogous to an arboreal canopy). This suggests that flight could have originated in almost any non-aquatic setting on earth, obviating the need for a simple ground-up/trees-down dichotomy. The origin of avian flight therefore remains an open question.

In any case, the focus here is not on the origin of the flight stroke itself, but the origin and evolution of the major anatomical components that power avian flight. Most of these components were assembled piecemeal throughout theropod evolution. Feathers originated in much more primitive theropod dinosaurs, or perhaps even more primitive archosaurs, and had already developed into the large, complex, vaned aerodynamic structures required for flight before the origin of birds (Xu and Guo, 2009). Although the forelimbs of theropods such as tyrannosaurids and abelisaurids were comically atrophied, there is a general trend in forelimb elongation along the line to birds, so much so that the most primitive birds have forelimb proportions essentially indistinguishable from those of derived non-avian theropods such as dromaeosaurids

(Middleton and Gatesy, 2000). Similarly, the degree to which the wrist can fold against the body also progressively increased along the line to birds. Wrist folding in living birds is due to two anatomical features: the wedge-like shape of one carpal bone (the radiale) and the highly convex proximal end of the fused carpal–metacarpus complex. This deep convexity is present in all coelurosaurian theropods, whereas the radiale became incrementally more wedge-shaped across theropod phylogeny (Sullivan et al., 2010). Finally, the range of motion of the theropod forelimb also underwent several important changes on the line to birds. The proximal joints, including the joints between the pectoral girdle and humerus and humerus and antebrachium, exhibited progressively increased flexibility in more derived theropods (Senter, 2006a, 2006b). However, the individual joints of the hand became less flexible, suggesting that the manus was departing from its ancestral function as a primary grasping organ in predation (Carpenter, 2002; Senter, 2006a, 2006b). The hands of living birds are essentially useless in grasping, but rather anchor primary flight feathers that provide much of the thrust during the flight stroke. Because of all these gradual changes along the line to birds, it is currently very difficult to draw a line between birds and non-avian theropods, or to determine exactly where on the theropod family tree flight evolved.

The unusual knee-driven hindlimb system of birds was also assembled piecemeal throughout theropod evolution. Several individual components are integral to this style of locomotion: a horizontal femur, which is largely immobile and must be resistant to strong torsional forces, extremely long distal hindlimb segments, large knee flexor and extensor muscles but reduced femoral retractors, and a more forward-positioned center of mass. Most theropods have cursorial limb proportions, so the elongate distal hindlimbs of birds are inherited from a distant theropod ancestor (Gatesy and Middleton, 1997; Carrano, 1999). No non-avian theropods possess the short, stout, robust femur of living birds, so it seems as if the immobile horizontal femur so characteristic of extant avians developed after the origin of birds (Carrano, 1998). However, the size of various hindlimb muscles and the position of the center of mass underwent demonstrable changes

during theropod evolution. The femoral retractor muscles, which stretch from the tail to the posterior surface of the femur, were clearly reduced along the line to birds, as more derived theropods have incrementally shorter tails and a less prominent muscular attachment site on the femur (Gatesy, 1990). These changes, along with the increasing size of the forelimb, would have progressively shifted the center of mass further forward, culminating in the posture of living birds, which balance their weight over their knee and not their hips (Gatesy, 1990, 2002; Farlow et al., 2000).

Gatesy and Dial (1996) summarized all these changes, in both forelimb and hindlimb anatomy and function, by discussing so-called "locomotor modules" (Fig. 6.8). These modules are defined as integrated portions of the muscular and skeletal systems that act as functional units during locomotion, and are a conceptually useful way to think about the myriad major changes that occurred as non-flying predatory theropods evolved into small volant birds. Their main idea is that basal non-avian theropods possessed a single locomotor module: they used their hindlimbs for propulsion and their tail for balance, and their forelimbs played no role in locomotion. The origin of avian flight, however, involved the superposition of a second locomotor module onto this primitive bauplan, as well as changes to the initial module itself. This second module consisted of the elongate and feathered arms that were capable of powered flight, whereas the modifications to the cursorial module involved decoupling into separate pelvic and caudal modules, which involved changes to the tail and pelvis. In primitive theropods the tail was a counterbalance to the large front of the body (including the big pelvic muscles), and was thus intimately associated with bipedal locomotion. In birds, however, the elongation of the forelimbs moved the center of mass forward, over the knee, and the tail shortened as a result. Therefore, in birds, the size and shape of the tail is closely affiliated with the size and position of the forelimbs, not the pelvis. These major changes did not happen overnight, but were assembled bit by bit across the theropod family tree and over tens of millions of years of evolution. This is the single greatest revelation of decades of research on theropod locomotion.

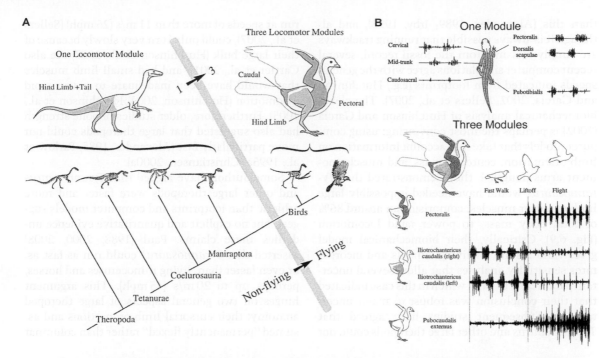

Figure 6.8 Locomotor modules in theropods, and changes in these modules during the dinosaur–bird evolutionary transition. (A) Schematic showing the change from one to three locomotor modules during the evolution of birds. (B) Comparison of muscle activity in a salamander with one locomotor module (note that various muscles from all over the skeleton are firing at once) and a pigeon with three locomotor modules (note the orderly progression of muscle activity divided into three regions, with pelvic muscles active during walking and pectoral and caudal muscles active during flight). Modified from Gatesy and Middleton (1997) and Gatesy and Dial (1996), and reproduced with permission.

Some Major Questions about Dinosaur Locomotion

Could large theropods run fast?

An enduring debate in both scientific and popular circles is whether *Tyrannosaurus rex* and other colossal theropods were capable of fast speeds. Old reconstructions commonly depicted large theropods as stocky lumbering creatures whose tails drooped onto the ground, but newer portrayals often present these 5-tonne carnivores as active energetic dynamos capable of great speed and agility. This trend reached its climax in the movie *Jurassic Park*, in which *Tyrannosaurus* is famously shown chasing down a jeep moving at highway speeds. Is this landmark movie scene plausible? This is an interesting question to answer, not mere-

ly for the novelty of proving Hollywood producers wrong or the excitement of studying a celebrity species, but also because it may reveal general insights about how enormous terrestrial bipeds functioned as living animals. This question, obviously, is intractable by studying modern species only, because nothing on the scale of *Tyrannosaurus* exists today.

All explicit evidence, ranging from studies of trackways to sophisticated statistical and computer modeling analyses, agrees that *Tyrannosaurus* and other large theropods could not run at high speeds. Maximum speed was probably in the ballpark of about 5–11 m/s (11–25 mph) and *Tyrannosaurus* probably could not run in the strict sense (i.e., it probably did not have an aerial phase in its step cycle, even at its maximum speed). No known trackways of large theropods indicate speeds greater than this (Alexander, 1989; Irby, 1996), and although it is always possible that running trackways are simply missing from the fossil record, several recent computer simulations agree with the general speeds indicated from footprints (e.g., Hutchinson and Garcia, 2002; Sellers et al., 2007). The clever biomechanical analysis of Hutchinson and Garcia (2002) is perhaps the most convincing: using computer models that take into account information on limb orientation, center of mass, and muscle moment arms (lengths), they demonstrated that *Tyrannosaurus* would have needed impossibly large limb extensor muscles, comprising an absurd 86% of total body mass, to power rapid locomotion (Fig. 6.9). Critically, their biomechanical model gives valid results for living animals and incorporates sensitivity analyses that allow several uncertain parameters to vary, which in this case indicated that their conclusion was robust to major uncertainties. Subsequent studies have agreed that *Tyrannosaurus* and other large theropods could not

run at speeds of more than 11 m/s (25 mph) (Sellers et al., 2007), could only turn very slowly because of their large bulk (Hutchinson et al., 2007; see also Carrier et al., 2001), and had small limb muscles that would have been inadequate to power rapid locomotion (Hutchinson, 2004; Hutchinson et al., 2005). Furthermore, older studies of bone strength had also suggested that large theropods could not move particularly fast (Alexander, 1985; Farlow et al., 1995a; Christiansen, 2000a).

Some authors have argued that *Tyrannosaurus* and other large theropods were faster and more athletic than footprints and computer models suggest, but no explicit and quantitative evidence underlies these claims. Paul (1988, 2000, 2008) asserted that *Tyrannosaurus* could run as fast as, or even faster than, living rhinoceroses and horses, perhaps up to 20 m/s (45 mph). This argument hinges on two general aspects of large theropod anatomy: their cursorial limb proportions and assumed "permanently flexed" rather than columnar

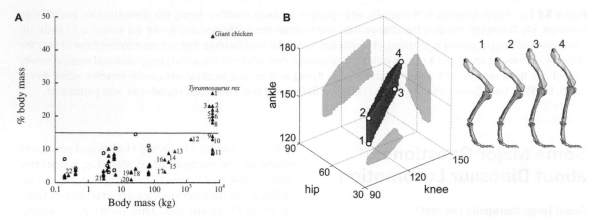

Figure 6.9 Biomechanical evidence suggesting that *Tyrannosaurus* was not capable of running fast. (A) Plot comparing the muscle mass and body mass of *Tyrannosaurus* and living animals. Open circles denote the percentage of body mass taken up by the leg muscles in living animals and the triangles denote how much muscle mass is needed in all theoretical digital models of animals running at fast speeds. The horizontal line (15% of body mass occupied by limb muscles) shows the largest known percentage of body mass occupied by the limb muscles (in living ostriches and emus). Note that all digital models of fast-running *Tyrannosaurus* require a greater proportion of body mass incorporated into the limb muscles than is known in any living animal, rendering it unlikely that tyrannosaurids could run rapidly. (B) Plot showing potential mid-stance running poses for *Tyrannosaurus*. Slow running poses, depicted by the dark black cloud with numbers 1–4, are restricted to a narrow wedge of configuration space, which is defined by the angles at various joints (the gray regions are projections of this wedge of configuration space onto two-dimensional planes). Four extreme slow running poses are depicted and plotted on the graph: 1 (50° hip, 108° knee, 132° ankle), 2 (50°, 109°, 149°), 3 (70°, 141°, 142°), 4 (61°, 139°, 168°). Modified from Hutchinson (2004) and Gatesy et al. (2009), and reproduced with permission.

limbs (the latter of which are seen in large living animals that cannot move fast). However, as shown by Carrano (1999), the limb proportions of dinosaurs and mammals span a vast continuum and have no clear relationship with absolute speed. Even a "cursorial" animal, especially a colossal one, need not run at a sprinter's pace. Furthermore, large theropods like *Tyrannosaurus* evolved from smaller, clearly fast-running ancestors, so it is likely that their cursorial limb proportions reflect evolutionary baggage rather than adaptation. It is also simplistic, and speculative, to conclude that *Tyrannosaurus* had a permanently flexed limb. Any limb can undergo a wide range of poses and motions, and there is no clear reason why a flexed posture is any more reasonable than other positions. Indeed, the recent quantitative study of Gatesy et al. (2009), which assessed millions of possible poses for *Tyrannosaurus*, found that a permanently flexed posture was unlikely (Fig. 6.9). The geometry of individual joints, in concert with the size and position of hindlimb muscles, simply would not have allowed *Tyrannosaurus* to maintain such a posture, despite the fact that it seems plausible when observing the dry bones of a skeleton in isolation. Instead, it is apparent that *Tyrannosaurus* had something of an intermediate posture, between a crouched bird-like stance and a straight-legged columnar gait.

Therefore, it is simply not possible to make a general argument about large theropod locomotion armed with a few quick measurements, superficial consideration of joint motion based solely on bones, and vague comparisons to modern analogues. As forcefully argued by Hutchinson and Allen (2009), in a summary of this debate which they describe as "tiring," the question of how large theropods moved is ultimately a quantitative, not a qualitative, issue. Explicit quantitative techniques, which take into account a range of parameters that can be subjected to sensitivity analyses, are the only proper methods for answering this question. Importantly, all such studies agree that large theropods were not Olympian sprinters capable of outrunning speeding vehicles.

Could some non-avian theropods fly?

Birds evolved from theropod dinosaurs, so the origin of flight itself must have occurred somewhere along the theropod–bird evolutionary lineage. Did flight originate in birds, or could some non-avian theropod dinosaurs themselves fly? If so, could they achieve powered flight as in living birds or were they restricted to passive gliding? These questions are difficult to answer, but exciting new fossil discoveries over the past decade have revealed some startling surprises. Indeed, the possibility that some non-avian theropods could glide or fly has been bolstered by the discovery of four-winged theropods and several non-avian species that are remarkably similar in anatomy and body size to early birds.

The biggest bombshell was the discovery of *Microraptor gui*, a small dromaeosaurid theropod from the Early Cretaceous of China that bears large, vaned, pennaceous feathers on both the forelimb and hindlimb (Xu et al., 2003). This discovery was completely unexpected: up until this point, complex aerodynamic feathers were only known to be present on the forelimbs of true birds and their closest dinosaurian relatives, as in living birds these types of feathers comprise an airfoil that is used to generate lift during the flight stroke. The presence of such feathers in *Microraptor* suggested two provocative hypotheses. First, *Microraptor* itself – a tiny, 80-cm, 1–3 kg non-avian theropod – was perhaps able to fly or glide, probably by using its limbs as gliding airfoils. Second, the presence of a dense network of complex hindlimb feathers suggests that the hindlimb was also used as a lift-generating airfoil, raising the intriguing possibility that birds passed through a transitional "four-winged" stage during the evolution of flight (Fig. 6.10).

Testing these hypotheses, however, has proven more difficult. Although it is almost certain that the size, complexity, and overlap of the hindlimb feathers would have conferred some aerodynamic benefits, it is unclear if these feathers actually formed a coherent airfoil that could generate large amounts of lift (Padian, 2003). Two biomechanical investigations that have subjected models of *Microraptor* to computer and catapult launch studies have both concluded that gliding was possible, with the hindlimbs held either obliquely (Chatterjee and Templin, 2007) or straight laterally (Alexander et al., 2010). However, because all known specimens of *Microraptor* are smashed flat on limestone slabs, it is quite difficult to reconstruct the orientation of the hindlimb feathers and the range of motion of the hindlimb joints. Although perhaps unappreciated,

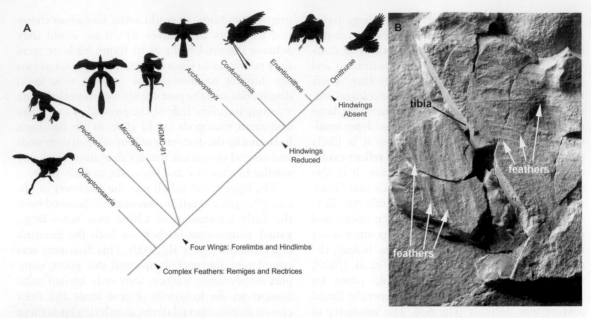

Figure 6.10 Theropods with four wings (wings on the forelimbs and hindlimbs). (A) Simplified schematic of wing evolution in birds and their closest non-avian theropod relatives showing that four wings may be the primitive condition for birds and was a common morphology among bird-like theropods. (B) Photograph of large feathers attaching to the hindlimb (tibia) of the oldest known bird, *Archaeopteryx*. Image (A) modified from Longrich (2006) and used with permission from the Paleontological Society. Image (B) courtesy of Dr Nick Longrich.

these uncertainties render any modeling study tenuous at best. For instance, the study of Alexander et al. (2010) is fatally flawed because the hindlimbs are modeled as projecting laterally, but this is simply impossible given the presence of various bony processes (supracetabular crest and antitrochanter of the ilium) that prevent splaying of the hindlimbs in theropods (Norell and Makovicky, 1997). Even though the known fossil material of *Microraptor* is crushed, these features were clearly visible but ignored by the authors of the study (Brougham and Brusatte, 2010).

It is also difficult to assess the idea that avian-style flight evolved through a four-winged stage. Assuming for the moment that *Microraptor* actually could fly or glide, it is not clear that its aerial abilities were homologous with those of birds. In other words, *Microraptor* may have evolved its gliding and/or powered flight habits independently of birds, which would mean that locomotor evolution in birds and their closest relatives was much more complicated than often considered. Ultimate-

ly, the question of whether avian flight was homologous to dromaeosaurid flight is a phylogenetic question: how are flight and flight-related characters optimized onto theropod phylogeny? There are nearly 40 known dromaeosaurid and troodontid dinosaurs – the closest relatives of birds – and these exhibit great variation in body size, feathery integument, and morphology. Just because *Microraptor* was able to glide/fly and had four wings does not mean that the avian ancestor would have possessed these qualities. Perhaps *Microraptor* and other small-bodied coelurosaurs whose skeletons seem primed for flight, such as *Rahonavis* and *Anchiornis*, may have evolved their volant capabilities independently of birds (Paul, 2002).

That being said, recent discoveries are strengthening the hypothesis that birds and their closest relatives inherited a homologous set of four wings (or at least large pennaceous forelimb and hindlimb feathers, whether or not they comprised airfoils) from a more distant theropod ancestor (Fig. 6.10). Large hindlimb feathers are now also known in other specimens of non-avian theropods, including the troodontid *Anchiornis* (Hu et al., 2009) and the indeterminate maniraptoran *Pedopenna* (Xu and Zhang, 2005), as well as some basal birds, such as *Confuciusornis* (Zhang et al., 2006), some enantiornithies (Zhang and Zhou, 2004), and even *Archaeopteryx* itself (Christiansen and Bonde, 2004; Longrich, 2006) (Fig. 6.10B). It now seems credible that primitive birds did inherit four wings from their theropod ancestors, but it is still an open question as to how the hindlimb feathers actually related to aerial locomotion, whether they formed a coherent airfoil, and whether they could generate much lift (Padian, 2003; Hutchinson and Allen, 2009). Clearly, this is an area that demands further study, using explicit biomechanical modeling approaches.

One issue that is not in doubt is the great similarity between primitive birds and some of their closest dinosaurian cousins. As has been reiterated throughout this chapter, there is a morphological and functional continuum between non-avian theropods and birds, and it is truly difficult to draw a line between "non-avian theropods" and "birds" and theropods that could fly and those that could not. This is dramatically, and somewhat ironically, illustrated by the fact that contemporary scientists sometimes mistakenly identify non-avian dromaeosaurids and troodontids as birds (e.g., *Rahonavis*, *Jingfengopteryx*, *Anchiornis*, and perhaps *Archaeopteryx* itself: Xu et al., 2011a). Simply put, it is extremely difficult to tell apart a primitive bird from other derived theropods, and if Mesozoic zoologists were observing these creatures as living animals they would probably fail to make any conceptual distinction between species like *Archaeopteryx* (a bird) and *Anchiornis* (a troodontid).

We now know that several non-avian dinosaurs possessed complex pennaceous feathers, which sometimes even attached to the ulna via quill knobs as in living birds (Forster et al., 1998; Norell et al., 2002; Xu et al., 2003, 2009a; Turner et al., 2007b; Hu et al., 2009; Xu and Guo, 2009) (see Fig. 3.11). Some of these species were remarkably small-bodied, which is usually a requirement for volant animals because flight becomes metabolically expensive at large sizes (Turner et al., 2007a) (Fig. 6.11). It is a reasonable conclusion that these small, aerodynamically feathered, non-avian theropods, especially genera such as *Rahonavis* and *Anchiornis*, were capable of

some form of gliding or flight (regardless of whether their flight capabilities were homologous to those of birds or derived independently). They may have even had aerial capabilities similar to those of primitive birds such as *Archaeopteryx*, which were probably not strong fliers (Nudds and Dyke, 2010). These hypotheses require testing with additional fossil discoveries, phylogenetic analyses, and biomechanical studies, but most reasonable paleontological gamblers would probably place their bets on flight having evolved in non-avian theropod dinosaurs.

Could some dinosaurs burrow?

Most dinosaurs were ground-dwelling animals, although birds and their closest non-avian relatives may have roosted in the trees. The functional habits of Mesozoic dinosaurs therefore seem somewhat restricted when compared with those of living mammals, which live in most every conceivable environment (Sereno, 1999). Recent discoveries, however, indicate that some dinosaur species were capable of burrowing. Varricchio et al. (2007) described a spectacular association between the skeletal fossils of the basal ornithopod *Oryctodromeus* and a tubular meandering fossilized burrow (Fig. 6.12). The bones of an adult and two juveniles were found within the expanded terminal chamber of the burrow, and the fact that *Oryctodromeus* has several skeletal features present in living burrowers supports its identification as the burrow-maker. For instance, the scapula and coracoid are fused and expanded, which would support large muscles necessary for excavating a burrow, and the pelvis and hindlimbs are especially robust, which would have helped brace the animal against the violent forces generated during digging. More recent discoveries from the Late Cretaceous of North America indicate that some derived theropods were also capable of burrowing, perhaps as a strategy for hunting small fossorial mammals (Simpson et al., 2010). Interestingly, both types of burrowing dinosaur would have possessed cursorial limb proportions, suggesting that some dinosaurs were able to both dig and run rapidly at the same time. These discoveries have expanded the known range of Mesozoic dinosaur locomotory and functional behaviors, and it is likely that additional examples of burrowing and digging dinosaurs will continue to be discovered.

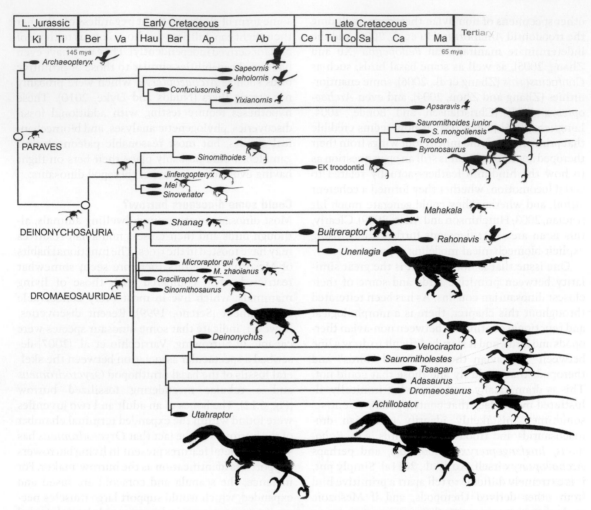

Figure 6.11 Body size evolution in the earliest birds and their closest theropod relatives (including troodontids and dromaeosaurids). Silhouettes denote relative body size; those facing left represent the body size of the species in the phylogeny, whereas those facing right are the estimated body sizes of the hypothetical common ancestors (the nodes) on the phylogeny, as predicted by a parsimony algorithm. Note that although some dromaeosaurids were large, basal dromaeosaurids and troodontids, as well as primitive birds, were mostly small-bodied animals. It is predicted that the common ancestor of birds, troodontids, and dromaeosaurids was a small-bodied animal approximately the same size (or smaller) than primitive birds such as *Archaeopteryx*. Modified from Turner et al. (2007a) A basal dromaeosaurid and size evolution preceding avian flight. *Science* 317, 1378–1381. Reprinted with permission from AAAS and the author.

How did sauropods hold their necks?

The elongate necks of sauropods, some of which were up to 10 m in length, have generated continued bewilderment and speculation since the discovery of the first well-preserved sauropod skeletons over a century ago. One central question concerns neck posture: did most sauropods hold their neck erect, allowing it to reach high into the treetops, or at a more horizontal or even downturned posture? This puzzle has important implications for understanding sauropod feeding, as an animal that could stretch its neck into the highest reaches of the

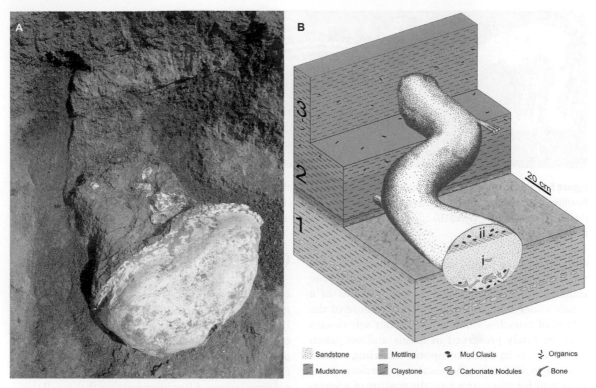

Figure 6.12 Burrowing dinosaurs. Photo and illustration of burrows belonging to the small ornithischian *Oryctodromeus*. Images courtesy of Dr David Varricchio and modified from Varricchio et al. (2007).

canopy had access to very different food sources than a species restricted to browsing at or near ground level. Unfortunately, it is difficult to directly assess the neck posture of sauropods based on their bones, as the neck is composed of several individual vertebrae, each of which is huge, heavy, fragile, and awkward to physically move and articulate with other vertebrae. Therefore, computer modeling techniques, comparisons to living analogues, and arguments based on physiology have played a central role in this debate.

Disagreements continue to linger over whether some, or any, sauropods could hold their necks in an erect high-browsing position. Computer modeling studies suggest that *Apatosaurus*, *Diplodocus*, and other sauropods were physically incapable of twisting their necks into an erect pose, but rather employed a more horizontal neck posture (Stevens and Parrish, 1999, 2005a, 2005b) (Fig. 6.13). These studies aligned the individual cervical vertebrae of

several sauropods into articulated digital models of the neck, with vertebrae contacting at that position in which overlap between the zygapophyses is maximal. This pose is referred to as the "osteological neutral pose," and has been argued to represent the most common posture for these sauropods. Arguments for a more horizontal posture have also been supported with physiological data, as Seymour and Lillywhite (2000) and Seymour (2009) calculated that some sauropods would need to expend half their energy, and employ absurdly high blood pressures, just to pump blood to an erect neck. Furthermore, Ruxton and Wilkinson (2011) calculated that a long neck would provide substantial energy savings for a giant sauropod, which must expend a great deal of effort to simply move its body, even if it was only used for feeding at or near the ground.

Other workers, however, have disagreed with these conclusions. Dzemski and Christian (2007)

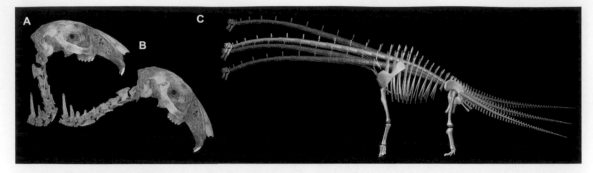

Figure 6.13 Range of motion in sauropod necks. (A, B) Extant rabbit in osteological neutral pose (B) but also in a much more flexible pose that has been observed in living animals (A). (C) Neck of *Brachiosaurus* in a variety of poses, based on a digital model. Images (A) and (B) courtesy of Dr Michael Taylor and modified from Taylor et al. (2009); image (C) courtesy of Dr Kent Stevens.

and Taylor et al. (2009) showed that living animals are rarely restricted to the osteological neutral pose implied by their vertebrae, but are capable of a much wider range of neck posture because of the action of muscles, tendons, and other soft tissues that are rarely preserved in fossils and not taken into account in most computer modeling studies (Fig. 6.13A,B). Study of bones alone, therefore, is not sufficient for reconstructing the posture of a vertebrate neck. Furthermore, although the arguments about the extreme blood pressures and energy expenditure needed to maintain an erect neck are persuasive, these would only preclude a high-browsing posture if the neck was permanently held this way. It is possible that the neck was held erect only intermittently, perhaps during feeding (Sander et al., 2009; see also Upchurch and Barrett, 2000). Most provocative, however, is Christian's (2010) biomechanical analysis showing that large sauropods would have required more energy to walk a short distance than raise their neck high. It seems as if sauropods were truly strange creatures, with energy requirements completely unlike those of living animals, and their energy expenditures for any activity may seem enormous by our standards. Precluding high-browsing neck postures based on energetics aruguments may therefore be a misleading exercise. In sum, it seems plausible that the necks of most sauropods were capable of a wide range of positions and postures, but that extreme poses may have been limited by physiological constraints.

Conclusions

Understanding how dinosaurs stood and moved is integral to understanding how they functioned as living animals. Dinosaur locomotion can be studied using several strands of data, including qualitative and quantitative aspects of skeletal anatomy, fossil footprints, and computer modeling and mathematical simulations. All evidence agrees that small theropods were likely the fastest of dinosaurs, whereas bulky sauropods were the slowest. Large theropods such as *Tyrannosaurus rex* were probably incapable of running at rapid speeds and may have been slower than some of their prey. All theropods were bipedal and all sauropods quadrupedal, but "prosauropods" and ornithischians had a wider range of locomotor and postural abilities, and it is likely that many species were capable of walking on two or four legs depending on the situation. Both the weight-bearing, columnar, quadrupedal posture of sauropods and the fleet-footed, long-armed, flying bauplan of birds were acquired piecemeal over millions of years of evolution. The morphological and functional distinction between birds and their closest non-avian theropod relatives is becoming increasingly blurred, and it is likely that some derived small-bodied coelurosaurian theropods were capable of gliding or even powered flight. Whether flight evolved through a four-winged stage or perhaps originated multiple times on the theropod family tree is uncertain, but promise to be among the most active areas of future research.

7 Feeding and Diet

Is there any more iconic image in paleontological lore than the toothy jaws of *Tyrannosaurus* chomping into the flesh of its prey? There is a reason why this image crops up so often in museum exhibits, movies, and other artistic renderings of dinosaurs. Procuring food is one of the basic biological functions of any animal, and depicting a dinosaur in the process of feeding helps bring a pile of dusty prehistoric bones to life. A battle scene between *Tyrannosaurus* and *Triceratops* injects some drama and intrigue into what otherwise could be a yawner of a museum exhibit, and the aerial views of *Brachiosaurus* feeding high in the treetops in *Jurassic Park* are much more inspirational than watching these animals stand or sleep.

It is not only museum goers and cinema fans who are fascinated by the dietary strategies and feeding habits of dinosaurs, but scientists as well. Because living animals spend much of their time feeding, understanding how dinosaurs fed and what they ate is integral to understanding how they functioned as biological organisms. Diet is also strongly intertwined with many other aspects of organismal biology, such as reproduction, growth, behavior, habitat preferences, and ecosystem dynamics. With this in mind, it is no surprise that dinosaur diet has generated considerable interest, speculation, and study since the discovery of the first dinosaur fossils over 150 years ago. Contemporary vertebrate paleontologists view studies of feeding and diet as an integral component of any rigorous understanding of dinosaur biology. In particular, scientists are keen to understand the major feeding strategies employed by different dinosaur subclades, how features of the anatomy relate to specific feeding strategies, how anatomical character complexes linked to certain feeding strategies were assembled over evolutionary time scales, and whether diet may help explain why certain species were so successful.

As with questions relating to locomotion explored in the preceding chapter, these and other questions relating to diet are often difficult to answer.

Dinosaur Paleobiology, First Edition. Stephen L. Brusatte.
© 2012 John Wiley & Sons, Ltd. Published 2012 by John Wiley & Sons, Ltd.

Dinosaurs cannot be observed directly, so scientists must piece together information from skeletal anatomy, rare trace fossils (such as bite marks and fossilized dung), and quantitative biomechanical modeling studies in order to understand potential feeding and dietary strategies. There are great uncertainties involved with each of these lines of evidence, and often paleontologists are only able to broadly categorize a certain species or group as carnivorous or herbivorous. In other cases, however, a wealth of information allows scientists to firmly hypothesize that certain species repeatedly ate a specific type of food, had specific anatomical adaptations that enabled a certain feeding style, or even hunted in packs or cannibalistically preyed on members of their own species. This chapter will outline the toolkit of methods that scientists use to study dinosaur feeding, explore the major dietary strategies employed by the fundamental dinosaur subgroups, discuss the evolution of dietary strategies in dinosaurs, and examine some major questions about dinosaur diet that have spawned significant debate.

Methods for Studying Feeding and Diet

Gross skeletal anatomy

As with many aspects of dinosaur behavior, study of dinosaur feeding should begin with a general consideration of gross skeletal anatomy. Even fragmentary bones are often enough to illustrate whether a certain dinosaur was carnivorous or herbivorous, as both of these diets are associated with distinctive anatomical features (Hildebrand and Goslow, 1998; Norman, 2001; Barrett and Rayfield, 2006).

Carnivores usually possess large heads in proportion to the body, jaws studded with serrated and recurved teeth akin to steak knives, and sharp claws on the hands and feet used to subdue and process prey. Many carnivores have a large brain and strongly developed sense organs, especially large and forward-pointed eyes and/or an acute sense of smell, as well as a skeleton designed for fast and agile movement. The jaws are usually optimized for quick snaps to snatch prey rather than prolonged chewing, and the jaw joint is usually positioned at approximately the same level as the tooth row. This latter feature allows the jaws to snap shut in a scissors-like action in which the brunt of the bite force is focused on one portion of the tooth row, rather than evenly distributed across the jaw, making it easier to grab and subdue struggling prey. All these features are present in most theropod dinosaurs, conclusively indicating that these species were carnivores.

Herbivores, on the other hand, usually have small heads, broader and often more chisel-shaped teeth for cropping and grinding vegetation, and bulkier bodies that accommodate the expansive gut needed to digest large amounts of vegetation. The jaw joint is usually located either above or below the tooth row, enabling the jaws to meet like two guillotine blades. This configuration allows all teeth along the jaw to shear vegetation simultaneously, rather than bite force being concentrated in one portion of the skull. Understandably, such a system is optimal for herbivores that must constantly process enormous quantities of vegetation: the entire tooth row acting together at once permits more vegetation to be chewed at one time than would a concentrated bite force. All these features are present in all sauropods, some basal sauropodomorphs, and most ornithischians, indicating that these animals were obligatory herbivores.

Differentiating obligate carnivores and herbivores based on gross anatomy is usually straightforward. It is more difficult, however, to use gross anatomy to identify exactly what types of prey species a carnivore may have targeted or what types of plant a herbivore would have ingested. Furthermore, obligate carnivory and herbivory are simply two ends on a spectrum of possible diets. Some species are omnivorous, and others may subsist on one or a few specific food items, such as insects, nuts, fruits, eggs, or shellfish. Recognizing omnivores or those species with a specialized diet is exceedingly difficult based on gross morphology alone. Omnivores do not simply possess a mixture of carnivorous and herbivorous characters and, frustratingly, come in a wide array of shapes and sizes with no consistent set of omnivory-related skeletal characters. Some discrete features of the anatomy may be related to a particular specialized diet, such as a sharp beak for cracking eggs or an elongate finger for prodding insects out of their nests, but most of these features

could conceivably be related to other functions. A beak, for instance, could also be used to crack nuts or crop plants. Therefore, it is clear that gross morphology is an important initial source of evidence for identifying the types of food a dinosaur may have eaten, but is rarely sufficient for a detailed understanding of dinosaur feeding and diet.

Trace fossils

In rare cases, trace fossils offer direct evidence of dinosaur diet (Fig. 7.1). If available and identifiable as belonging to a specific dinosaur, trace fossils can obviate the need for general dietary speculations based on gross morphology and allow paleontologists to recognize exactly what type of food a dinosaur was ingesting. The most common types of trace fossils that are useful in analyzing diet are

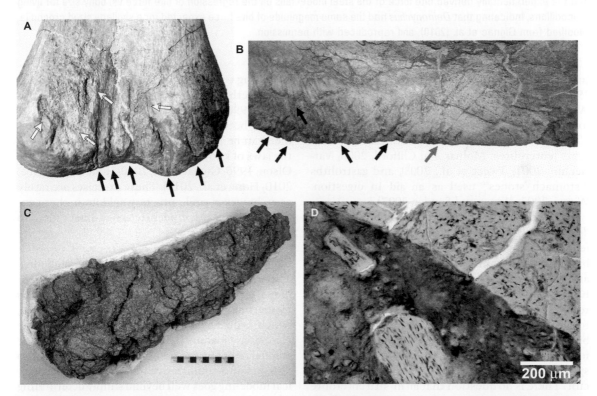

Figure 7.1 Dinosaur trace fossils that record feeding behavior. (A, B) Bite marks made by a tyrannosaurid theropod on the bones of the hadrosaurid *Saurolophus*. (C, D) An enormous coprolite from the Late Cretaceous of North America that has been assigned to *Tyrannosaurus* (C), with a close-up photomicrograph of the specimen showing pieces of bone (lighter colored blocks) in the ground mass of the coprolite (D). Images (A, B) courtesy of Dr David Hone; (C, D) courtesy of Dr Karen Chin.

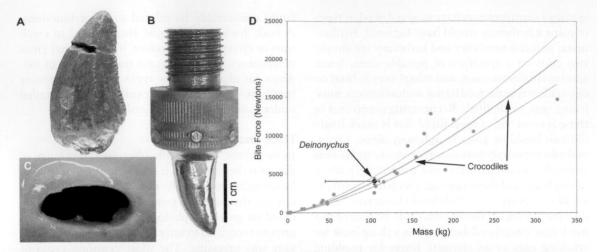

Figure 7.2 Using bite marks to study bite force and feeding style. A fossil tooth of the dromaeosaurid theropod *Deinonychus* (A), a steel model of the same tooth (B), and bovine bone punctured by the steel model (C). The plot (D) shows that the experimentally derived bite force of the steel model falls on the regression of bite force vs. body size for living crocodilians, indicating that *Deinonychus* had the same magnitude of bite force expected for a similarly sized crocodile. Modified from Gignac et al. (2010), and reproduced with permission.

tooth marks on prey bones (Erickson and Olson, 1996; Jacobsen, 1998; Carpenter, 2000; Hone and Watabe, 2010; Hone et al., 2010; Paik et al., 2011), fossilized feces (coprolites: Chin and Gill, 1996; Chin et al., 1998; Chin 2007), fossilized gut contents (enterolites: Molnar and Clifford, 2000; Varricchio, 2001; Tweet et al., 2008), and gastroliths ("stomach stones" used as an aid in digestion: Wings, 2007; Wings and Sander, 2007).

In some circumstances trace fossils can provide "slam dunk" evidence for specific dietary and feeding strategies in dinosaurs (Fig. 7.2). However, bite marks, coprolites, enterolites, and gastroliths are, on the whole, very rare discoveries. Only a handful of examples are known from the fossil record, and it is foolhardy to rely on such rarities to provide primary evidence for dinosaur feeding behavior. Furthermore, as with any trace fossil, identifying which species, or even broad group of species, produced these specimens is usually difficult. There is no question that an in-situ enterolite or gastrolith mill belonged to the animal it was found within, but there is great uncertainty in linking a specific bite mark or coprolite to the species that made it. Sometimes individual bite marks can be used as a mold to produce a model of

the biting tooth, and these can be compared with fossil teeth from the same locality (Erickson and Olson, 1996; Gignac et al., 2010). In other cases, the spacing between individual bite marks on a specimen can be matched with the spacing of teeth in the jaws of known predatory species (Erickson and Olson, 1996; Gignac et al., 2010; Hone and Watabe, 2010; Hone et al., 2010). These exercises are rarely easy, however, and only very seldom is there enough evidence to confidently assign a trace fossil to a specific dinosaur.

Quantitative studies and biomechanical modeling

The most explicit, rigorous, and powerful technique for studying dinosaur feeding and diet is biomechanical modeling. Modeling approaches, which utilize advanced mathematics and various computer programs, allow dinosaur feeding to be studied in detail and permit explicit hypotheses to be tested in a comparative framework. Biomechanical modeling goes well beyond simply determining whether a dinosaur was a carnivore or herbivore, and may help reveal specific food types that a certain species was optimized to exploit, or certain feeding styles that an organism was well designed to employ.

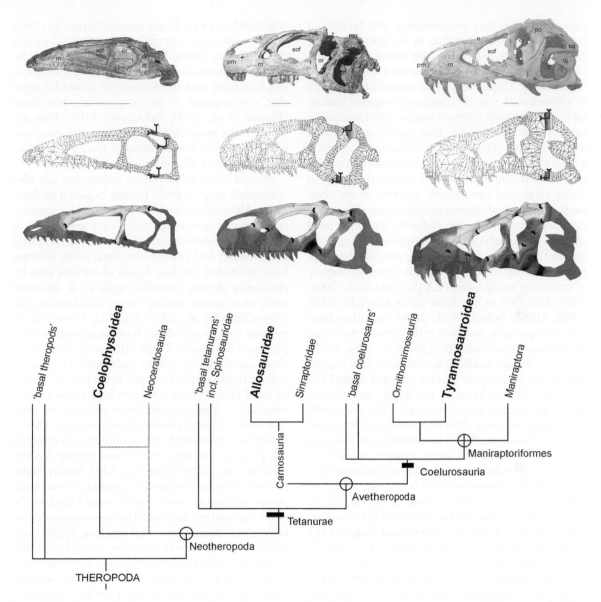

Figure 7.3 Finite element analysis in studies of dinosaur feeding. Skull photos, finite element skull models, and finite element results showing stress and strain (different shades of gray and black) for three representative species from across theropod phylogeny (*Coelophysis*, *Allosaurus*, *Tyrannosaurus*). Images modified from Rayfield (2005a), and reproduced with permission.

One method in particular, finite element analysis (FEA), has swept through the field of dinosaur paleobiology in recent years (Rayfield, 2007; Dumont et al., 2009) (Fig. 7.3; see Plate 14). FEA is a computerized technique, commonly used by engineers, that calculates the stress and strain distributions in a modeled structure after it is subject to an applied load. Engineers will often build digital models of bridges or buildings and, using a computer program, subject them to a series of loads that approximate the types of situations the real-life structures may be faced with. In this way, engineers can be confident that the bridge they are designing will be able to handle the weight of traffic, or the building being planned will be able to endure strong winds. The same analytical protocols can be used to study dinosaur feeding. Digital models of a dinosaur skull can be constructed, usually based on computed tomography (CT) or laser scanned fossils, and these can be subjected to various loads that may be encountered during feeding (Fig. 7.3). The researcher can then assess whether the skull was strong enough to process certain food items or perform a certain type of biting style, and can identify portions of the skull such as individual sutures or robust bones that were especially important for enduring feeding-related stresses. Although used for many years in engineering and medicine, FEA has only recently been applied to dinosaurs. Theropods have been the subject of extensive study using FEA (Rayfield et al., 2001, 2007; Mazzetta et al., 2004, 2009; Rayfield, 2004, 2005a, 2005b; Snively et al., 2006), but other dinosaur groups have received only cursory attention (Witzel and Preuschoft, 2005; Bell et al., 2009; Reichel, 2010). Future studies promise to greatly expand the sample of dinosaur species analyzed using FEA.

Other biomechanical modeling approaches, which are usually less computationally and mathematically intensive than FEA, have also been used to study the strength and construction of dinosaur skulls (Figs 7.4 and 7.5). Beam analysis models the complex three-dimensional skull of a dinosaur as a simple beam, whose strength (i.e., ability to resist forces generated during biting) is calculated by simple mathematical equations that take into account the length and cross-sectional shape of the beam, the magnitude of the estimated force on the beam (analogous to the force on the jaw due to feeding), and the distance of that force from the end of the beam (analogous to the jaw joint or the attachment site of jaw musculature). Because beam analysis requires only a handful of straightforward measurements and not a complex digital model or advanced computer programming, it is useful as a first approximation of skull strength and can reveal differences among a group of taxa. In some cases, these differences may be related to differences in feeding style or food choice. This technique has been used to study both theropods and ceratopsians (Henderson, 2002, 2010; Therrien et al., 2005). A similar method, frame and truss analysis, which reduces a skull to a series of struts, has been used to assess the strength of the *Tyrannosaurus* skull (Molnar, 2000). Yet other studies model the jaws of dinosaurs as simple levers in order to calculate bite force and the amount of muscular force that was required to impart a certain bite force (Tanoue et al., 2009; Sakamoto, 2010). This approach is the most simplistic of all such biomechanical models, but is still useful in identifying major differences in skull construction among taxa.

Several other quantitative approaches are also commonly used to study feeding behavior in dinosaurs. Although these do not require the construction of a representative model, they are similar to the above modeling techniques in their explicit methodology and quantitative rigor. Some workers have estimated the bite forces of various taxa by physically driving metallic replicas of dinosaur teeth into extant bovine bones (Erickson et al., 1996; Gignac et al., 2010) (Fig. 7.2). These studies indicate that theropods such as *Tyrannosaurus* and *Deinonychus* required high bite forces, larger than those employed by extant mammals, to produce indentations similar in depth to bite marks found on fossil bones. Other scientists have measured the orientation of microscopic wear patterns on dinosaur teeth and used statistical techniques to identify distinct, consistent sets of scratches that record the primary motions of the jaws during feeding (Williams et al., 2009) (see Fig. 7.17). Finally, theoretical ecological studies have predicted how much energy certain dinosaur species would need to obtain from food per day, based on assumptions about body size, metabolism, and foraging ability (i.e., speed) of the species in question. Armed with this prediction, it can then be hypothesized that a certain food source (such as different types of plants for herbivores, or carrion vs. live prey for carnivores) either was or was not a tenable source of nutrition (Colbert, 1993; Ruxton and Houston, 2003; Barrett, 2005; Hummel et al., 2008).

Quantitative techniques are clearly the most promising method for studying dinosaur feeding and diet in any sort of rigorous detail, but they do have some drawbacks. The construction of idealized models sometimes reduces a complex structure such as a skull into a much too simplified abstraction. Most published quantitative studies are also taxon specific: they assess the skull strength,

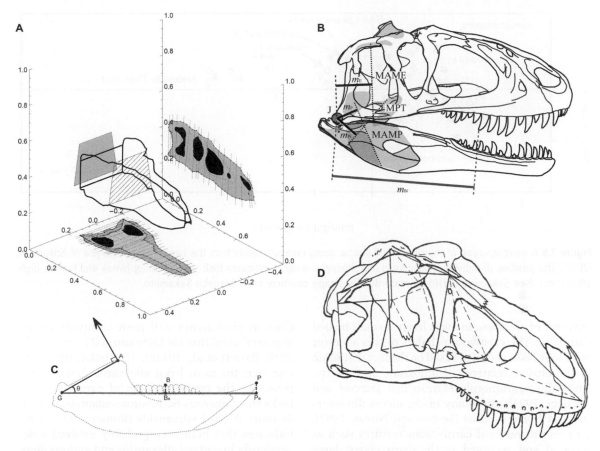

Figure 7.4 The use of biomechanical modeling in studies of dinosaur feeding. (A) Skull of a theropod dinosaur plotted into three-dimensional space, used to build a model to study bite force and feeding habits. (B) Theropod skull showing various measurements that can be input into equations, derived from beam theory, indicating the relative bite force and biting style of theropods. (C) Similar beam theory schematic of the lower jaw of a ceratopsian. (D) Frame and truss model of the skull of *Tyrannosaurus*. Image (A) modified from Henderson (2002), and reproduced with permission; image (B) courtesy of Dr Manabu Sakamoto and modified from Sakamoto (2010); image (C) modified from Tanoue et al. (2009), and reproduced with permission; image (D) modified from Molnar (2000), and reproduced with permission.

bite forces, tooth wear, or energy requirements in a single dinosaur species only. Some recent studies, however, have begun to explore these topics in a phylogenetic framework, by analyzing skull strength or bite forces across a range of taxa whose genealogy is well established (Rayfield, 2005a; Therrien et al., 2005; Sakamoto, 2010). Future work must follow this path, as only this type of study can reveal major differences in inferred feeding habits among close relatives or contemporaries, permit correlation of certain anatomical characters or other organismal attributes with feeding modes across a wide range of

taxa, and help uncover the major evolutionary changes in dinosaur diet and feeding during the Mesozoic.

The Evolution of Feeding and Diet in Dinosaurs

Dinosaurs boasted a wide range of diets and feeding styles, and diversified to fill many ecological niches throughout the Mesozoic, but the common

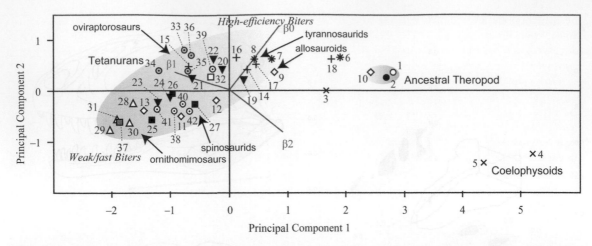

Figure 7.5 A morphospace of theropod skull shape, using measurements from the beam theory analysis of Sakamoto (2010). The position of individual species in morphospace roughly indicates their style of biting (weak and fast vs. high efficiency). See Sakamoto (2010) for more details. Image courtesy of Dr Manabu Sakamoto.

ancestor of all dinosaurs was likely a small-bodied carnivore. The carnivorous habits of this ancestor have long been assumed based on the fact that some of the closest relatives of dinosaurs, such as the basal dinosauromorph *Marasuchus* (Sereno and Arcucci, 1994), and many of the oldest dinosaurs such as *Herrerasaurus* (Sereno and Novas, 1992), possess unequivocal carnivorous features such as serrated and recurved teeth, sharp claws, large heads, and skeletons designed for speed.

New discoveries, however, have muddied this picture (Langer et al., 2010; Barrett et al., 2011a). The very closest relatives of dinosaurs, the speciose clade Silesauridae, are primarily herbivorous or omnivorous, as indicated by their leaf-shaped teeth with coarse denticles and beak-like anterior tip of the lower jaw (Dzik, 2003; Nesbitt et al., 2010). Furthermore, the basal theropod *Eoraptor* and basal sauropodomorph *Panphagia* have remarkably similar dentitions, in which the anterior teeth are sharp and recurved as is common in carnivores, but the posterior teeth are broader and more leaf-shaped as in herbivores (Sereno et al., 1993; Martinez and Alcober, 2009). It is likely that these animals were omnivorous, and not either strict carnivores or herbivores as were more derived members of their respective subclades (however, recent work indicates that *Eoraptor* may in fact be a sauropodomorph: Martinez et al., 2011).

Current phylogenies still posit carnivory as the ancestral condition for Dinosauria (Nesbitt et al., 2010; Barrett et al., 2011a). This is largely due to one fact: the most basal silesaurid, *Lewisuchus*, possesses the recurved teeth of carnivores and lacks the numerous herbivorous adaptations seen in more derived silesaurids (Romer, 1972). This indicates that herbivory probably evolved independently in derived silesaurids and various dinosaur groups. This being said, any conclusion about the ancestral dietary modes of dinosaurs or their constituent subclades is liable to change with the discovery of new fossils. In short, it is clear that dietary evolution near the base of the dinosaur family tree is much more complicated than previously thought.

Regardless of the exact diet of the dinosaur ancestor, more derived dinosaurs developed a remarkable array of feeding strategies and dietary modes. Most theropods were carnivorous, probably retaining the primitive condition of the dinosaur ancestor, but some species likely subsisted on fish, insects, and even plants. Other theropods, most notably the colossal tyrannosaurids, were hypercarnivores that could crunch through the bones of prey much larger than themselves. All ornithischians, with the exception of some possible omnivorous basal forms, seem to have been herbivorous. Derived ornithischians, the ornithopods and ceratopsians, developed sophisticated chewing mechanisms for cropping and processing large amounts of food at once. Sauropods were also obligate herbivores, and almost certainly developed their plant-eating diet independently of ornithischians. More basal sauropodomorphs, however, may have been capable of eating a great diversity of food items, ranging from meat and plants to insects and fruits. Therefore, it appears as if the ancestral dinosaur was carnivorous, that most theropods retained this habit, and that herbivory evolved at least three times within dinosaurs: at least once apiece in theropods, sauropodomorphs, and ornithischians (Barrett and Rayfield, 2006; Barrett et al., 2011a).

Theropod Feeding and Diet

Carnivorous theropods

Most theropods clearly ate meat, although some derived subclades closely related to birds possess leaf-shaped teeth, edentulous beaks, barrel-shaped guts, and other features indicative of herbivory, omnivory, or a specialized diet. These aberrant forms will be discussed later in this section, but the focus first is on the vast majority of theropods that were obligate carnivores. These species are united by a consistent bauplan, as they all possess large skulls with bone sutures optimized for enduring feeding-related stresses (Rayfield, 2005b), a kinetic joint between the bones of the lower jaw that allowed the mandible to gape outward when feeding on large or struggling prey (Sereno and Novas, 1994), sharp teeth with serrations for puncturing and gripping meat (Abler, 1992; D'Amore, 2009), piercing claws on the hands and feet for slashing and grasping prey (Carpenter, 2002; Manning et al., 2006), and relatively rapid bipedal locomotion for chasing down their victims.

The relatively conservative nature of the theropod body plan, among those species that were clearly carnivorous, contrasts with the greater variety of skull and tooth shapes seen in living carnivoran mammals (Van Valkenburgh and Molnar, 2002). It is likely that carnivorous theropods exploited a narrower set of dietary strategies than extant carnivoran mammals, and specifically it seems as if they did not experiment with the bone-cracking

and more omnivorous ecotypes of hyenas and ursids and canids, respectively. Although large tyrannosaurids could crush through the bones of their prey, there is little evidence that they, or other theropods, regularly cracked, gnawed on, or consumed large bones as a primary source of nutrition. The teeth of most species were simply too weak to allow bone crunching, theropod coprolites rarely include traces of bone and, most importantly, fossilized dinosaur bones rarely exhibit the characteristic gnawing traces that hyenas make when chewing the ends of long bones (Farlow et al., 1991; Fiorillo, 1991; Jacobsen, 1998; Hone and Rauhut, 2010).

It seems, therefore, as if the great majority of unequivocally carnivorous theropods were "hypercarnivorous," in the sense that they obtained most of their nutrition from the flesh of prey. Like most living hypercarnivores, such as lions, it is plausible that theropods preferentially hunted small, young, and weak individuals, although this is difficult to conclusively support using the fossil record (Hone and Rauhut, 2010). It is not straightforward, at least based on fossil evidence, to shoehorn carnivorous theropods into different guilds as is often done for living mammals, and it is likely that ecological separation among coexisting theropods was enabled more by differences in body size and prey size than overall dietary strategy (Van Valkenburgh and Molnar, 2002).

Although the theropod bauplan is quite conservative generally, it is still possible to recognize differences between species based on gross morphology, sensory abilities, and biomechanical analyses of skull strength. These differences are probably indicative of subtly different feeding styles among species, which could have helped enable niche partitioning if all these species were hypercarnivorous. Differences in gross morphology are often obvious: there is great variability in theropod body size and inferred speed, and species frequently differ in the size and robusticity of their skulls and the length and range of motion of their arms. A large-bodied tyrannosaurid, for instance, has an enormous muscular skull but puny arms, whereas a man-sized dromaeosaurid has a relatively smaller skull but enormous forearms tipped with large claws and powerful hindlimbs with a hyperextensible "killer claw" on the second digit. It is no great stretch to imagine that these animals, which did live side by side in some ecosystems, probably targeted different prey and hunted using different styles.

More sophisticated and quantitative studies also reveal general differences among theropod hypercarnivores. Biomechanical analyses using beam theory and lever mechanics indicate that skull strength and bite force varied greatly among theropods (Henderson, 2002; Therrien et al., 2005; Sakamoto, 2010). Therrien et al. (2005) divided carnivorous theropods into five basic morphotypes based on the shapes and inferred strength of their jaws. Abelisaurids were found to be similar to Komodo dragons in their skull strength, and were posited as ambush predators that dispatched their prey with weak bites. Dromaeosaurids were described as slashers, whereas spinosaurids had a strong anterior mandible that would have been useful in capturing small prey and allosauroids were well suited for delivering hatchet-like blows to large prey (see also Rayfield et al., 2001). Tyrannosaurids were found to be the strongest of theropods, as the nearly circular cross-section of their anterior mandible was well adapted for enduring extreme bite forces and high stresses from struggling prey. A similar comparative analysis by Sakamoto (2010) used lever mechanics to identify dromaeosaurids and spinosaurids as possessing weak but fast bites, and tyrannosaurids and allosauroids as possessing stronger and slower bites (Fig. 7.5).

Comparative assessments of theropod skull strength and sensory abilities using FEA and CT are still in their infancy. In a particularly illuminating analysis that compared the feeding-induced stress patterns in three phylogenetically disparate theropods, Rayfield (2005a) demonstrated that peak feeding stresses in the tyrannosaurid skull are concentrated over the snout, whereas those in coelophysoids and allosauroids are focused over the orbits (Fig. 7.3). Mazzetta et al. (2009) compared FEA models for *Allosaurus* and the abelisaurid *Carnotaurus*, and argued that the latter had a weaker and faster bite than the former, which agrees with the findings from beam theory and lever mechanics described above. Finally, in one of the few comparative studies of sensory abilities across theropods, Zelenitsky et al. (2009) used high-resolution CT scans to show that tyrannosaurids and dromaeosaurids had significantly larger olfactory bulbs in relation to body mass than other theropods, indicating that they had a particularly strong sense of smell. As is clear, there are many recognizable differences among species that share the generally conservative theropod body plan, which is likely indicative of different feeding strategies and prey choices.

Tyrannosaurids: megapredatory theropods

Among the carnivorous theropods, the colossal Late Cretaceous tyrannosaurids are set apart by their extreme adaptations for feeding on large prey. As such, these animals are commonly referred to as dinosaurian "megapredators" and are recognized as among the largest and most formidable predators to ever live on land. Although they evolved from smaller and more gracile ancestors, almost all members of the derived clade Tyrannosauridae are characterized by a unique set of anatomical features, including a deep skull with enormous jaw-closing muscles, thickened teeth that are nearly circular in cross-section, a fused set of nasals along the snout, deeply interlocking sutures between individual skull bones, large olfactory bulbs, and puny forelimbs of the same absolute size as a human's (Holtz, 2004; Brusatte et al., 2009a, 2010d; Sereno et al., 2009). The largest tyrannosaurids, *Tyrannosaurus* and *Tarbosaurus*, reached lengths of 13 m and masses of up to 6 tonnes, and they were undoubtedly the only colossal predators in their ecosystems.

The feeding habits, bite force, and prey selection of tyrannosaurids, especially *T. rex* itself, have been studied using a great diversity of evidence, including observations of gross morphology, study of bite marks and coprolites, and biomechanical modeling (Barrett and Rayfield, 2006). The diet of no other single dinosaur species has been studied in such extensive detail, and this legacy of research is a prime example of how a toolkit of varied techniques and evidence can confidently illuminate the biological habits of long-extinct animals. All lines of evidence agree that tyrannosaurids were truly unique among theropods. They were capable of bite forces of more than 13,400 N, an order of magnitude greater than lions and probably substantially greater than the strongest living chompers, alligators (Erickson et al., 1996; Meers, 2002). The jaw muscles were obscenely large to produce such bite forces and the skull bones were optimized to endure stress from both the bite and struggling prey. In particular, the fused and vaulted nasals of tyrannosaurids were a stress sink (Rayfield, 2005a; Snively et al., 2006), the thick bars of bone around the orbit provided strength and rigidity (Molnar, 2000; Henderson, 2002), and the robust lower jaw was almost circular in cross-section so that it could endure high stresses from all directions (Therrien et al., 2005). None of these features are present, at least to this degree, in other theropods, even allosauroids, ceratosaurs, and spinosauroids that are nearly as large as *Tyrannosaurus* and in some cases even larger.

Such extreme bite forces, and the strong skull that enabled them, allowed tyrannosaurids to employ an unusual method of feeding called "puncture–pull feeding" (Erickson and Olson, 1996). Tyrannosaurids would have bit deeply and powerfully into their prey, often through bone itself, and then pulled back, creating long cuts in the flesh and drag marks on the bones. Biomechanical models indicate that the tyrannosaurid skull was strong enough to produce the bite forces, and endure the stress, required of this activity (Fig. 7.3). However, unequivocal evidence is preserved in trace fossils. Tyrannosaurid bite marks have been found on the bones of a wide diversity of species, and these usually take the form of deep punctures that grade into elongate furrows (Erickson and Olson, 1996; Carpenter, 2000; Hone and Watabe, 2010) (Fig. 7.1A,B). A gargantuan coprolite from the Late Cretaceous of North America, which by its massive size and bone-filled interior could only belong to the sole large predator in its fossil assemblage (*Tyrannosaurus*), contains chunks of bone that were broken during feeding (Chin et al., 1998) (Fig. 7.1C,D). Finally, characteristic "spalled" wear facets on the robust teeth of tyrannosaurids could only be produced by consistent contact with hard food, such as bone (Farlow and Brinkman, 1994; Schubert and Ungar, 2005) (Fig. 7.6). It is possible that tyrannosaurids targeted bone as a source of sustenance, but it is equally likely that they incidentally plowed through bone with brute force while stripping flesh (Hone and Rauhut, 2010). Bone-crunching behavior is currently unknown in other theropods, and likely would have been very difficult

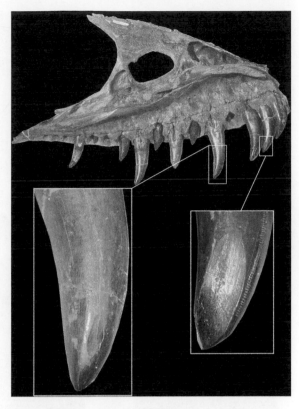

Figure 7.6 Wear facets on the teeth of *Tyrannosaurus*, which reflect strong tooth-on-bone bite forces. Image courtesy of Dr Blaine Schubert and modified from Schubert and Ungar (2005).

in species that were not endowed with such strong bites or robust skulls.

Tyrannosaurids: predators, scavengers, or both?

One specific question about tyrannosaurid feeding has generated substantial speculation: were *Tyrannosaurus* and kin active predators or primarily scavengers? Horner (1994) and others have argued that tyrannosaurids were not tenacious megapredators as often envisioned, but merely 5-tonne collectors of dead carcasses. This argument is based on general observations of tyrannosaurid anatomy: it has been asserted that a predator would not possess characteristic tyrannosaurid features such as small eyes, large olfactory bulbs, tiny forelimbs, and stocky hindlimbs incapable of great speed. Rather, it is said that some of these features,

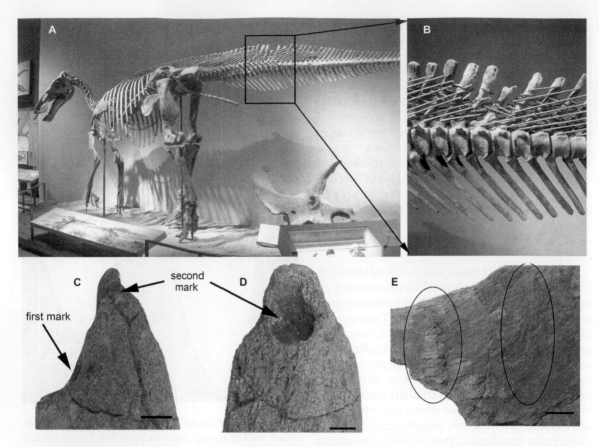

Figure 7.7 Bite marks made by tyrannosaurids, with evidence that the prey survived and healed after the attack. (A, B) Bite marks on the tail of the hadrosaurid *Edmontosaursus*. (C, D, E) Bite marks on the skull of *Triceratops*; (C) is a left postorbital horn core in left lateral view and (D) is the same specimen in dorsal view, with arrows denoting two marks that match the spacing and size of *Tyrannosaurus* teeth and show evidence of healing; (E) left squamosal in dorsal view with two parallel score marks circled, which also match the spacing of *Tyrannosaurus* teeth and show evidence of healing. Images (A) and (B) courtesy of Dr Ken Carpenter; images (C–E) modified from Happ, J.W. in Larson, P., Carpenter, K. (eds), *Tyrannosaurus rex*: The Tyrant King, pp. 355–368. Copyright © 2008, Indiana University Press. Reprinted with permission of Indiana University Press.

especially the large olfactory bulbs, would be ideal for a carrion-sniffing lifestyle.

These arguments were thoroughly dismantled by Holtz (2008) in a lively and passionate review of the subject. The eyes of tyrannosaurids are no smaller than the eyes of other large theropods, and in fact small eyes seem to be a required component of a strong skull that is capable of sustaining high forces during feeding (Henderson, 2002). The olfactory bulbs of tyrannosaurids are indeed large, but so are those of the obviously predatory dromaeosaurids (Zelenitsky et al., 2009). Hindlimbs are known

to become stockier and locomotion generally slower in larger animals (see review in Holtz, 2008), and although the forelimbs of tyrannosaurids are comically small, their extreme robusticity and large muscle attachments suggest some functional utility (Carpenter and Smith, 2001; Lipkin and Carpenter, 2008). Finally, and most persuasively, healed bite marks attributed to *Tyrannosaurus* have been found on the bones of *Triceratops* and *Edmontosaurus*: clearly, these animals must have survived a predatory attack (Carpenter, 2000; Happ, 2008) (Fig. 7.7). Therefore, in sum, none of

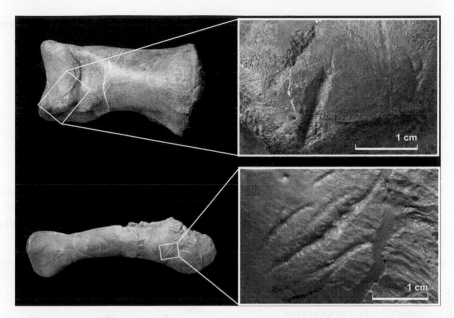

Figure 7.8 Evidence that *Tyrannosaurus* was a cannibal. The images depict bite marks, which match the size and spacing of *Tyrannosaurus* teeth, on the bones of *Tyrannosaurus*. Images modified from Longrich et al. (2010b). Reproduced with permission from Elsevier.

the features proposed as evidence of a scavenging lifestyle hold up to scrutiny, and explicit fossil evidence proves that tyrannosaurids hunted, at least on occasion. As is true of most predators, it is likely that tyrannosaurids both hunted and scavenged, depending on the availability of food.

Theropod cannibals

Some carnivorous dinosaurs were cannibals, a fact trumpeted with sophomoric zeal by the media but which is unsurprising given that cannibalism is extremely common in living mammalian, crocodilian, and reptilian predators (Fox, 1975). Bite marks matching the size, shape, and spacing of tyrannosaurid teeth have been described on several *Tyrannosaurus* specimens from the Late Cretaceous of North America, and it is logical to consider *Tyrannosaurus* itself as the culprit given that it is the only large theropod in this ecosystem (Longrich et al., 2010b) (Fig. 7.8). Similarly, bite marks matching the size and spacing of *Majungasaurus* teeth are found on bones of conspecifics from the Late Cretaceous of Madagascar (Rogers et al., 2003). However, one celebrated case of dinosaur cannibalism, parroted for decades in popular accounts and museum ex-

hibits, turns out to be fictional. It was thought that some specimens of the common Late Triassic theropod *Coelophysis* contained bones of juvenile conspecifics in their stomach region (Colbert, 1989). Recent reexamination of the specimens, however, reveals that the small bones belong to crocodylomorphs, not juvenile *Coelophysis* (Nesbitt et al., 2006). Regardless of this single instance of falsification, the commonality of cannibalism among living carnivores, and its known occurrence in two species of dinosaurs, suggests that it was a usual occurrence for theropods.

Spinosaurids: fish-eating theropods

Not all carnivorous dinosaurs may have subsisted primarily on a diet of terrestrial vertebrate flesh. The aberrant spinosaurids, a subclade of large-bodied basal tetanurans, probably fed mostly on fish, a feeding strategy that seems to be unique among theropods (Fig. 7.9). The most conclusive evidence for a piscivorous diet is the presence of partially digested fish scales in the gut region of one of the most complete spinosaurid specimens, the holotype of *Baryonyx* (Charig and Milner, 1997). It is likely that fish were not merely a circumstantial

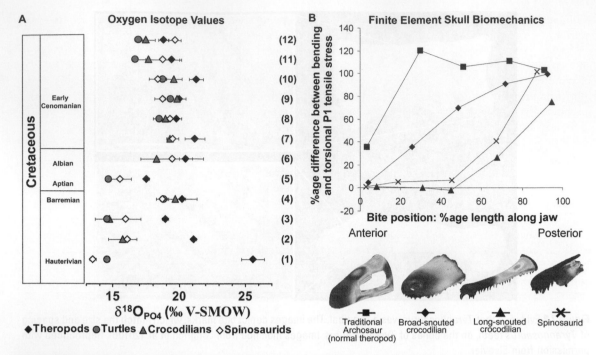

Figure 7.9 Evidence that spinosaurids had a semiaquatic habitat. (A) Oxygen isotopic data showing that spinosaurid teeth have an isotopic signature similar to that of unambiguously aquatic contemporaries such as crocodilians and turtles, but a different isotopic signature from contemporary theropods with more traditional skulls and teeth (i.e., hypercarnivorous theropods). Image modified from Amiot et al. (2006). Reproduced with permission from Elsevier. (B) Finite element biomechanical skull modeling of bending and torsional stresses across the skulls of crocodilians, traditional theropods, and spinosaurids. Note that the spinosaurid has a similar stress profile to the long-snouted crocodilian, but a quite different stress profile from the broad-snouted crocodilian and traditional theropod. Modified from Rayfield et al. (2007), and reproduced with permission.

component of this one individual's diet, because spinosaurids display a range of morphological features expected in a piscivore. The skull is long, narrow, and studded with conical teeth – eerily similar to the condition in living crocodiles – and the forelimbs are remarkably robust for a basal tetanuran and capped with enlarged claws that may have been used as "hooks" for catching fish (Charig and Milner, 1997; Sereno et al., 1998; Sues et al., 2002).

More sophisticated biomechanical studies also support crocodile-like feeding behaviors for these bizarre dinosaurs. Rayfield et al. (2007) used FEA to show that the skull of *Baryonyx* was functionally similar to that of the living gharial, as the tubular snouts of both species are well optimized to resist torsional stresses that are regularly caused by snatching and subduing large fish (Fig. 7.9B). The

beam analysis of Therrien et al. (2005) found that spinosaurids had a unique mandibular morphology among theropods, as they were particularly well adapted for catching small and agile prey due to their reinforced mandibular symphysis and terminal rosette of teeth that would have functioned as a toothy net. Sakamoto's (2010) approximation of the spinosaurid mandible as a lever indicates that it had an unusually fast bite, which would be expected in an animal that needed to snatch small slippery prey such as fish (Fig. 7.5). Both Sakamoto (2010) and Henderson (2002) found that the spinosaurid skull was weak and incapable of strong bites, which contrasts with the skull architecture in most other large theropods and would be truly unexpected if spinosaurids were targeting large-bodied prey.

Not only do spinosaurids exhibit anatomical tools and skull shapes ideal for fish-eating, but geological evidence indicates that they lived primarily in semiaquatic habitats. Spinosaurid fossils are usually found in rocks deposited on floodplain and coastal environments (Milner, 2003). More convincingly, Amiot et al.'s (2010) clever study of tooth enamel composition suggests that spinosaurids, but not other theropods, spent a large amount of their daily time in the water, as do living crocodiles and hippos (Fig. 7.9A). These authors measured the oxygen isotope composition of phosphate in the tooth enamel of spinosaurids and made comparisons with the enamel composition of contemporary species that were unequivocally terrestrial (theropods) and semiaquatic (crocodilians and turtles). Any differences in composition among these groups would likely be due to environment, because the oxygen isotope composition of tooth enamel in living animals is known to be controlled primarily by the composition of drinking water and food. Amiot et al. (2010) found that spinosaurids had a significantly different tooth enamel composition compared with contemporary terrestrial theropods, but were indistinguishable from the contemporary semiaquatic species.

In sum, Amiot et al.'s (2010) isotopic work provides prime evidence that spinosaurids occupied a semiaquatic habitat, and were something of a dinosaurian equivalent to a hippo. Fish was almost certainly the primary component of their diet, as indicated by gut contents, gross anatomical features, and quantitative biomechanical analyses. That being said, gut contents and bite marks also indicate that spinosaurids fed on herbivorous dinosaurs and pterosaurs, suggesting that these unusual theropods may have been something of an ecological generalist (Charig and Milner, 1997; Buffetaut et al., 2004).

Dromaeosaurids: pack-hunting theropods?

The mostly small, sleek, bird-like dromaeosaurids exhibit some of the most remarkable, and unusual, predatory adaptations among dinosaurs. Their lithe but muscular skeletons are clearly optimized for agility and speed, their enormous and clawed forelimbs were ideal for grasping and snatching their victims, and they could procure prey with brains (keen senses) and brawn (arsenal of sharp teeth and claws). However, their most salient anatomical feature is certainly the hyperextensible "killer claw" on the second digit of the foot, whose enormous size and wide range of motion would have allowed a dromaeosaurid to disembowel or latch onto prey with a kickboxer's vigor (Ostrom, 1969, 1990; Manning et al., 2006) (Fig. 7.10). One particular dromaeosaurid, *Balaur* from the Late Cretaceous of Romania, even had a double set of "killer claws" on its feet (Csiki et al., 2010) (Fig. 7.10D). These animals were clearly "killing machines," to use the evocative terminology of Ostrom (1994), despite the fact that most species were barely larger than a human.

Dromaeosaurids are often depicted in movies, museum exhibits, and popular accounts as pack-hunting theropods that cooperated in groups to take down prey much larger than themselves. A pack-hunting lifestyle for *Velociraptor*, *Deinonychus*, and other dromaeosaurids was passionately put forward by Ostrom (1990) and Maxwell and Ostrom (1995). Their argument was almost completely based on fossils from one site, the type locality for *Deinonychus* in the Middle Cretaceous Cloverly Formation of Montana, in which the remains of four dromaeosaurids are found associated with the skeleton of a larger herbivorous *Tenontosaurus*. The large number of shed *Deinonychus* teeth found at the site, as well as the presence of several individuals, suggested to Maxwell and Ostrom (1995) that a *Deinonychus* pack was actively feeding on *Tenontosaurus* when the five individuals died. They suggested that the larger *Tenontosaurus* was able to kill the four dromaeosaurids, but then later succumbed to its injuries.

Although plausible, there is little unequivocal evidence for this scenario. Maxwell and Ostrom (1995) admit that this fossil site could record scavenging by several individual dromaeosaurids, not active pack hunting. Furthermore, based on comparisons with the social and feeding behavior of living crocodiles and birds, which comprise the extant phylogenetic bracket for dinosaurs, Roach and Brinkman (2007) suggested that it was more plausible that *Deinonychus* was not a pack hunter, because this behavior is rare in living archosaurs. Yet this conclusion is also tenuous: just because a behavior is not common among living archosaurs does not mean that it was impossible, or even implausible, for dinosaurs (because, remember, it

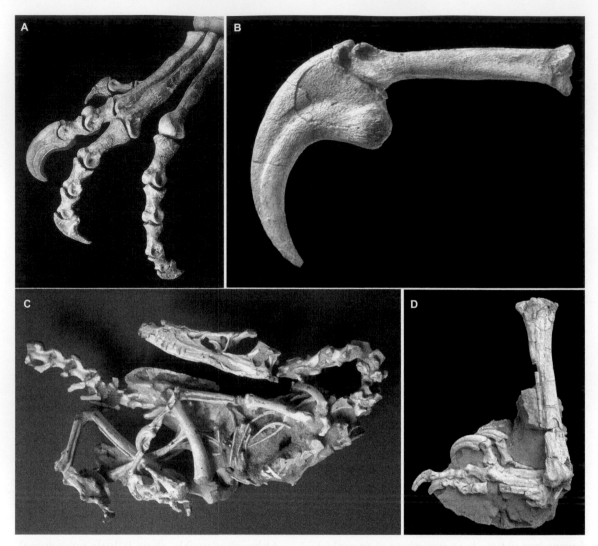

Figure 7.10 Dromaeosaurid theropods, the "killer-clawed" predators that are among the closest relatives of birds. (A) The foot of *Deinonychus*, showing the enlarged and hyperextensible second digit (second from left, with largest claw in foot); (B) the enlarged "killer claw" of the second digit of the foot of *Velociraptor*; (C) an articulated, well-preserved, and nearly complete specimen of *Velociraptor* from the Late Cretaceous of Mongolia; (D) the articulated foot of *Balaur*, a dromaeosaurid with hyperextensible "killer claws" on both the first and second digits of the foot (top two digits), from the Late Cretaceous of Romania. All photographs © Mick Ellison.

is present in some living archosaurs). Therefore, there is a lack of current consensus or clear evidence for the pack-hunting abilities of dromaeosaurids. What is undoubted is that these animals possessed a sophisticated armory of predatory weapons and would have been formidable carnivores.

Non-carnivorous theropods

Theropods have the reputation of being the biggest and baddest predators in earth history, and megapredators such as *Tyrannosaurus* and "killing machines" such as *Deinonychus* certainly fulfill the stereotype. However, not all theropods were well adapted for taking down large prey or processing

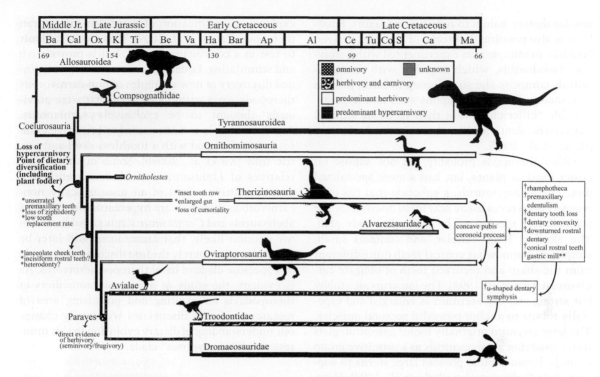

Figure 7.11 Non-carnivorous theropods. A phylogeny of coelurosaurian theropods indicating the inferred diets of individual subgroups, based on anatomical evidence (including the various features listed in the figure) and direct evidence of diet (such as gut contents and coprolites). Image courtesy of Dr Lindsay Zanno and modified from Zanno et al. (2009).

flesh with sharp teeth and claws. Several species, including a number of entire subgroups that are closely related to birds, were not carnivores at all, but rather ate plants or subsisted on a more specialized diet (Kobayashi et al., 1999; Barrett, 2005; Barrett and Rayfield, 2006; Zanno et al., 2009; Zanno and Makovicky, 2011) (Fig. 7.11). It is even possible, and perhaps likely, that obligate carnivory was not the ancestral condition for the derived theropod clade Maniraptora, which includes birds and their closest relatives (Zanno et al., 2009; Zanno and Makovicky, 2011) (Fig. 7.11).

Recognition that not all theropods ate meat, but rather had diverse and often specialized diets, has followed from the discovery of new fossils. The roster of non-carnivorous theropods is constantly expanding with new discoveries, and now it seems as if the majority of maniraptorans were not obligate carnivores. The peculiar therizinosauroids, perhaps the most inconceivable dinosaurs of all,

have small skulls with leaf-shaped teeth and enormous pot-bellies for digesting plant matter, both of which are unequivocal indicators of herbivory (Paul, 1984; Zanno et al., 2009; Zanno, 2010). The ostrich-mimic ornithomimosaurs were probably the fastest of all theropods, but the skulls of most species are toothless and capped by a beak, and their gut housed a mill of gastroliths (Kobayashi et al., 1999; Norell et al., 2001b; Barrett, 2005). These features are expected in a herbivore, or perhaps an omnivore that ingested a large amount of plant material, but not in animals that are obligate carnivores. Oviraptorosaurs, the alien-like theropods closely related to birds, are also mostly toothless and have been found with gastric mills. The most primitive members of the group possess teeth, but these are procumbent and reduced to the front of their snout, an arrangement suboptimal for killing and processing large prey (Ji et al., 1998; Xu et al., 2002b). Instead, it is likely that oviraptorosaurs had similar dietary habits to ornithomimosaurs. Finally, it is also possible that some of the most derived bird-like maniraptorans experimented with herbivory. Troodontids, which together with dromaeosaurids comprise the sister group to birds, possess convincing predatory weapons such as a hyperextensible "killer claw," but their teeth are studded with coarse denticles like those seen in herbivores (Holtz et al., 2000).

Other theropods probably did not subsist on either meat or plants, but had a more specialized diet. The alvarezsaurids, a subclade that has only recently been revealed by new fossil discoveries, are immediately recognized by a peculiar body plan. The skull is long, tubular, and contains small, simple, and somewhat conical teeth quite different from the sharp and recurved teeth of obligate carnivores (Clark et al., 1994). The forearms are stubby but strong and the sternum is enlarged and especially robust to anchor powerful pectoral muscles. The large sternum originally fooled paleontologists into classifying alvarezsaurids as a primitive group of birds, because birds possess large sterna to support their flight muscles (Perle et al., 1993). However, it is now thought that these large muscles of alvarezsaurids, as well as their robust and small forelimbs, were optimized for a very different behavior: insect-eating (Senter, 2005; Longrich and Currie, 2009). Not only would the massive muscles have powered an efficient digging stroke for ripping into insect nests, but the small and conical teeth were ideal for puncturing insect carapaces. Furthermore, and perhaps most convincingly, the range of motion of the alvarezsaurid forelimb was different from that of other theropods. The humeri sprawled laterally, the forearm was positioned vertically, and the palms of the hands could face ventrally (and were not restricted to a medially facing position as in other theropods) (Senter, 2005). This configuration simply would not have allowed grasping or prey capture, but was perfect for delivering strong blows to an insect nest or performing the "hook-and-pull" movements that living anteaters use to open nests (Senter, 2005).

Non-carnivorous theropods are clearly quite diverse (Fig. 7.11). Carnivory was not the only dietary strategy among theropods, and it has been hypothesized that the development of herbivory, omnivory, and specialized diets may have helped drive the explosive diversification of derived maniraptoran theropods (Zanno et al., 2009). This idea is difficult to test in a convincing fashion, but is provocative and stimulating. Equally provocative is the continued discovery of new examples of non-carnivorous theropods, even within subclades that were previously thought to be exclusively carnivorous. A great example is *Limusaurus*, a primitive ceratosaurian theropod with a toothless skull and gastric mill (Xu et al., 2009b). Some of the closest relatives of *Limusaurus*, which is currently the most primitive record of an unequivocally non-carnivorous theropod, are hypercarnivores such as abelisaurids and *Ceratosaurus*. It is possible, and in my opinion likely, that *Limusaurus* will later be remembered as merely the first discovered example of a speciose clade of basal theropod herbivores. As is evident, the study of non-carnivorous diets in theropods is an exciting and promising area of research, and new discoveries will surely change our understanding of dietary evolution in this quintessential "carnivorous" clade.

Sauropodomorph Feeding and Diet

Basal sauropodomorphs

The most primitive sauropodomorphs – the evolutionary grade commonly referred to as "prosauropods" – have been said to possess nearly every conceivable type of diet (see reviews in Galton, 1985; Barrett, 2000; Galton and Upchurch, 2004b). Many early workers regarded these moderately large, bipedal or quadrupedal, long-necked species as carnivores or scavengers of carrion, largely due to the discovery of recurved serrated teeth near their skeletons (Swinton, 1934). However, the discovery of more complete specimens revealed that the skulls of prosauropods never contained these types of teeth, which almost certainly belonged to predatory theropods that lost their teeth while hunting or scavenging prosauropod carcasses. It is now generally agreed that most prosauropods, which exhibit a relatively conservative skeleton and set of feeding-related cranial features, were primarily herbivorous, although they probably dabbled in omnivory as well (Barrett, 2000; Galton and Upchurch,

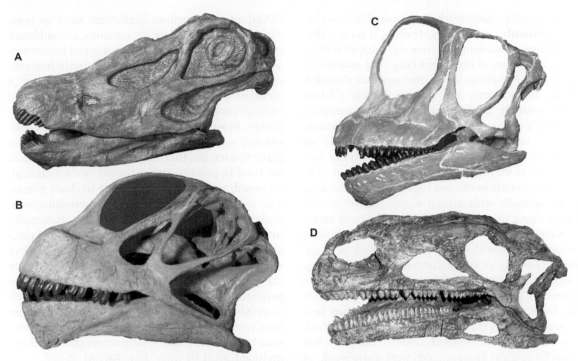

Figure 7.12 Sauropodomorph skull montage illustrating major skull shapes that were associated with different feeding styles: (A) *Diplodocus*; (B) *Camarasaurus*; (C) *Europasaurus*; (D) *Plateosaurus*. Also note the thin pencil-like teeth of *Diplodocus*, which are restricted to the anterior end of the snout, and contrast these with the broader, more spatulate teeth of *Camarasaurus* and *Europasaurus* that extend across much of the jaw. *Plateosaurus* also has an extensive tooth row, but simpler leaf-shaped teeth similar to those of living iguanas, which were perhaps useful for an omnivorous diet. Images (A) and (B) courtesy of WitmerLab at Ohio University.

2004b; Barrett and Upchurch, 2007). As such, prosauropods represented the first major radiation of herbivorous dinosaurs, because they were remarkably common during the earliest phase of dinosaur history during the Late Triassic. They were also the first high-browsing vertebrates in the history of terrestrial life that were routinely able to access plants higher than a meter or two off the ground.

The general body plan and specific features of the skull and dentition support a primarily herbivorous lifestyle for most prosauropods (Fig. 7.12D). The jaw joint is depressed below the dentary tooth row, allowing the two jaws to come together simultaneously like a guillotine. The vast majority of teeth in most taxa are broad and spatulate, with coarse denticles, remarkably similar to the teeth in living herbivorous reptiles such as iguanas (Barrett, 2000). The expansion of the tooth crowns relative

to the root, and the en echelon arrangement of individual teeth along the jaw, resulted in a closely packed dentition, forming something of a continuous cutting edge that would have been ideal for shearing plants (Galton and Upchurch, 2004b). In contrast, the teeth of carnivorous or insectivorous reptiles, including theropod dinosaurs, are not usually tightly packed, because they need to function as individual cutting blades to slice through flesh (Abler, 1992). Furthermore, the moderately long necks of prosauropods, which are admittedly shorter than those of true sauropods but much longer than the necks of other Late Jurassic to Early Cretaceous terrestrial herbivores, would have allowed access to higher vegetation. Several prosauropod skeletons have also been found with gastrolith mills, which in living birds are used to aid digestion of tough plant material that cannot be processed in the mouth alone (Shubin et al., 1994).

Although prosauropods did possess features that are commonly present in herbivores, it is clear that their plant-processing abilities were not as sophisticated as those of ornithischians, true sauropods, and other Mesozoic herbivores such as rhynchosaurs, dicynodonts, and procolophonids (Galton and Upchurch, 2004b). The skulls of prosauropods are lightly constructed and were probably incapable of enduring high feeding stresses, although this has yet to be adequately tested by FEA or other biomechanical methods. Most sutures between individual bones are loose, not rugose and interlocking as in animals with strong skulls. However, the bones along the snout midline were firmly interdigitating, which would have prevented kinetic motions of individual bones that would have enabled chewing or other complex oral processing. Furthermore, prosauropods only had small cheeks restricted to the posterior end of the jaw, not the large and fleshy cheeks of ornithischians that retained food in the mouth during chewing (Paul, 1984; Barrett and Upchurch, 2007). Finally, the jaw muscles of most prosauropods were quite weak, as evidenced by their shallow and small attachment sites on the bones, and the quite simple and generalized teeth lack wear facets that would indicate complex occlusion between the upper and lower tooth rows.

In sum, various features of the skeleton suggest that prosauropods ate plants, but probably simply stripped leaves and ingested them without chewing or other refined oral processing. Their unsophisticated cranial masticatory apparatus, along with the presence of gastric mills to break down food, imply that prosaropods probably employed an unusual feeding strategy. Leaves, branches, and other high-reaching plant material that was difficult for other herbivores to collect could easily be procured by the long neck of prosauropods. This vegetative matter was then minimally processed in the mouth and delivered to the barrel-shaped gut, where the repetitive action of the gastric mill would facilitate digestion. Ingested plants would remain in the gut for long periods of time, enabling prosauropods to ingest poor-quality vegetation that was nutritionally unfeasible for more sophisticated herbivores to feed on.

It is likely that many prosauropods supplemented their diets with meat. As noted by Barrett (2000), extant reptilian herbivores such as iguanas, whose teeth share an uncanny resemblance to those of prosauropods, are not strict herbivores as often considered. Although primarily herbivorous, these reptiles include a large amount of meat in their diet, and will often succumb to malnutrition if they are restricted to eating only plants. Most teeth of iguanas are broad and leaf-shaped, with coarse denticles ideal for cropping plant matter, but their anterior-most teeth, which are used to procure and cut meat, are subconical, recurved, serrated, and terminate in sharp points. The same is also true of many prosauropods, making it plausible that they also habitually included meat in their diet. Even more convincing is the discovery of a heavily worn mandible of a small reptile, seemingly etched and pitted by the activity of digestive enzymes, within the gastric mill of a well-preserved prosauropod skeleton (Shubin et al., 1994). It is plausible therefore that most prosauropods were primarily herbivorous but also facultative omnivores that regularly included meat in their diet (Barrett, 2000).

The evolution of "sauropod" feeding characters

Prosauropods, of course, are not a single clade of animals but rather an evolutionary grade of species along the evolutionary line toward true sauropods (Fig. 7.12A–C). Sauropods are unmistakably herbivorous to an extreme degree, and must have processed enormous quantities of vegetation in order to fuel their large bodies (Coombs, 1975; Farlow, 1987; Sander et al., 2011). Several anatomical features enabled sauropods to ingest and digest plants in bulk amounts: their long necks may have enabled them to reach vegetation inaccessible to other herbivores, their large body size allowed slower passage time through the gut and longer periods of gut fermentation to break down low-quality plant matter, and their broad muzzles and spatulate or pencil-shaped teeth would have formed an efficient apparatus for gathering plants en masse. In short, sauropods employed a highly unusual "megaherbivorous" feeding style that set them apart from all other dinosaurs, and may have enabled their evolutionary success (Farlow, 1987; Sander et al., 2011).

Until recently, however, it was very difficult to study the evolution of this novel dietary strategy.

Did all the various megaherbivorous adaptations develop simultaneously, perhaps in the earliest sauropods? Alternatively, were these features assembled piecemeal throughout millions of years of sauropodomorph evolution, only eventually (and perhaps serendipitously) culminating in the unique dietary apparatus of sauropods? New fossil discoveries have helped bridge the gap between the characteristic "prosauropods" that are so common in the Late Triassic and Early Jurassic and the enormous, unequivocally megaherbivorous sauropods that dominated the Middle Jurassic and beyond. This new evidence, interpreted in an increasingly robust phylogenetic framework, conclusively indicates that the sauropod feeding system evolved incrementally, over millions of years (Barrett and Upchurch, 2007; Upchurch et al., 2007). There was a general increase in neck length, body size, and muzzle width across basal sauropodomorph phylogeny (Barrett and Upchurch, 2007). Furthermore, individual characters that relate to skull and muzzle shape, such as the width of the lower jaw, the robusticity of the jaw symphysis, and the depth of the skull, did not develop at the same time, but seem to follow each other in a general series of progressive changes.

That being said, there was not a simple and clean progression of dietary evolution on the line to sauropods. There was substantial variability in neck length, body size, and skull shape of basal sauropodomorphs, and optimizing these characters onto a cladogram reveals great amounts of homoplasy (Barrett and Upchurch, 2007). Furthermore, even closely related basal sauropodomorphs may have differed in their tooth shapes and cranial adaptations for herbivory (e.g., depth of the jaw joint below the tooth row, size of the coronoid eminence for jaw muscle attachment). What is clear, however, is that dietary evolution on the line to sauropods was more complex than previously thought, and many features thought to be unique to sauropods or specifically related to sauropod megaherbivory first appear in older, smaller, more generalized prosauropods.

Sauropods

All sauropods possessed the aforementioned megaherbivorous diet and suite of anatomical features that enabled bulk feeding at large size (Figs 7.12

and 7.13). There is no question that sauropods were among the most extreme and unusual herbivores in earth history, and scientists have long been puzzled as to exactly how such animals could fuel themselves. This mystery has attracted the attention, and in recent years the collaborative teamwork, of paleontologists, functional biologists, animal behaviorists, and nutrition scientists (see review in Sander et al., 2011). Using a breadth of information, ranging from observations of gross anatomy to comparative studies of digestive physiology in living analogues, these workers have assembled a convincing picture of what sauropods ate, how they fed, and how they sustained their massive frame. In sum, it appears as if sauropods were megaherbivores that ingested enormous amounts of plant matter, perhaps up to 40 kg per day, and relied on their large gut rather than advanced chewing mechanisms or a gastric mill to break down their food. In other words, sauropods simply guzzled massive quantities of vegetation, much of it probably low in nutritional quality, and let their large size take care of the rest (Hummel et al., 2008; Franz et al., 2009; Sander et al., 2011). No other herbivores, living or extinct, functioned in this way.

A megaherbivorous bulk-feeding strategy, without any mechanical manipulation of food, is supported by several lines of evidence. First, it is clear that sauropods were incapable of sophisticated oral processing of their food. Although wear facets on their teeth indicate that tooth-on-tooth occlusion was present, the short tooth rows, lack of cheeks, and akinetic skulls would have made chewing impossible (Fiorillo, 1998; Upchurch and Barrett, 2000; Upchurch et al., 2004; Sander et al., 2011; Whitlock, 2011a). Instead, teeth concentrated at the front of the jaws, along with the possession of a wide muzzle and reinforced dentary symphysis, would have allowed sauropods to efficiently and quickly crop and ingest large amounts of foliage (Fig. 7.12A–C). Second, it seems as if sauropods did not employ other mechanical methods for breaking down food, namely a gastric mill. Although gastroliths are commonly discovered with sauropod skeletons, Wings and Sander (2007) argued that these were probably only incidentally swallowed and served little, or no, mechanical function. Compared with the sizes of gastric mills in living organisms, masses of gastroliths preserved with sauropod skeletons were simply too small to mechanically process large amounts of vegetation, and their surfaces are polished rather than covered with networks of scratches that indicate repeated use in grinding food. Therefore, sauropods did not seem to employ any mechanical mechanism for breaking down food. How, then, could they process such large quantities of plant matter? The answer probably lies in their great bulk: because the size of the gut increases with body mass, the colossal size of sauropods would have guaranteed long digestion times (Hummel et al., 2008; Sander and Clauss, 2008; Franz et al., 2009). If food is digested long enough, enzymes can chemically degrade even very hard plants or large amounts of foliage, without any need for mechanical mechanisms.

Determining exactly what types of plants sauropods ate is a more complicated matter. Direct evidence of sauropod food choice, such as coprolites and fossilized stomach contents, are remarkably rare (Sander et al., 2010). It is difficult to assign a particular coprolite to a sauropod source, because there were many different dinosaurian herbivores living in most Mesozoic ecosystems. One set of coprolites from India, which are tenuously assigned to sauropods, contain remnants of small silica inclusions (phytoliths) that match those known from extant grasses, providing remarkable evidence that these advanced flowering plants were present during the Cretaceous and formed a major component of some dinosaurian diets (Prasad et al., 2005). These and other coprolites from India also contain fragments of many other types of plants, including conifers and palms, as well as fungi and algae (Ghosh et al., 2003; Prasad et al., 2005). Therefore, if sauropods were the culprits, then they must have fed on a wide variety of plants. Neontological experiments, which measure the energy content of several living plants whose close relatives may have been eaten by dinosaurs, also suggest that a diversity of plants could have fulfilled the predicted energy requirements of most sauropods (Hummel et al., 2008). This is not surprising, given the many lines of evidence supporting an indiscriminate bulk-feeding lifestyle, as opposed to more selective food choice. Nonetheless, it is not likely that sauropods haphazardly ate any plant that was available. Some plants, such as the energy-rich but slow-fermenting *Araucaria*, were ideal for the languidly digesting sauropods, whereas other plants such as cycads probably could not provide enough energy to power a massive 25-tonne animal even if eaten in bulk (Hummel et al., 2008).

Although all sauropods were probably bulk feeders that subsisted on a variety of plants, not every species fed in the exact same way. Important differences in body size, skull morphology, neck length and mobility, and dental features probably allowed coexisting sauropods to target different food sources and feed in distinct ways. It has long been recognized, for instance, that coexisting Late Jurassic diplodocoids (such as *Diplodocus*) and macronarians (such as *Camarasaurus* and *Brachiosaurus*) have distinctly different cranial and dental designs (Upchurch and Barrett, 1994, 2000; Christiansen, 2000b; Upchurch et al., 2004; Whitlock, 2011a). The skulls of *Brachiosaurus* and *Camarasaurus* are short and deep, the jaw muscles were oriented nearly vertically, the individual teeth are robust and spatulate, and the tooth row extends across most of the upper and lower jaws. *Diplodocus*, on the other hand, possesses a longer snout in which the jaw muscles were oriented at a more anteroventral inclination and thinner cylindrical teeth that are restricted to the front of the skull and project at a procumbent angle. Furthermore, the articular surface between the upper and lower jaws is more anteroposteriorly extensive in *Diplodocus* than *Brachiosaurus*, and *Camarasaurus* has more severe tooth microwear, consisting of deep pits and scratches, than *Diplodocus* (Fiorillo, 1998).

Based on these features, several authors have hypothesized that *Diplodocus* and kin fed by "raking" softer plant material, using their anterior tooth arcade and ability to move the jaw in a fore–aft fashion (propaliny) to efficiently strip leaves from branches (see summary in Upchurch et al., 2004). *Brachiosaurus* and *Camarasaurus*, on the other hand, were better suited for eating harder foods, due to their more robust teeth, deeper skulls, and jaws constrained to an up-and-down motion (orthal shearing, although *Camarasaurus* itself could probably move its jaws back and forth to some degree). The more extensive microwear in *Camarasaurus* would agree with this hypothesis (Fiorillo, 1998), as would the inclination of wear scratches on the teeth of all three genera (Christiansen, 2000b). It is likely that the stronger, orthal chewing stroke of these macronarians was primitive for sauropods, as it seems to be present in basal sauropodomorphs, whereas the propalinal "raking" capabilities of diplodocids were unusual and highly derived. It is also plausible that diplodocids were better suited for browsing on lower vegetation (Upchurch and Barrett, 2000). Their necks may not have been able to reach as high as those of macronarians (Stevens and Parrish, 1999), and the orientation of the semicircular canals of the inner ear indicate that the head was habitually held in a downturned position, ideal for feeding near the ground and not high in the treetops (Sereno et al., 2007). Interestingly, broad-toothed sauropods such as *Brachiosaurus* disappear from the fossil record by the dawn of the Late Cretaceous, at which point macronarians convergently acquired the narrow-crowned morphologies

of diplodocoids. Although skulls of these late-surviving macronarians are poorly known, it is likely that they also fed on softer foods and browsed closer to the ground (Chure et al., 2010).

Aside from the general differences between diplodocoids and many macronarians, which may have enabled niche partitioning when they coexisted, some individual sauropod genera or small subclades of sauropods are just plain bizarre (Fig. 7.13). The dicraeosaurids, a subgroup of diplodocoids that is currently only known from a few genera, had the shortest necks of all sauropods. One genus in particular, *Brachytrachelopan*, was only about the size of *Tyrannosaurus* and had a neck only a few meters in length (Rauhut et al., 2005). It is likely that these sauropods filled a unique ecological niche, perhaps by employing a low or mid-browsing feeding strategy similar to that of some

Figure 7.13 Aberrant sauropods. (A) Digital skull model and skeletal reconstruction of *Nigersaurus*, a small sauropod with a tooth battery at the front of its jaws (images courtesy of Dr Paul Sereno and Carol Abraczinskas and modified from Sereno et al., 2007). (B) Skull illustration and skeletal reconstruction of *Bonitasaura*, a small sauropod with a guillotine-like beak at the posterior end of its jaws (directly below the eye) (illustrations by Jorge Antonio Gonzalez and courtesy of Dr Sebastián Apesteguia; see Apesteguia, 2004 for additional information).

ornithopods. The derived macronarian *Bonita-saura*, from the Late Cretaceous of Argentina, had a squared-off lower jaw that sported a sharp beak behind the teeth, probably used to crop plants like a guillotine (Apesteguía, 2004). It was also quite small for a sauropod, of the same general size as *Brachytrachelopan*, and may have been adapted for feeding on harder vegetation close to the ground.

However, the most sublime sauropod of all is undoubtedly *Nigersaurus*, a member of another small and poorly known diplodocoid subclade called the Rebbachisauridae (Sereno et al., 1999; Sereno and Wilson, 2005; Sereno et al., 2007; Whitlock, 2011a, 2011b). This sauropod was also small, about the same size as its distant relatives *Bonita-saura* and *Brachytrachelopan*, but its most extraordinary features are in its skull. All teeth are restricted to the front of the squared-off jaws and form a tightly packed battery composed of more than 30 individual columns. Each tooth was replaced at the astounding rate of approximately one per month and preserved teeth are decimated by extreme amounts of wear. The individual bones of the skull are paper thin, so much so that they are translucent when held up to light. And the semicircular canal orientation indicates that the head was habitually held perpendicular to the ground, such that *Nigersaurus* would literally look downward. All these features persuasively suggest that *Nigersaurus* was adapted for slicing large amounts of soft vegetation that grew near the ground (Sereno et al., 2007). This "extreme" lifestyle is substantially different from the usual image of sauropods browsing high in the treetops, and may have been a more common feeding strategy among sauropods than is currently indicated by the handful of rare rebbachisaurid fossils.

Ornithischian Feeding and Diet

The ornithischian body plan

The third and final major radiation of dinosaurian herbivores, the ornithischians, had a sophisticated array of cranial, dental, and postcranial characters finely tuned to their plant-eating diet (Fig. 7.14). They were clearly the most refined of all herbivo-rous dinosaurs, and probably subsisted on a diet of higher-quality vegetation that was extensively processed in the mouth before swallowing (Weishampel and Norman, 1989; Norman and Weishampel, 1991; Norman, 2001). Their feeding style therefore differed from that employed by the bulk-feeding sauropods, which simply guzzled large amounts of low-quality vegetation en masse, and the aberrant herbivorous theropods, which seemed to have lacked complex oral processing mechanisms and relied primarily on gastric mills to break down their food.

Several features of the anatomy convincingly demonstrate that ornithischians were sophisticated herbivores that used advanced techniques, relative to those of other dinosaurian herbivores, to procure and process their food (Galton, 1986; Weishampel and Norman, 1989; Norman and Weishampel, 1991; Norman, 2001) (Fig. 7.14). The teeth at the front of the snout are absent and replaced by a sharp keratinous beak that would have grown continuously through life, therefore allowing ornithischians to crop abrasive plants (Fig. 7.14C–F). A bone unique to ornithischians, the predentary, supports the beak on the lower jaw, and the ceratopsians further developed a second novel bone, the rostral, to form much of the beak on the upper jaw (Fig. 7.14D). The teeth of ornithischians are large, robust, and shaped like a leaf, with coarse denticles ideal for grinding plant matter (Fig. 7.14A,B). Importantly, these teeth are inset relative to the external surfaces of the jaws, which acts to form a deep depression that almost certainly housed muscular cheeks in life (Fig. 7.14D). Although not present (or extremely small) in sauropodomorphs and herbivorous theropods, cheeks would have prevented food from haphazardly falling out of the mouth during feeding, therefore enabling it to be chewed over and over before swallowing. Finally, the most salient character of ornithischians, their backward-projecting pubis, provided more area for a large gut, a necessary feature of herbivores. Aside from being larger, the gut was now positioned below the center of mass of the animal, allowing at least some ornithischians to stand bipedally and move at fast speeds. Sauropods, whose large gut was restricted to a more anterior position in the abdomen, were limited to a quadrupedal posture and slow movement.

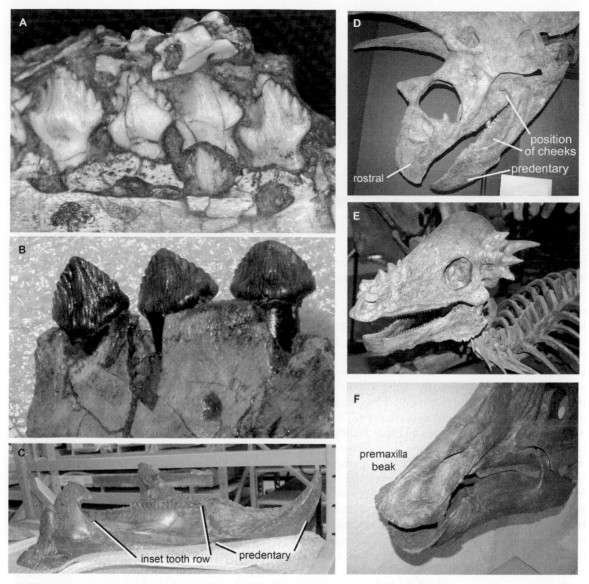

Figure 7.14 Features of ornithischian cranial anatomy related to feeding: (A) leaf-shaped teeth of a basal ornithischian; (B) leaf-shaped teeth of a pachycephalosaurid; (C) lower jaw of a ceratopsian; (D) the skull of the ceratopsian *Titanoceratops*, showing the sharp rostral and predentary bones that would have supported a beak and the inset fossa on the lateral surface of the jaws that would have supported a cheek; (E) the skull of *Pachycephalosaurus*; (F) the anterior portion of the snout, showing the broad "duck-like" bill, in the hadrosaurid *Corythosaurus*. Image (A) courtesy of Dr Richard Butler; images (B–F) courtesy of Dr Thomas Williamson.

Basal ornithischians

Most of the aforementioned features are present in every ornithischian, even the oldest and most primitive members of the group from the Late Triassic.

The most primitive of all ornithischians, the heterodontosaurids, may have supplemented their diet with some meat. Although heterodontosaurids possess all the aforementioned skeletal features

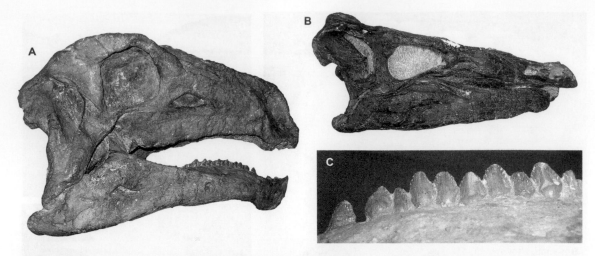

Figure 7.15 Features of stegosaur anatomy relating to feeding: (A) skull of *Huayangosaurus*; (B) skull of *Stegosaurus*; (C) leaf-shaped teeth of *Huayangosaurus*. All images courtesy of Dr Susannah Maidment.

optimized for high-level herbivory, they also boast large hands that were capable of grasping and capped with enormous claws (Butler et al., 2007). Furthermore, several species had enlarged "canine-like" teeth at the front of their jaws, which were almost certainly not sexually dimorphic or ornamental features as long assumed because they are present even in small juveniles (Butler et al., 2008b). Together, the powerful hands and sharp teeth may have been used to procure and dismember prey. Alternatively, Norman et al. (2004a) hypothesized that the large hands and sharp claws may have functioned as digging implements, perhaps to excavate burrows or uproot tubers. It is difficult to assess both of these dietary hypotheses at present, and clearly more work is required to understand the feeding habits and functional morphology of the most primitive ornithischians.

The thyreophorans, which include stegosaurs and ankylosaurs, seem to have employed a somewhat simple and primitive style of food processing, at least compared with more derived ornithischians such as ornithopods and ceratopsians (Norman, 2001). The teeth were small, simple in shape, and not arranged into complex dental batteries, and only a limited range of jaw motion was possible during the feeding stroke (Fig. 7.15). In particular, stegosaurs had a relatively unsophisticated masticatory apparatus. Their teeth lack extensive wear, their bite forces were

probably lower than those of extant dogs, and most of the forces incurred during feeding were probably endured by the beak (Reichel, 2010). The skull of most thyreophorans was essentially akinetic: the quadrate fit tightly into the glenoid of the lower jaw, preventing any substantial fore–aft propalinal movement, and the individual bones of the upper jaw could not move against each other to facilitate complex chewing motions (Barrett, 2001). Therefore, most thyreophorans orally processed food by simply bringing their jaws together orthally to produce a vertical shearing motion between the maxillary and dentary teeth (Norman and Weishampel, 1991; Barrett, 2001; Norman, 2001). There is some evidence, however, that at least one ankylosaur had more complex jaw motions, in which some propalinal shifting at the quadrate–glenoid contact and a small degree of medial pivoting of the lower jaws were permitted (Rybczynski and Vickaryous, 2001). However, it is unclear whether this style of cranial kinesis was present more broadly among ankylosaurids, and regardless it paled in comparison to the extreme cranial kinesis and extensive dental batteries of the more refined ornithopods and ceratopsians.

Because of their relatively simple feeding styles, and inability to process food in the more sophisticated manner of derived ornithischians, thyreophorans are usually posited as something of an

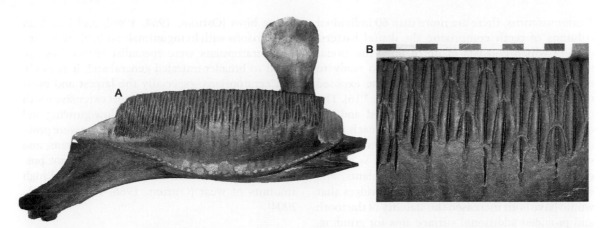

Figure 7.16 Features of ornithopod anatomy relating to feeding: the dental battery. Lower jaw of a lambeosaurine hadrosaurid, shown in medial view (A) and with a close-up of the tightly packed teeth of the dental battery (B). Images courtesy of Dr Albert Prieto-Márquez and the University of Alberta Vertebrate Paleontology Laboratory.

ornithischian equivalent to sauropods, albeit with a greater capacity to process food in the mouth (Norman, 2001; Norman et al., 2004b). It is likely that stegosaurs and ankylosaurs were bulk feeders and needed to ingest large quantities of vegetation, relying on mechanical or chemical processing of plant matter in the gut more so than complex oral processing. It is no stretch to imagine the tank-like ankylosaurs, which were probably only capable of slow locomotion and which were restricted to walking on four legs, as something of an ornithischian version of a sauropod. Because of their short legs and stout necks, ankylosaurs were probably limited to feeding on plants near ground level. Stegosaurs, on the other hand, may have had access to a greater diversity of plants. In general, stegosaurs have much shorter forelimbs than hindlimbs, so their neutral limb posture would bring the skull close to the ground during feeding. This has led most authors to consider them as specialized for feeding on ground-level plants (Bakker, 1978). However, the recent discovery of the aberrant long-necked stegosaur *Miragaia* indicates that some species were able to browse at greater heights (Mateus et al., 2009).

Sophisticated feeding mechanisms in derived ornithischians

Two major clades of derived ornithischians, the ornithopods and ceratopsians, had even more spe-cialized dietary habits than basal species (Figs 7.16 and 7.17). Members of these clades were capable of sophisticated oral processing of their food prior to swallowing, due largely to the presence of ex-pansive dental batteries and, in ornithischians, complex chewing motions enabled by the kinetic nature of the skull (Norman, 1984; Weishampel, 1984; Norman and Weishampel, 1985, 1991; Weishampel and Norman, 1989). These advanced feeding styles evolved independently in the two groups, because their common ancestor did not possess tightly packed dental batteries. Moreover, marked differences in skull architecture clearly demonstrate that the elaborate dentition of or-nithopods and ceratopsians functioned in distinct-ly different ways.

The dental batteries of ornithopods and ceratop-sians are among the most fantastic feeding adapta-tions ever developed in a herbivorous animal (Fig. 7.16). The ancestors of both groups possessed a traditional array of large leaf-shaped teeth that were widely spaced along the jaws (Norman et al., 2004a, 2004b, 2004c). Within the ceratopsian and ornithopod clades, however, the teeth progressively increased in number and became ever more tightly packed together, forming a single functional cutting surface that could efficiently shear even the toughest of plant matter (Dodson et al., 2004; Horner et al., 2004; Norman, 2004). In the most derived ornithopods, such as the duckbilled

Edmontosaurus, there are more than 60 individual columns of teeth comprising the dental battery, each of which consists of up to three functional teeth and up to five replacement teeth ready to erupt in a conveyer-belt fashion as the exposed teeth are worn down during feeding (Fig. 7.16). The individual teeth are tightly interlocked and the enamel is restricted to only one side of each tooth so that the cutting surface self-sharpens as the softer dentine wears away. Each tooth is strong, coarsely denticulated for pulverizing plants, and strengthened with robust longitudinal ridges that would have both increased the stability of the tooth and provided additional surface area for grinding. Nearly identical dental batteries are present in the most derived ceratopsians, although the number of tooth columns is usually lower (Ostrom, 1966; Dodson et al., 2004). In both clades, these remarkable structures would have allowed a great degree of oral processing prior to food ingestion (although only ornithopods could probably chew in the strict sense, due to their cranial kinesis), and permitted the ingestion of tough plants that caused individual teeth to wear down quickly. And, in both clades, the size and complexity of the dental battery progressively increased during the evolution of more derived species.

Ceratopsians
Although ornithopods and ceratopsians both possess extraordinary dental batteries and sophisicated oral processing capabilities, their feeding styles differed in detail. The tooth batteries of ceratopsians are arranged in a nearly vertical manner (Ostrom, 1964, 1966; Dodson et al., 2004). That is, the individual teeth erupted straight vertically and the entire interlocking array of teeth is oriented perpendicular to the floor of the mouth. This arrangement, along with the akinetic nature of the skull and the limited range of motion at the jaw joint, restricted the jaws to moving in a simple up-and-down orthal fashion. Therefore, during the biting stroke the upper and lower tooth rows would have sheared past each other like opposing guillotine blades. Additional cutting ability was conferred by the beak, which was the largest, most robust, and probably sharpest of any dinosaur. The narrow profile of the beak in anterior view was probably better suited for plucking rather than

strong bites (Ostrom, 1964, 1966), and based on comparisons with living animals probably indicates that ceratopsians were specialist feeders (as opposed to broader-muzzled generalists). It is likely that ceratopsians, especially the largest and most derived forms that possess the most extensive tooth batteries, subsisted primarily on low-growing and abrasive vegetation. These plant sources were probably inaccessible to sauropods, thyreophorans, and herbivorous theropods, because they did not possess complex dentition that could withstand high amounts of wear (Ostrom, 1966; Dodson et al., 2004).

Ornithopods
Derived ornithopods, on the other hand, were capable of much more complicated jaw motions than the simple up-and-down shearing of ceratopsians. Like living mammals, ornithopods were able to chew their food by repeatedly grinding it in the mouth using a multidimensional array of feeding strokes. Mammalian chewers are able to move their jaws in a wide range of motion, including orthal (up–down), propalinal (fore–aft), and transverse (sideways) directions. The primitive ancestors of ornithischians were limited to a simple orthal stroke, like that of ceratopsians and most other dinosaurs, which restricted their degree of oral processing. These animals, as well as more derived ornithopods, simply could not slide their lower jaws sideways relative to the upper jaws, because the tight articulation between the quadrate and articular at the jaw joint prevented any substantial motion. Derived ornithopods, however, seemingly developed an ingenious solution to this problem: their "pleurokinetic" skulls permitted a wide range of motion between some individual bones of the snout, such that a passive transverse stroke occurred as the result of the primitive orthal stroke (Norman, 1984; Weishampel, 1984; Norman and Weishampel, 1985, 1991; Weishampel and Norman, 1989). This novel functional system entailed the rotation of a unit of bones, including much of the side of the face and the palate, relative to the snout and skull roof via a loose "hinge" at the premaxilla–maxilla suture. Because of this moveable articulation, the maxilla and associated bones would literally swing outward as a passive consequence when the upper and

lower jaws came together. When this occurred, the diagonally oriented tooth batteries of the two jaws slid past one another, with the maxillary cutting surface shearing against the outer surface of the dentary tooth battery. In effect, this provided a transverse power stroke, in addition to the orthal stroke. Derived ornithopods were therefore able to employ a range of jaw motions when feeding, which allowed them to chew their food in a manner analogous to mammals.

Originally, the evidence for this unusual feeding mechanism was based on gross observation of the ornithopod skull, especially careful observation of the degree of motion between various sutures (Norman, 1984; Weishampel, 1984). More recent studies, relying on a toolkit of analytical techniques unavailable to earlier scientists, lend credibility to the pleurokinetic hypothesis. Analysis of tooth wear on hadrosaurid teeth revealed four discrete sets of scratches arranged in consistent, statistically significant orientations (Williams et al., 2009) (Fig. 7.17C,D). Some of these could only be produced via transverse motion of the lower jaw relative to the upper jaw, as would occur during pleurokinetic skull movement. Furthermore, FEA demonstrates that the lower jaws of hadrosaurids were remarkably robust and well suited to withstand high torsional forces caused by medial rotation of the lower jaw relative to the maxilla during pleurokinetic feeding (Bell et al., 2009).

However, some recent work has cast doubt on some aspects of the pleurokinetic hypothesis. Rybczynski et al. (2008) constructed a digital model of the *Edmontosaurus* skull and used a computer animation program to simulate the motion of skull bones during a pleurokinetic bite (Fig. 7.17A,B). This exercise found that large amounts of secondary motion between other cranial bones was required to accommodate the motion of the maxilla and prevent the skull from systematic dislocation. Some of these motions required large separations of more than 1 cm between individual bones, including between the jugal and quadratojugal and pterygoid and palatine. This may indicate that the skull was not pleurokinetic to the extreme degree posited by gross morphological studies, but it is important to remember that the digital model only studied the relative motions of bones during

pleurokinetic feeding. Muscles, tendons, ligaments, and cartilage were not incorporated into the model, and in a living animal these soft tissues would surely have constrained bone motion and limited disarticulation. Holliday and Witmer (2008) criticized the pleurokinetic hypothesis from the standpoint of sutural geometry: they could not find sufficient evidence for extreme mobility at the premaxilla–maxilla suture, as well as other bone contacts that simply must be kinetic in order for pleurokinesis to work. On the contrary, they suggested that the broad contact areas between these bones, and in some cases tighter and more interlocking sutural relationships than previously appreciated, seriously argued against the notion of pleurokinesis. However, Holliday and Witmer (2008) acknowledged that tooth microwear data does indicate the presence of a transverse chewing stroke in some ornithopods, and therefore more work is needed to harmonize observations from gross anatomy, sutural relationships, and tooth wear. It is possible that some ornithopods had more pleurokinetic skulls than others, meaning that large comparative studies of many species are needed to help resolve questions about ornithopod feeding.

Regardless of how kinetic their skulls were, ornithopods were probably well suited for feeding on "mid-level" plants growing several meters above the ground. Their moderate size and ability to rear up on two legs would have given ornithopods access to higher-blossoming plants not easily accessible to most thyreophorans and shorter-necked sauropods. Only the very largest ornithopods, such as the 15-m *Shantungosaurus* from the Late Cretaceous of Asia, could browse at heights accessible to longer-necked sauropods. Therefore, most ornithopods probably exploited a feeding envelope between that of low-browsing thyreophorans and high-browsing sauropods. Fossilized stomach contents and coprolites prove that hadrosaurs ate a variety of plants, including conifers and angiosperms, as well as specific plant structures such as fruits and even decaying wood (Chin, 2007; Tweet et al., 2008; Sander et al., 2010). Because ornithopods had a wide range of muzzle shapes, ranging from narrow to extremely broad, it is likely that some were more specialist feeders and others generalists.

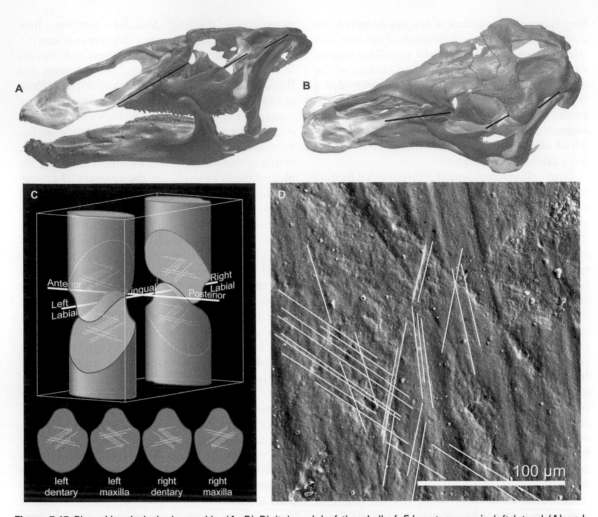

Figure 7.17 Pleurokinesis in hadrosaurids. (A, B) Digital model of the skull of *Edmontosaurus* in left lateral (A) and dorsolateral oblique (B) views, with the straight lines indicating the hypothesized position of the pleurokinetic hinge between bones of the skull roof and those of the side of the face. (C, D) Tooth microwear in hadrosaurids, including a scanning electron microscope image of wear on the surface of a hadrosaurid tooth (D) and a schematic illustration showing the major patterns of wear in hadrosaurids in general (C). These wear patterns indicate the presence of a transverse chewing stroke, which could be explained by pleurokinesis, although the model shown in (A) and (B) suggests that the classic pleurokinesis hypothesis may be incorrect (see text for details). Images (A) and (B) courtesy Alex Tirabasso © Canadian Museum of Nature (see Rybczynski et al., 2008 for further details); image (C) modified from Williams, V.S. et al. (2009) Quantitative analysis of dental microwear in hadrosaurid dinosaurs, and the implications for hypotheses of jaw mechanics and feeding. *Proceedings of the National Academy of Sciences (USA)* 106, 11194–11199, and reproduced with permission; image (D) courtesy of Dr Vince Williams.

Did dinosaurs "invent" flowers?

Angiosperms, the "flowering plants" that are so exceptionally diverse in the modern world, originated sometime during the reign of dinosaurs, probably in the Early Cretaceous (Sun et al., 1998). Bakker (1978, 1986) hypothesized that the rise of angiosperms from a marginal clade to their modern dominance was driven by a coevolutionary arms

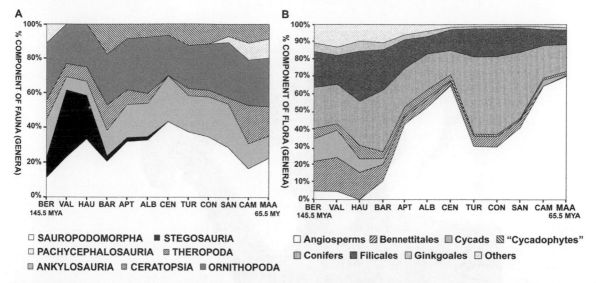

Figure 7.18 Plots showing the diversity of dinosaurs and plants across the Cretaceous. The plots show the proportion of the total fauna/flora of the time in question that is comprised of each individual dinosaur subgroup/major plant group. Images modified from Butler et al. (2009b), and reproduced with permission.

race with dinosaurs. More specifically, he suggested that the sophisticated oral processing capabilities of ceratopsians and ornithopods, which also became exceptionally abundant and diverse in the Cretaceous, developed in concert with the diversification of flowering plants. A careful look at the fossil record, however, indicates that there is no strong correlation between the diversity patterns of dinosaurs and angiosperms (Sereno, 1997; Weishampel and Jianu, 2000; Barrett and Willis, 2001; Lloyd et al., 2008; Butler et al., 2009b) (Fig. 7.18). In other words, angiosperm diversity does not rise in close association with the diversity of any herbivorous dinosaur groups, and major pulses in angiosperm diversification do not correspond to any major radiations of dinosaurian herbivores. Furthermore, if ornithopods coevolved in harmony with angiosperms, then angiosperm fossils should regularly be found in association with ornithopods, but only rarely with the less dietarily refined sauropods. However, a comprehensive database of fossil dinosaur and plant occurrences recovers no statistically significant correspondence between ornithopod and angiosperm fossils (Butler et al., 2010b). Therefore, there is little evidence at present that dinosaurs and angiosperms evolved together in a dynamic waltz throughout the Cretaceous.

Conclusions

One of the primary functions required of all living organisms is the acquisition of nutrients. Therefore, understanding how dinosaurs fed and what they ate is essential for understanding how they functioned as biological species (Fig. 7.19). Information from gross anatomy, trace fossils (coprolites, bite marks, gastroliths, stomach contents), and computational biomechanical analyses can shed light on dinosaurian diets and feeding strategies. The common ancestor of all dinosaurs was probably carnivorous, although the realization that several of the closest dinosaurian relatives were herbivorous indicates that dietary evolution in early dinosaurs was more complex than previously thought. Most theropods were carnivores, and some were specialized for piscivorous diets, megacarnivorous feeding strategies in which large prey were dispatched by haphazardly crunching through bone, and perhaps even pack hunting. Other theropods, such as the herbivorous therizinosauroids and insect-eating alvarezsaurids, had distinct diets. All sauropods were obligate bulk-feeding herbivores that guzzled enormous quantities of vegetation in order to fuel their colossal bodies. More primitive sauropodomorphs, however, were probably omnivores.

Figure 7.19 A hypothetical illustration of a carnivorous theropod (the carcharodontosaurid *Tyrannotitan*) feeding on the carcass of a sauropod. Illustration by Brett Booth.

All ornithischians were herbivorous, and the derived ornithopod and ceratopsian subclades developed sophisticated feeding mechanisms in which extensive dental batteries were used to efficiently crop tough vegetation. Some ornithopods were probably capable of chewing, enabled by their pleurokinetic skulls, although this is an active subject of debate. There is no evidence, however, that dinosaurs and angiosperms coevolved during the Cretaceous.

8 | Reproduction, Growth, and Physiology

A 25-tonne sauropod or a 13-m *Tyrannosaurus* did not begin life as such colossal majestic creatures. Like any animal, they were conceived, underwent embryonic development, were born, and then matured into a fully grown adult capable of reproducing yet another new generation. How dinosaurs reproduced and grew are some of the most fascinating aspects of their biology. It was no quick or easy task for a sauropod hatchling to develop into a multi-ton adult that was capable of shaking the earth as it walked. Even smaller dinosaurs went through substantial changes in their transition from embryo to adult. Understanding how dinosaurs reproduced and grew provides yet more critical information, along with data on locomotion and feeding, necessary to comprehend dinosaurs as living animals. Not only that, but some grasp of how dinosaurs grew should help reveal exactly how some species were able to achieve such enormous sizes and dominate terrestrial ecosystems for millions of years.

Studies of dinosaur reproduction and growth, as well as the physiological attributes that underpinned their development and fueled their lifestyles, have moved to the forefront of dinosaur paleobiological research in recent years. Some of the most exciting new research techniques are being used to address several important questions about dinosaur life history (Horner, 2000; Chinsamy-Turan, 2005; Erickson, 2005; Varricchio, 2011). Did dinosaur parents care for their young and, if so, did some species exhibit bird-like behaviors? Did dinosaurs grow extremely rapidly like modern birds, or slower like living crocodiles and other reptiles? Were faster growth rates or prolonged periods of growth the key to attaining gigantic size? What major changes in body proportions, feeding habits, and locomotor capabilities occurred as dinosaurs matured from embryos into adults? Did many quintessentially "avian" characters, such as a fast growth rate

Dinosaur Paleobiology, First Edition. Stephen L. Brusatte.
© 2012 John Wiley & Sons, Ltd. Published 2012 by John Wiley & Sons, Ltd.

and high levels of egg brooding and parental care, first evolve in birds or originate in non-flying, non-avian dinosaurs? And, perhaps most fascinating of all, were some or all dinosaurs capable of high metabolic activity enabled by internal temperature control (endothermy)?

Like many questions explored in previous chapters, these mysteries are actively being addressed with a combination of evidence from skeletal fossils, trace fossils, and experimental and quantitative analytical techniques. Discoveries of eggs, nests, and embryonic skeletons have helped scientists understand the reproductive mechanisms of dinosaurs and the level of parental care invested by some species. Histological techniques have opened a new frontier for studying dinosaur growth, as growth line counts allow individual specimens to be aged and microscopic bone texture imparts information on how fast or slow an animal was developing. The ever-expanding inventory of dinosaur fossils now includes relatively complete growth series for several species, allowing researchers to pinpoint how body size, skeletal proportions, and discrete features (such as those relating to feeding or locomotion) changed during ontogeny. Finally, a range of evidence has been used to study dinosaur physiology, including features of gross anatomy, bone texture and inferred growth rates, preserved soft tissues, and information on the isotopic composition of dinosaur bones and teeth. The overall picture that has emerged from this research is provocative: all dinosaurs laid eggs, many species grew into adults remarkably fast and began reproducing long before they were mature, several species exhibited advanced "avian-style" egg brooding and parental care, and most species probably had high metabolisms and were capable of elevated activity levels.

Reproduction: Eggs, Nests, and Parental Care

Dinosaur eggs

All dinosaurs, as far as we know, laid eggs (Figs 8.1–8.3). Fossilized eggs containing embryos have been discovered for all three of the major dinosaurian subgroups: theropods, sauropodomorphs, and ornithischians (Carpenter, 1999; Horner, 2000; Varricchio, 2011) (Fig. 8.1). More precisely, within these major clades, embryos are known from a range of taxa: basal tetanurans, therizinosauroids, oviraptorosaurs, and troodontids among theropods (Norell et al., 1995, 2001c; Mateus et al., 1997; Varricchio et al., 2002; Kundrát et al., 2008); "prosauropods" and large-bodied sauropods among sauropodomorphs

(Chiappe et al., 1998; Reisz et al., 2005, 2010); and hadrosaurs and ceratopsians among ornithischians (Horner and Weishampel, 1988; Horner and Currie, 1994; Horner, 1999; Balanoff et al., 2008). Although some workers such as Bakker (1986) have speculated that some dinosaurs may have given birth to live young, this is nothing more than conjecture based on no solid evidence. The prevalence of fossilized eggs for a range of dinosaur species, as well as the fact that the closest living relatives of dinosaurs (birds and crocodiles) are exclusively oviparous, is more than enough evidence to infer that most (if not all) dinosaurs laid eggs.

As is true of the eggs of living animals, dinosaur eggs would have provided protection and nutrition for the growing embryo (Carpenter, 1999). The

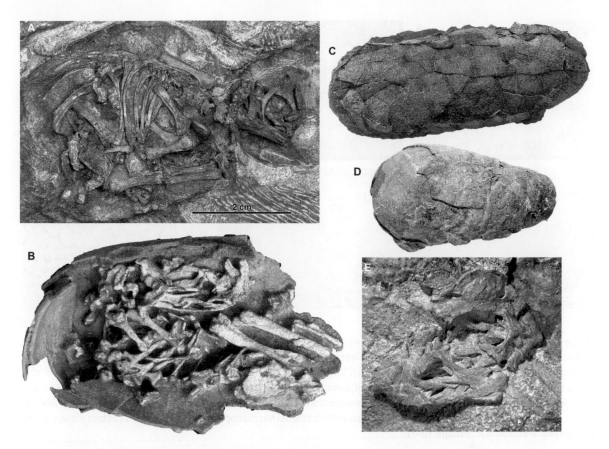

Figure 8.1 Dinosaur eggs and embryos: (A) embryo of the basal sauropodomorph ("prosauropod") *Massospondylus* from the Early Jurassic of South Africa; (B) embryo of the oviraptorosaurian theropod *Oviraptor* from the Late Cretaceous of Mongolia; (C) the egg of a large theropod (probably an oviraptorosaur); (D) the egg of a troodontid theropod; (E) the skull of a sauropod embryo from the Cretaceous of Argentina. Image (A) modified from Reisz et al. (2010), and reproduced with permission; (B) photograph © Mick Ellison; (C) courtesy of Dr Gerald Grellet-Tinner; (D) courtesy of Dr David Varricchio; (E) courtesy of Dr Luis Chiappe.

eggshell was both a buffer from the outside world and a source of calcium for the embryo's developing bones. All dinosaurs, as far as is known, laid hard-shelled eggs composed of one or more layers of calcite crystals (Fig. 8.2), although it is possible that some primitive dinosaurs may have produced softer eggs like those of their close pterosaurian cousins (Unwin and Deeming, 2008). Calcitic eggshell fragments, which are remarkably common in many Jurassic and Cretaceous dinosaur assemblages, are incredibly rare in Triassic sediments, and currently represented only by a few specimens associated with sauropodomorph embryos (Bonaparte and Vince, 1979; see also Early Jurassic eggs and embryos described by Reisz et al., 2005, 2010). It is possible that this rarity reflects the fact that many Triassic dinosaurs did not produce hard-shelled eggs, but alternatively it may stem from a preservational bias. Future discoveries, or the lack thereof, should help determine which of these two explanations is correct.

Different types of dinosaurs produced different eggs, which is not unexpected given the great diversity of egg size and shape among living birds (Carpenter, 1999) (Figs 8.1 and 8.3). Dinosaur eggs run the gamut from spherical to elliptical, and some

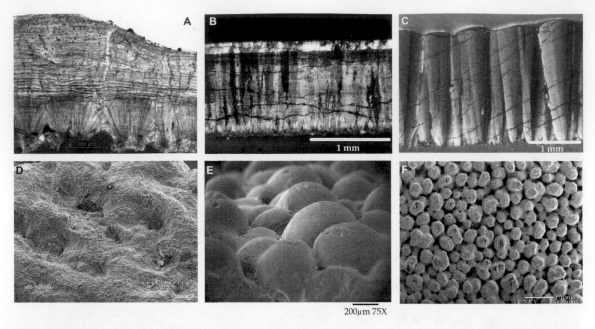

200μm 75X

Figure 8.2 The microstructure of dinosaur eggs: (A) histological thin section of a maniraptoran theropod egg (most likely *Deinonychus*) illustrating the two calcite crystal layers that are diagnostic of most theropod eggs; (B) thin section of a troodontid egg, also exhibiting two calcite layers; (C) thin section of a sauropod egg showing the primitive condition of a single calcite layer; (D) surface ornamentation of a maniraptoran theropod egg (most likely *Deinonychus*); (E) surface ornamentation of a sauropod egg; (F) surface ornamentation of a sauropod egg. All images courtesy of Dr Gerald Grellet-Tinner and include histological thin sections viewed with traditional microscopy (A–C) and surface structure viewed using a scanning electron microscope (D–F).

even possess the classic asymmetrical shape of most extant bird eggs, which exhibit blunt and pointed ends. The recovery of embryonic remains inside some eggs indicates that sauropods largely produced spherical eggs, whereas theropods had more elliptical eggs which were asymmetrical in the most derived bird-like species (Norell et al., 1995; Varricchio et al., 1997, 2002; Chiappe et al., 1998; Grellet-Tinner et al., 2006; Zelenitsky and Therrien, 2008a). Dinosaur eggs also vary greatly in size, although remarkably no known examples are larger than about 10 liters in volume (Carpenter, 1999; Horner, 2000). Instead, the largest dinosaur eggs are approximately the same size as those produced by the extinct 3-m "elephant bird" *Aepyornis*. This strongly suggests that even the most colossal dinosaurs came into the world as pint-sized hatchlings and attained their giant bulk almost entirely after birth. There seems to be no relationship whatsoever between egg size and adult size: some of the

smallest dinosaur eggs on record, with a volume of less than 1 liter, belong to sauropods, whereas closely related and similarly sized hadrosaurs have egg volumes that differ threefold or fourfold (Horner, 2000).

Because they are not components of a dinosaur's skeleton, eggs themselves are classified according to their own "parataxonomic" classification system. This is also true of other trace fossils, such as footprints and even, in some cases, bite marks. A dinosaur egg that contains an embryo inside will not be given the name of the embryo, but rather receives its own moniker. Therefore, for example, eggs containing embryos of the hadrosaur *Maiasaura* are referred to as their own "ootaxon," *Spheroolithus* (Horner, 2000). The parataxonomic classification of eggs is akin to the classification of biological species: egg species, genera, families, and higher taxa are named and described based on unique combinations of characters, which usually relate to size, shape, microstructural calcite crystal orientation, surface ornamentation, and surface pore geometry (Mikhailov, 1991, 1997; Carpenter, 1999). In recent years, dinosaur egg specialists have begun using cladistics to group together similar egg types and study the acquisition of primitive and derived eggshell features (Varricchio and Jackson, 2004; Zelenitsky and Therrien, 2008a).

Dinosaur nests

Dinosaur eggshells are often found broken and isolated, but numerous well-preserved sites have yielded organized assemblages of complete or nearly complete eggs (Horner and Makela, 1979; Norell et al., 1995; Varricchio et al., 1997, 1999; Chiappe et al., 1998, 2004; Zelenitsky and Therrien, 2008b; Grellet-Tinner and Fiorelli, 2010; Vila et al., 2010). In many cases, these assemblages clearly represent nests (Fig. 8.3). Known dinosaur nests come in a wide range of shapes, most commonly either elliptical or circular, and can contain few or many eggs, sometimes haphazardly gathered together, in other cases arranged in a neat order. In general, it seems as if sauropods built elliptical nests filled with a small number of eggs (~25) arrayed in no discernible pattern (Chiappe et al., 1998, 2004; Grellet-Tinner and Fiorelli, 2010; Vila et al., 2010) (Fig. 8.3D). Sedimentary evidence indicates that these nests were bowl-like depressions, actively dug out of the surrounding earth by the adult and later filled in during flooding episodes (Chiappe et al., 2004).

The nests of derived theropods, on the other hand, seem to have been mostly circular, and in some cases built on mounds surrounded by a rim of sediment (Varricchio et al., 1999; Zelenitsky and Therrien, 2008b) (Fig. 8.3A,B,E–G). Egg number varies depending on the species in question – *Oviraptor* nests contain 20–36 eggs whereas those of *Troodon* hold 12–24 – but the eggs are paired and the pairs are arranged together into a ring-like configuration (Norell et al., 1995; Varricchio et al., 1997; Zelenitsky and Therrien 2008b) (Fig. 8.3E–G). Moreover, the individual eggs are asymmetrical in these species, and in well-preserved nests the blunt end of each egg usually points inward (Fig. 8.1C,D). It is thought that the narrower end of the egg was exposed first during egg-laying, based on comparisons with living birds, which

suggests that the mother dinosaur sat at the center of the nest (which is usually flat) during egg deposition (Zelenitsky and Therrien, 2008b). In these nests, the eggs are often partially buried, which would have helped stabilize the eggs and maintained the circular shape of the nest, and this may have been particularly important in species that incubated their eggs by brooding (see below).

Although fossil nests are quite rare, the known evidence demonstrates that dinosaurs nested in a variety of settings. Dinosaur eggs and nests are known from nearly every conceivable terrestrial sediment, including limestones, mudstones, and sandstones (Horner, 2000). It seems as if some dinosaurs nested near streams and rivers, along the shores of lakes, lagoons, and seas, on tidal flats and beaches (Sanz et al., 1995; López-Martínez et al., 2000), and even in Yellowstone-like hydrothermal geyser settings (Grellet-Tinner and Fiorelli, 2010). There have been some suggestions that certain dinosaurs nested in more nearshore or upland environments, or in wetter or drier conditions, than others, but this deserves further study as more fossils become available for comparative research (Horner 1999).

Remarkably, some dinosaur nests are found associated with large numbers of other nests at what appear to be nesting sites, which in some cases are spread over hundreds of square meters and contain thousands of individual eggs (e.g., Horner 1982, 2000; Chiappe et al., 1998; Grellet-Tinner and Fiorelli 2010). Sometimes these sites are extensive laterally, which reflects a large colony of individuals reproducing at once, and also stacked one on top of the other in vertical stratigraphic section, probably representing hundreds or thousands of years of continued nesting activity. Both of these conditions characterize some of the famous nesting sites studied by John Horner and colleagues in Montana (e.g., Horner and Makela, 1979; Horner, 1982). Here, individual nesting horizons are repeatedly stacked upon each other vertically, and individual horizons contain more than 10 nests, which are evenly spaced from each other at large enough distances so that full-grown adults could walk between them. This is stunning and convincing evidence for "site fidelity": the continued usage of a nesting area by a communal group over an extended time period.

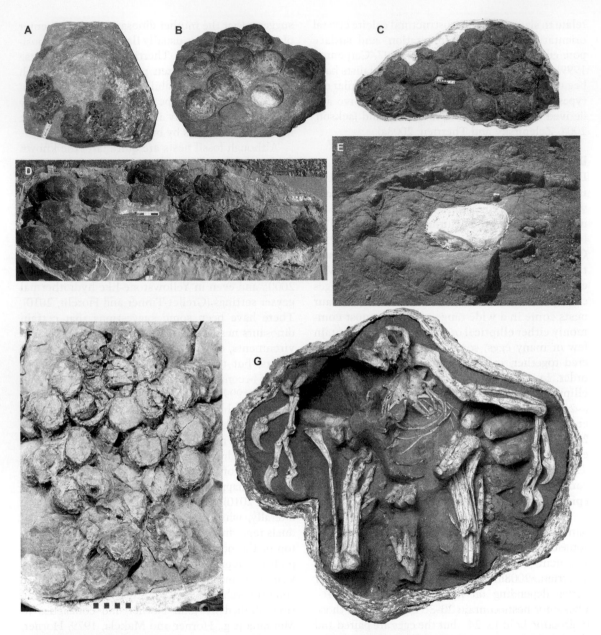

Figure 8.3 Dinosaur nests: (A) nest of a maniraptoran theropod with a central platform where the mother (and perhaps father) sat during egg-laying (and brooding); (B) the egg clutch of a therizinosauroid; (C) the egg clutch of a hadrosaurid; (D) the egg clutch of a sauropod; (E) the nest of a troodontid (covered with a plaster jacket), surrounded by a rim of sediment; (F) the egg clutch of a troodontid; (G) a spectacular fossil of an oviraptorosaur parent preserved brooding a nest of eggs. Images (A–C) courtesy of Dr Darla Zelenitsky; (D) modified from Vila et al. (2010); (E) and (F) courtesy of Dr David Varricchio; (G) photograph © Mick Ellison.

Parental care

Living birds and crocodiles exhibit varying degrees of parental care, so it is reasonable to hypothesize that dinosaurs did as well (Carpenter, 1999; Horner, 2000; Varricchio, 2011). However, "parental care" is an imprecise umbrella term that can refer to many different behaviors. It is more informative to ask three specific questions about dinosaur parental habits. Did dinosaurs invest substantial time and energy in making nests? Did some dinosaurs actively incubate their eggs with brooding behaviors? Did dinosaur parents actively care for their offspring after hatching? In each case, it is also possible to ask whether dinosaurian behaviors were similar, and perhaps even directly homologous, to those of their living avian descendants.

First, there is solid evidence that various dinosaur species devoted large amounts of time and energy to construction of their nests, rather than haphazardly depositing their eggs in a hastily built structure. Nests belonging to troodontids are commonly surrounded by a rim of sediment 15–40 cm beyond the edge of the egg clutch, which probably functioned as something of a barrier against flooding and predation (Varricchio et al., 1997) (Fig. 8.3E). A maniraptoran nest described by Zelenitsky and Therrien (2008b) was carefully constructed on a mound of sediment, and built at the precise size needed for the mother to sit in the center and lay eggs without disturbing those eggs that were already present (Fig. 8.3A). This level of parental care is not characteristic only of derived theropods, but is also present in other dinosaurs. The communal hadrosaurs described by Horner and Makela (1979) and Horner (1982) keenly arranged their nests so that there was enough room for large adults to navigate around the nesting ground, whereas some sauropods clearly excavated their bowl-like nests by using their hind feet (Chiappe et al., 2004; Vila et al., 2010).

Second, several discoveries firmly demonstrate that derived bird-like theropods sat on their nests and actively incubated their eggs. Spectacular fossils preserve adults of *Troodon* and the oviraptorids *Oviraptor* and *Citipati* in contact with their eggs and, in the case of the oviraptorids, in the stereotypical brooding posture of living birds in which the wing is extended over the eggs to provide heat and protection and the abdomen is in contact with the clutch (Norell et al., 1995, 2001c; Dong and Currie, 1996; Varricchio et al., 1997; Clark et al., 1999; Horner, 2000; Varricchio, 2011) (Fig. 8.3G). Clearly, these stimulating fossils are the most direct evidence possible of brooding, active incubation, and intentional parent–egg interaction in extinct dinosaurs. Not all dinosaurs, however, likely behaved in this way. There is no fossil evidence that sauropods or hadrosaurs sat atop their nests and incubated their eggs, and indeed this would be physically difficult given the enormous sizes of most of these animals. Many, if not most, dinosaur species probably incubated their eggs passively, either by burial in sediment or covering the nests with vegetation. The former is more likely, because there is little evidence from dinosaur nesting grounds of fossilized plants in association with eggs or nests (Carpenter, 1999; Horner, 2000). Furthermore, Deeming (2006) showed that the substantial thickness and high porosity of many dinosaur eggs indicated that they were buried in humid conditions. Even some eggs of derived theropods, such as those of oviraptorids, possess these features, suggesting that the eggs may have been partially buried and received much of their incubation heat from sediment and not the brooding parent. Brooding may therefore have functioned more in protection than incubation in these bird-like species, although this is difficult to test with certainty.

Third, there is some evidence that some dinosaur parents may have cared for their offspring long after hatching, but this is debatable and open to interpretation because of the uncertainty of taphonomic and sedimentological evidence. There are many instances in the fossil record of adult dinosaurs preserved in association with juveniles, but these records are not sufficient to demonstrate active parental care. On the contrary, they may represent chance associations of fossils that were brought together after death. One good example, which more than any other may truly represent parental care of offspring, is the discovery of adult and juvenile *Oryctodromeus* bones within their burrows. Because there is evidence that this species actively burrowed, and that the burrow it was found within was its dwelling, it is no great stretch to conclude that *Oryctodromeus* adults at least interacted with their offspring (Varricchio et al., 2007).

Another classic and convincing case, based solely on the discovery of juvenile remains, is Horner and Makela's (1979) discovery of hatchling and larger juvenile (~1 m long) *Maiasaura* preserved within their nests. Histological evidence indicates that the bones of these juveniles were poorly ossified and likely incapable of powering active locomotion (Horner and Weishampel, 1988; Horner et al., 2000), and the prevalence of broken and trampled eggshells suggests that the juveniles were actively inhabiting the nests. Based on this information, it is reasonable to hypothesize that juveniles remained in the nest, where they were fed and cared for by adults, for quite some time after hatching. Of course, it is possible that the association of nests and larger juveniles was due to taphonomic happenstance or another behavior unrelated to parental care of offspring. Perhaps, for instance, a flock of juveniles retreated into an abandoned nest to escape a predator. A scenario like this cannot be ruled out, but what is important is that the association of nests, trampled eggshells, and juveniles poorly suited for gathering their own food is consistent with the hypothesis of active parental care in *Maiasaura*.

The evolution of "avian-style" reproduction

The ever-improving fossil record of dinosaur eggs and nests, along with the increasingly well understood phylogenetic relationships of birds and their closest dinosaurian relatives, allows scientists to study the evolutionary development of stereotypically "avian" reproductive features that are unique to birds among living animals. These features include asymmetrical eggs with blunt and pointed ends, which probably result from the physical constraints of the small avian oviduct; eggs with three distinct layers of calcite crystals, which probably lend strength that is necessary to bear the brunt of a brooding parent's weight; brooding and active incubation behaviors; and a single functional oviduct that lays only a single egg at one time. These and other reproductive features of dinosaurs, such as egg shape and ornamentation, have been explicitly studied in a phylogenetic framework (Grellet-Tinner et al., 2006), and in one case even using cladistic methodology (Zelenitsky and Therrien, 2008a). These studies convincingly demonstrate that most of these so-called "avian" features developed long before the evolution of birds or flight, as is the case with so many other stereotypically "avian" features of osteology, soft tissue anatomy, and locomotion that have been discussed in previous chapters.

Although the current fossil record of dinosaur eggs and reproductive behaviors is far from complete, there seems to have been an incremental development of avian-like characters during dinosaur evolution. Grellet-Tinner et al. (2006) argued that sauropod dinosaurs retained many primitive reproductive features that are present in living crocodiles. Sauropod eggs are usually spherical and symmetrical, composed of a single layer of calcite crystals, and arranged in haphazard piles in the nest. This latter feature suggests that there were still two functional oviducts, neither of which was restricted to producing one egg at a time. Furthermore, there is no evidence that sauropods incubated their young, much less provided much direct care or protection before hatching. Derived theropods, however, exhibit more derived reproductive behaviors shared with birds (Grellet-Tinner et al., 2006; Zelenitsky and Therrien, 2008a). Oviraptorosaurs and troodontids have asymmetrical eggs that were actively cared for, and perhaps incubated, by parents (Figs 8.1 and 8.3). Their eggs are composed of two calcite layers, and Varricchio and Jackson (2004) argue that some troodontid eggs may even exhibit three layers as in birds; however, Grellet-Tinner and Makovicky (2006) argue that the supposed third layer is a diagenetic artifact (Fig. 8.2). The eggs are not strewn about in the nest, but are rather paired together (Fig. 8.3). This indicates that there were still two functional oviducts, but that each could produce only one egg at a time. The discovery of two eggs within a female oviraptorosaur supports this hypothesis: clearly two functioning oviducts were present, but there simply was not enough room in the body cavity for more than two eggs (Sato et al., 2005).

Troodontids, which are the closest non-avian relatives of birds for which egg and nest fossils are known, are eerily bird-like in their reproductive traits. They are even more bird-like than oviraptorosaurs, due to their far more asymmetrical eggs, which are nearly conical as in many living birds (Fig. 8.1D). In these birds, the pointed tip of the conical egg houses an advanced "air cell" that helps the embryo breathe. The eggs of unequivocal birds from the Late Cretaceous of Argentina are almost indistinguishable from those of troodontids, aside from their unambiguous triple-layered calcite. Because of their remarkable similarities, Grellet-Tinner et al. (2006) hypothesized that troodontids had a "fully avian style" of incubating their eggs, and would have looked and behaved much like nesting birds. This is yet more convincing evidence that birds and their closest dinosaurian relatives were nearly identical anatomically and functionally. One important difference between birds and their closest relatives does remain: the presence of only a single functioning oviduct in birds. However, it is unknown exactly when this trait evolved. Fossil nests are unknown for primitive Mesozoic birds, so it is possible that the single oviduct developed much later during the evolutionary history of birds, long after their origination from theropods.

In summary, nearly every single egg-related character and reproductive behavior unique to modern birds is now known to have been present in non-avian, non-flying dinosaurs. No single character, with the possible exception of a single functional oviduct and three calcite crystal layers, distinguishes the eggs and reproductive biology of birds from those of their closest non-avian theropod relatives. Avian-style reproductive features are not present in all dinosaurs, however, and sauropods in particular seem to possess many primitive features and behaviors shared with crocodiles. It is difficult to study the pattern and tempo of the evolutionary transition between primitive "crocodile-like" and more derived "avian-like" reproductive features, because so few dinosaur groups are represented by well-preserved eggs, embryos, and nests. What we do know is that most of these avian-style features originated somewhere within theropod phylogeny, long before birds took to the skies.

Paternal care in dinosaurs?

It seems natural for humans to assume that most animals exhibit maternal care, based on our own anthropocentric experiences, but this is by no means universally true. Most crocodile mothers care for their young, but living birds are usually either biparental, with both parents involved in child care, or paternal, in which the male takes the lead role in caring for the offspring (Carpenter,

1999; Varricchio, 2011). It seems like this latter condition was characteristic of derived theropods. Varricchio et al. (2008a) collected data on clutch size and body size of the brooding parent in over 400 species of living archosaurs. They found individual diagnostic correlations between clutch size and body size that distinguish paternal, maternal, and biparental living archosaurs. When these two measurements were compiled for *Troodon*, *Oviraptor*, and *Citipati*, these maniraptorans were found to have the signature clutch size/body size relationship of parental birds (Fig. 8.4). More specifically, these theropods have clutch sizes that closely match those of paternally caring living ratites of similar body size, but clutches that would be far too large (for their body size) for a maternal or biparental bird. Furthermore, these scientists also examined the bone histology of the brooding adults found in association with the eggs of all three taxa. None of them exhibited medullary bone, the characteristic calcium-storing bony tissue that female birds use as a calcium source for

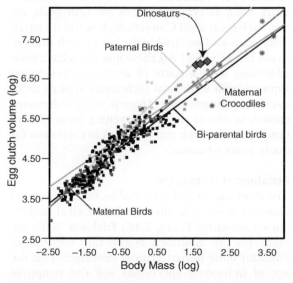

Figure 8.4 Plot of egg clutch size vs. body size in living birds and crocodiles and some extinct dinosaurs. Note that the dinosaurs (denoted by diamonds) plot out close to the regression line for paternal birds, suggesting that these dinosaurs also exhibited paternal care. Image courtesy of Dr David Varricchio and based on data from Varricchio et al. (2008a).

shelling eggs, nor extensive internal bone remodeling, which female crocodiles undergo in order to mine calcium from their bones. This indicates that the brooding adults are male, in agreement with the paternal care strategy predicted by the clutch size/body size comparison. It is unclear if more primitive theropods and non-theropods behaved this way, and more fossil discoveries are clearly needed.

Methods for Studying Dinosaur Growth

Once dinosaurs hatched from their eggs they needed to continue growing into an adult, a process that took considerable time and energy. Several intriguing questions about dinosaur growth have captured the imagination of scientists. How fast did dinosaurs grow and how do dinosaurian growth rates compare with those of living taxa? At what ages did dinosaurs become sexually mature, reach full size, and ultimately die? How did some dinosaurs, such as tyrannosaurids and sauropods, attain such giant size? Conversely, how did the small body sizes and astonishingly fast growth rates of living birds evolve? All these questions have been addressed with a spate of recent studies, using sophisticated analytical techniques applied to an ever-expanding sample of fossils. Study of dinosaur growth is one of the most exciting areas of contemporary research and promises to remain so for many years to come.

Histological techniques

The most useful and powerful means for studying dinosaur growth is through histological techniques (Chinsamy-Turan, 2005; Erickson, 2005; see Plate 15). As described below, thin sections of dinosaur bone impart critical information on the age of individual specimens and the tempo of growth at the time bony tissue was deposited. Before interpreting thin sections, however, these samples must be extracted from dinosaur bone. Two general techniques are commonly used by researchers (Erickson, 2005). First, bones can be sectioned transversely, usually at midshaft, using a diamond-tipped saw. The section is thin affixed to a glass slide and slowly polished until it is thin enough to be viewed under polarized and/or reflected light microscopy. Second, Sander (2000) developed an alternative method that uses a diamond-tipped drill to extract a cylinder of bone from a specimen, analogous to how geologists sample cores of rock or ice. This core, which slices through each successive bone layer from the external surface to the inner cortex, is then polished into a thin section using the same methods as the transverse section described above.

Although these methods can be applied to any specimen, scientists should carefully choose which bones to sample. In order for a thin section to give any meaningful information on specimen age or growth rates, the sampled bone should show minimal remodeling. Bones are not static elements but are dynamically remodeled throughout life as an animal grows, reproduces, and recovers from injuries. Heavily remodeled bones may no longer contain bony tissues deposited during earlier stages of development, preventing a complete understanding of the growth history of the animal and hindering our ability to age a specimen (Horner and Padian, 2004; Chinsamy-Turan, 2005). In many cases it seems as if bones that do not bear the weight of the animal, such as ribs, gastralia, and the fibula, are not as extensively remodeled as those that do (Erickson et al., 2004; Erickson, 2005). Therefore, these bones are often preferentially sampled. However, if possible, researchers try to sample multiple bones from the same individual and look for consistent patterns. If many bones from across the skeleton indicate the same age and growth rate, then scientists can be confident that what they are seeing is not biased by remodeling.

Bone texture

The organization and texture of the mineral crystals in a bone gives information on the rate of bone deposition and the general style of growth (Chinsamy-Turan, 2005). There are two common textures of compact bone, which define the end points on a spectrum of slow-growing to rapidly deposited bone (Fig. 8.5). Slow-growing compact bone is typically lamellar, with the mineral crystals arranged in discrete concentric rows called laminae that are stacked on top of each other in

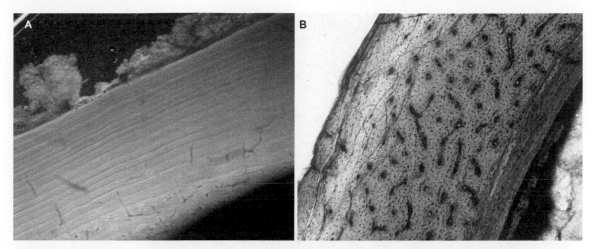

Figure 8.5 Histological thin sections of two major types of bone, slow-growing lamellar bone (A) and faster-growing fibrolamellar bone (B). Note that lamellar bone is characterized by numerous individual orderly layers, whereas fibrolamellar bone has a more woven, random texture. Images courtesy of Dr Gregory Erickson.

thin section (Fig. 8.5A). Fast-growing bone is described as fibrolamellar, because the collagen fibers in the original bone (and hence the mineral crystals in a fossil bone) are more randomly and haphazardly arranged to form a fibrous, or woven, texture (Fig. 8.5B). Fibrolamellar bone is deposited so quickly that collagen and minerals literally entrap blood vessels, much like a person being engulfed in a mosh pit at a rock concert. Later, as the bone begins to be remodeled, slowly growing lamellar bone forms around each vessel, producing a characteristic structure called a primary osteon. Often, these primary osteons are further remodeled by the addition of more lamellar bone, forming secondary osteons. Dense patchworks of several generations of secondary osteons, which chaotically overlap each other like sheets of paper on a messy desk, define a characteristic fibrolamellar bone texture called Haversian bone. As was first recognized by pioneering paleohistologist Armand de Ricqlès (1980), dinosaur bone typically consists of fibrolamellar bone with a Haversian texture, which is highly vascularized and indicative of rapid growth rates (see also Erickson et al., 2001; Padian et al., 2001; Chinsamy-Turan, 2005; Erickson, 2005). Similar bony tissues are common in extant mammals and birds, both of which exhibit fast growth, whereas slower-growing lamellar bone is typical of living reptiles.

Aging dinosaur bones

As well as indicating whether an animal was growing relatively fast or slow, histological thin sections sometimes record how many years old an individual was at the time of death (Fig. 8.6). Three general techniques, each with its own benefits and limitations, are commonly used to age individual dinosaur specimens. The most common, useful, and reliable technique entails counting growth lines, technically termed "lines of arrested growth" (abbreviated as LAGs) (Reid, 1981, 1997). Much like the rings of a tree trunk, these lines represent periods when the deposition of bone tissue slowed or stopped completely, probably because of seasonal shifts in climate or nutrient availability (Fig. 8.6). Conversely, the highly vascularized regions between the rings were formed during periods of active growth. In some specimens, several LAGs are tightly stacked together near the external surface of the bone, forming a structure called an "external fundamental system" that represents remarkably slow bone growth over a long period of time, probably because the animal is fully grown. It was long thought that dinosaurs did not exhibit growth rings in thin section, but this conclusion was based on only a handful of examples. However, extensive histological sampling by British paleontologist Robin Reid identified clear growth rings on nearly every dinosaur thin section he examined.

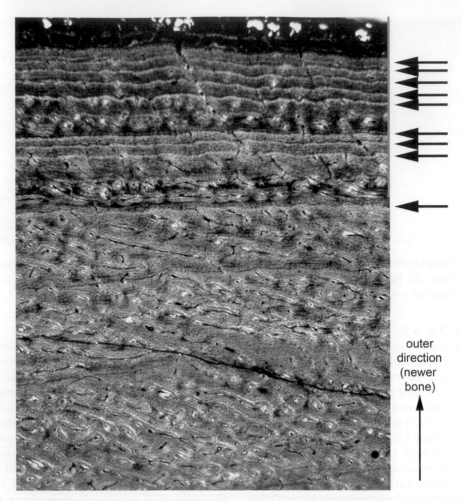

outer direction (newer bone)

Figure 8.6 Lines of arrested growth (LAGs, or "growth lines") in a histological thin section of *Tyrannosaurus*. Individual arrows denote individual LAGs and the arrow at bottom right indicates the direction of newer bone deposition toward the outer surface of the bone. Image courtesy of Dr Gregory Erickson.

Growth rings are therefore normal for dinosaurs, but some specimens (especially many sauropods) do genuinely lack these structures in most long bone elements traditionally used for aging (Erickson, 2005).

Growth rings can be used to age specimens if it is assumed that these structures formed annually. This is not mere wishful thinking, because the yearly spacing of growth lines is supported by a range of evidence, as reviewed by Erickson (2005). First, annual growth rings are known to form in a wide variety of living animals, including lepidosaurs and crocodiles. Second, in most dinosaur specimens the spacing between the lines is consistent with annual deposition: growth lines close to the center of the bone are spaced far apart, indicative of rapid juvenile growth, whereas those closer to the external surface are packed more closely together. If growth lines were formed by some other, non-cyclical process – say, completely unpredictable fluctuations in food supply – then random widths between the individual rings would be expected. Finally, the clever study of Tütken et al. (2004) measured the oxygen isotope composition across individual thin sections, from the center of the bone to the external surface. They reported a cyclical change in composition that corresponds to the alternation between LAGs and the vascularized regions of growth. Because oxygen isotope composition is closely tied to climate and drinking water sources, it is likely that these cyclical changes represent annual climate variations. If there was no annual change, or other clear cyclical change, random isotope variation across the bone would be expected. In sum, these three lines of evidence lend firm support to the annual nature of dinosaur growth lines.

Not all dinosaur specimens preserve growth rings, however, requiring that alternative methods be used to age these specimens. Two approaches in particular are used, one which can determine absolute age and the other relative ontogenetic stage, but each of these are subject to more uncertainty and bias than counts of growth lines. First, scientists can use the principles behind Amprino's Rule, which holds that there is a correlation between bone texture and depositional rate, to estimate the numerical age of a bone (see reviews and citations in Chinsamy-Turan, 2005 and Erickson, 2005). In other words, distinct bone textures are thought to form at their own unique rates, and therefore scientists can determine the age of a specimen by dividing the amount of bone present by the rate at which that type of bone forms. Unfortunately, studies of living animals show that there is no clear correlation between bone texture and rate of deposition: two textures that may look indistinguishable may be formed at rates that vary by up to 10-fold (Castanet et al., 2000; de Margerie et al., 2002). This is a clear problem with the reasoning behind Amprino's Rule, which renders it almost useless in determining an accurate age for individual specimens. That being said, Erickson (2005) accepted that these types of calculations are still somewhat useful in very roughly determining the age of specimens when no other information is available.

A potentially more useful system forgets about determining the precise numerical age of a specimen, but instead focuses on identifying relative ontogenetic stage (Fig. 8.7). Klein and Sander (2008) introduced the concept of "histological ontogenetic stages" (HOS), which comprise a relative scale indicating the level of histological maturity of a specimen. Identification of hundreds of sauropod thin sections allowed Klein and Sander (2008) to identify common bone textures and features usually associated with certain ontogenetic stages, such as juveniles, sub-adults, and fully grown adults (Fig. 8.7). The general idea is that growth rates slow but bone remodeling increases progressively throughout the life of an individual, and by identifying histological features associated with these ontogenetic changes the relative age of a specimen can be plotted on the HOS scale. Each discrete stage on the scale is defined based on the organization of bone tissue (fibrolamellar vs. more laminar), the degree of infilling and remodeling of vascular canals (no infill, primary osteons, secondary osteons), and the degree of bone remodeling (little remodeling up to dense accumulations of many secondary osteon generations). The scale ranges from HOS 1, representing embryonic bone, to HOS 14, representing individuals with completely remodeled bones (Klein and Sander, 2008; Stein et al., 2010). It is important to remember that an HOS does not indicate numerical age: it is not true, for instance, that an animal exhibiting the bone characteristics of HOS stage 8 is always 13 years old. However, Klein and Sander (2008) demonstrated that, in general, the histological stages exhibited throughout the growth series of an individual taxon correlate closely with body size, suggesting that ontogenetic stage and absolute age are closely related to each other (see plot in Fig. 8.7). These correlations do not necessarily hold true across taxa, however, because one species may become histologically mature earlier or later in its life relative to another species.

Growth curves

If scientists wish to understand the growth rates of dinosaurs or the mechanisms by which some species attained giant or miniature size, the construction of a growth curve is essential. These curves plot the age of individual specimens against some measure of body size, usually either mass or length (Erickson, 2005) (Fig. 8.8). Numerical ages are usually determined by counting growth lines, whereas body mass can be calculated using many of the methods described in Chapter 5. Alternatively, due to the uncertainties of body mass reconstruction, the length of a single standard bone (such as the femur) is often used as a proxy for body size. When

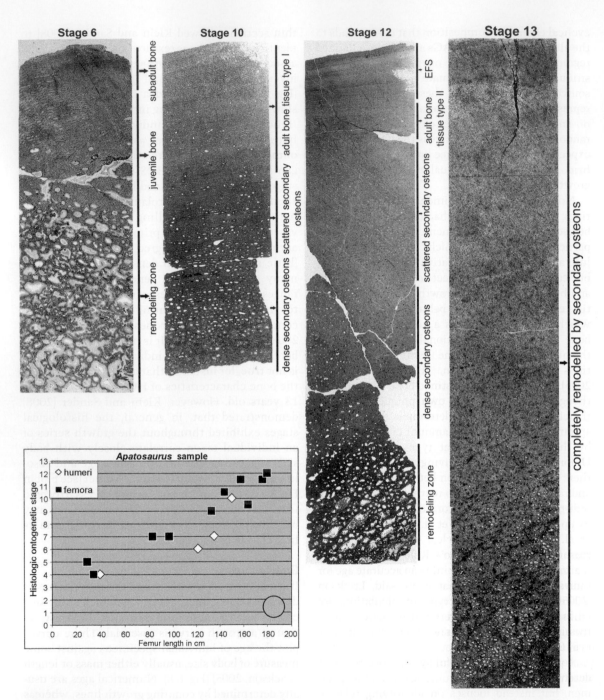

Figure 8.7 Histological ontogenetic stages. Histological thin sections from four specimens of the sauropod *Apatosaurus* showing how the form and extent of remodeling changes during ontogeny. These four sections represent histological ontogenetic stages (HOS) 6, 10, 12, and 13, respectively. The plot at the bottom left shows that body size and HOS are tightly correlated in *Apatosaurus*, which indicates that HOS is a good indication of specimen age (if ontogenetic age is assumed to be roughly correlated with body size). Modified from Klein and Sander (2008). Used with permission from the Paleontological Society.

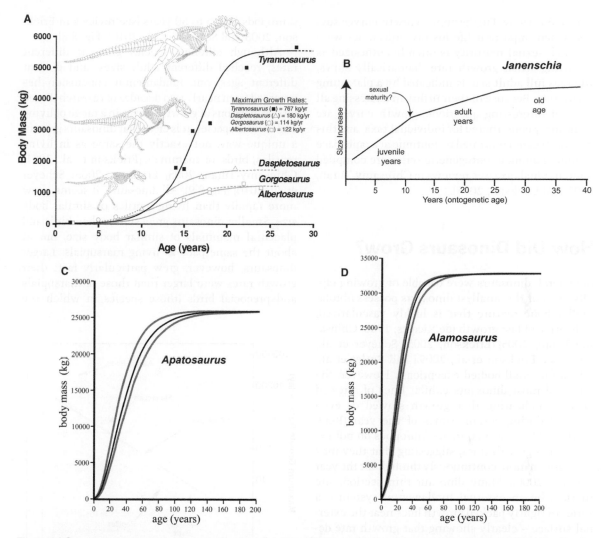

Figure 8.8 Growth curves showing the growth rates of various dinosaur species, constructed by plotting estimates of body size against ontogenetic age. (A) Tyrannosaurid curve showing that *Tyrannosaurus* grew to its enormous size by increasing its growth rate relative to close cousins. Image courtesy of Dr Greg Erickson and modified from Erickson et al. (2004) and Brusatte et al. (2010d). (B) Growth curve for the sauropod *Janenschia* showing major stages in the life history of the animal (juvenile, adult, old age) and the hypothesized onset of sexual maturity when fast juvenile growth rates transition to slower adult rates. Modified from Sander (2000). Used with permission from the Paleontological Society. (C, D) Growth curves for the sauropods *Apatosaurus* and *Alamosaurus* showing the faster growth rate and larger size attained by *Alamosaurus*. Modified from Lehman and Woodward (2008). Used with permission from the Paleontological Society.

constructing growth curves, it is essential to include specimens that span the ontogenetic series of an individual taxon. Measuring the ages and body size of several adults will not divulge much information, but being able to plot juveniles, sub-adults, and adults together will help reveal the overall growth trajectory of an individual species throughout its life. This permits the testing of explicit hypotheses, such as those outlined above, as well as the calculation of growth rates at any point during ontogeny. Furthermore, growth curves suggest when important life history milestones were reached: sexual maturity is often hypothesized as occurring when growth rate dramatically slows, whereas full adult size is indicated by a plateauing asymptote. Because of their utility in addressing all sorts of interesting questions, growth curves are commonly constructed for individual taxa, and this type of research is primed to continue at a rapid pace as more complete ontogenetic series are compiled for individual taxa (see reviews in Chinsamy-Turan, 2005 and Erickson, 2005).

How Did Dinosaurs Grow?

In general, dinosaurs were capable of growing rapidly. All but the smallest dinosaurs possess fibrolamellar bone texture that is highly vascularized, indicative of fast growth (de Ricqlès, 1980; Chinsamy-Turan, 2005; Erickson, 2005; Scheyer et al., 2010; see Erickson et al., 2009a and Butler et al., 2010a for small-bodied exceptions). However, the bones of most dinosaurs exhibit lines of arrested growth, indicating that growth slowed or even stopped during certain times of the year (Reid, 1981, 1997). However, many sauropods do not exhibit such growth lines, suggesting that they may have grown more continuously throughout the year (Sander, 2000). Many dinosaur thin sections are marked by an external fundamental system – a series of tightly packed growth lines near the external surface – clearly showing that growth rate decreased markedly and that any substantial growth of new bony tissue had ceased. This is firm evidence that dinosaurs stopped growing at a certain age and size, as is true of most animals. It is often said that crocodiles and other reptiles grow continuously throughout their lives, and therefore have different growth dynamics from dinosaurs, but this is incorrect. Living reptiles do in fact exhibit a slowdown in growth later in life and attain a stable adult body size (Pough et al., 2003). In dinosaurs, individual longevity was probably related to body mass: small theropods may have only lived for 3–4 years, moderate-sized dinosaurs such as larger theropods and prosauropods for 7–15 years, colossal theropods such as tyrannosaurids for 24–30 years, and

sauropods for up to 50 years (see reviews in Erickson, 2005 and Scheyer et al., 2010) (Fig. 8.8).

Although different species grew at different rates, reached different adult sizes, and died at different ages, one fundamental conclusion has emerged from nearly two decades of research. Based on comparisons of growth curves in a range of living and extinct species, it is clear that dinosaurs grew in a unique way, not exactly the same as in living reptiles, birds, or mammals (Erickson et al., 2001; Chinsamy-Turan, 2005; Erickson, 2005; Scheyer et al., 2010) (Fig. 8.9). All dinosaurs, it seems, grew more rapidly than living reptiles of similar body size. Smaller dinosaurs grew slower than birds and placental mammals of similar body size, but at about the same pace as living marsupials. Larger dinosaurs, however, grew particularly fast: their growth rates were larger than those of marsupials and precocial birds (those species in which the

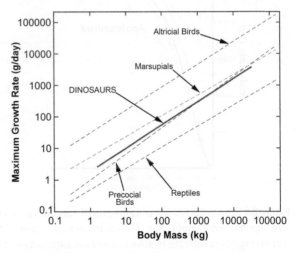

Figure 8.9 Dinosaur growth rates compared with those of living animals: regression line of non-avian dinosaur growth rates vs. body size, compared with regression lines for extant altricial birds (birds born small that need nourishment from parents), precocial birds (birds born well developed and requiring less or no parental care), marsupials, and reptiles. Non-avian dinosaurs grew faster than living reptiles of similar body size but slower than extant altricial birds. Their growth rates were similar to those of living marsupials and precocial birds. Image courtesy of Dr Gregory Erickson and modified from Erickson et al. (2009a).

young are born relatively mature) and about equivalent to those of placentals. No dinosaurs, in truth, grew at the breathtakingly fast rates of living altricial birds, which are born small and immature but develop into full-grown adults in just a couple of months (Erickson et al., 2009a). It appears as if dinosaurs had their own distinctive growth rates, intermediate between those of living reptiles on the one hand, and extant birds and placental mammals on the other (Fig. 8.9). Their closest living analogue may be marsupials, although large dinosaurs grew much faster than living pouched mammals (Erickson et al., 2009a). It must be remembered, therefore, that differences in growth rate between small and large dinosaurs indicate that there was no universal "dinosaur growth rate."

The evolution of dinosaur growth strategies

All dinosaurs were capable of growing fast, as indicated by their fibrolamellar bone textures and, in some cases, growth curves that quantify the speed of development (Fig. 8.8). This raises an important question: is rapid growth unique to dinosaurs among archosaurs, or did it first evolve in basal archosaurs and was merely retained in dinosaurs? Histological studies of other extinct archosaurs indicate that rapid growth was not a dinosaur innovation. Pterosaurs, which are among the closest relatives of dinosaurs, exhibit fast-growing fibrolamellar bone texture almost indistinguishable from that in dinosaurs (de Ricqlès et al., 2000, 2003a; Padian et al., 2004). Even many crocodile-line archosaurs, as well as basal archosauromorphs that are immediately outside the archosaur crown group, possess highly vascular fibrolamellar bone that is indicative of rapid growth (de Ricqlès et al., 2003a, 2008). Living crocodiles, which grow at a slow pace, are therefore unusual among archosaurs, and it is likely that their glacial growth rates are a secondary reversal to more primitive, non-archosaur growth strategies. This being said, extinct crocodile-line archosaurs probably did not grow quite as fast as dinosaurs and pterosaurs. Extinct crurotarsans, even taxa such as *Effigia* that were anatomically convergent on the posture and gait of dinosaurs, seem to have grown rapidly early in life but then switched to slower-growing and less vascular lamellar bone later in ontogeny (de Ricqlès

et al., 2003a, 2008; Nesbitt, 2007). Most dinosaurs, on the other hand, formed fibrolamellar bone throughout their lifetime. These growth differences may have underpinned other biological and functional differences between Triassic crurotarsans and dinosaurs, and potentially could help explain why the latter preferentially survived the end-Triassic extinction and subsequently dominated terrestrial ecosystems.

Ornithischians

Most ornithischians grew at fast speeds, including the basal taxon *Lesothosaurus*, whose bones are composed mostly of highly vascularized fibrolamellar tissue (Knoll et al., 2010). All ceratopsians and ornithopods that have been studied using histological techniques also possess extensive fibrolamellar bone and seem to have grown rapidly (Chinsamy, 1995; Horner et al., 1999, 2000, 2009; Erickson and Tumanova, 2000; Cooper et al., 2008; Woodward et al., 2011). The hadrosaurid *Maiasaura*, in particular, grew at extremely high rates after hatching, continued to develop at high rates as a juvenile and sub-adult, and gradually decreased its growth rate as an adult until full size was reached at 6–8 years (Horner et al., 2000). The high rates of post-hatching growth lend evidence to the hypothesis that juveniles were born in an immature precocial state and depended on caring parents for food and protection (Horner and Makela, 1979; Horner, 1982).

Not all ornithischians grew rapidly. Thyreophorans seem to be characterized by an unusual condition, probably diagnostic for the group, in which growth is exceptionally slow compared with other ornithischians. The basal thyreophoran *Scutellosaurus* exhibits very little fibrolamellar bone, but instead is characterized by poorly vascularized and parallel-fibered bone as is typical of slow-growing crocodiles (Padian et al., 2004). The long bones of a more derived thyreophoran, *Stegosaurus*, indicate that it probably grew faster than *Scutellosaurus*, due to the presence of more extensive fibrolamellar tissue toward the inner bone cortex (reflecting faster growth rates as a juvenile) (Redelstorff and Sander, 2009). However, poorly vascularized and parallel-fibered bone becomes more prevalent toward the outer surface of the bone, showing that growth decreased dramatically during ontogeny. Importantly, when compared with other dinosaurs of similar body size,

Stegosaurus clearly grew more slowly. Whether such slow growth rates characterize other thyreophorans, such as ankylosaurs, remains to be studied.

Sauropodomorphs

All studied sauropodomorphs were capable of rapid growth, based on the prevalence of fibrolamellar tissue and, for some species, the construction of growth curves (Chinsamy-Turan, 2005; Erickson, 2005; Scheyer et al., 2010) (Fig. 8.8B–D). This is true of both basal "prosauropods" (Chinsamy, 1993; Sander and Klein, 2005) and bona fide sauropods (Curry, 1999; Sander, 2000; Sander et al., 2004; Lehman and Woodward, 2008; Sander and Klein, 2008; Woodward and Lehman, 2009). However, the growth of sauropods generally seems to differ from that of most dinosaurs. Growth lines are only rarely observed, which suggests that sauropods deposited their fibrolamellar bone tissue continuously throughout the year (Sander, 2000; Erickson, 2005). This is clear evidence for rapid and sustained growth. Furthermore, comparisons between the growth curves of prosauropods and sauropods indicate that the latter had a much elevated growth rate, which probably enabled sauropods to attain such enormous sizes (Sander et al., 2004) (Fig. 8.10).

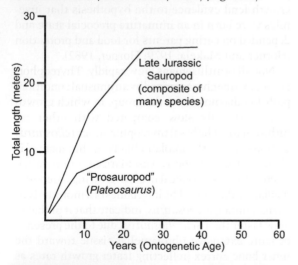

Figure 8.10 Growth rates in basal sauropodomorphs ("prosauropods") compared with those of larger, more derived sauropods. This plot shows that sauropods achieved their enormous sizes by accelerating their growth rate relative to their smaller ancestors. Image modified from data in Sander et al. (2004). Reproduced with permission from Elsevier.

Constructing growth curves for sauropods, as well as determining when species attained full size and ultimately died, is exceptionally difficult because most specimens do not preserve growth lines. Therefore, there is great uncertainty about the numerical ages of sauropod individuals, which led Klein and Sander (2008) to devise their relative system of histological ontogenetic stages (Fig. 8.7). Some specimens, however, do preserve either faint growth rings or so-called "polish lines," which physically stand out in bas relief on thin sections because of compositional differences between different bone layers (Sander, 2000). These lines are not equivalent to lines of arrested growth, but they do seem to represent annual decreases in growth, and therefore they can be counted to determine the age of a specimen (see Sander, 2000 for details). Growth curves based on counting either growth lines or polish lines indicate that sauropods grew very fast, and reached maximum size in only a couple of decades (Fig. 8.8B–D), but different studies disagree on the details. Curry (1999) suggested that *Apatosaurus* grew to full size in as little as 15 years, but another study using slightly different statistical methodology found that full size was attained only after 70 years (Lehman and Woodward, 2008). These results imply an order of magnitude difference in the annual average growth rate, which would be approximately 520 kg per year if Lehman and Woodward (2008) are correct and more than 2000 kg per year if Curry (1999) is correct. Regardless of these differences, however, it is clear that sauropods grew fast and did not take more than a century to slowly beef up to full adult size, as was long assumed (Case, 1978).

Comparisons of growth rates for different sauropods calculated using a single methodology indicate that there was great variability in growth dynamics within the sauropod clade. Lehman and Woodward (2008) found that *Apatosaurus* took 70 years to reach full size, but that the titanosaurs *Janenschia* and *Alamosaurus* attained adult size in 20–30 and 45 years, respectively (Fig. 8.8B,D). Some titanosaurs were therefore larger than *Apatosaurus* but achieved full size much earlier. This is suggestive evidence that different sauropod subclades experimented with different growth rates, in addition to the overall fast rates that were present in all sauropods. In particular, the large-bodied titanosaurs may have been the fastest growing of all sauropods (Lehman and Woodward, 2008; Woodward and Lehman, 2009).

Theropods

Only a few theropod species have been studied in extensive detail using histology and growth curves, largely because the fossils of these carnivores are generally rarer than those of the herbivorous ornithischians and sauropodomorphs. It is difficult, therefore, to compile extensive growth series for individual theropod species, although such data are available for the basal theropod *Syntarsus* (Chinsamy, 1990), the large-bodied *Allosaurus* (Bybee et al., 2006), and the colossal tyrannosaurids (Erickson et al., 2004). These theropods, along with *Troodon* (Varricchio, 1993), all exhibit the highly vascularized fibrolamellar texture typical of fast growth. *Troodon* seems to have required 3–5 years to reach full size, *Syntarsus* 7–8 years, and tyrannosaurids 16–20 years (Fig. 8.8A). No known tyrannosaurid individual is older than 30 years, indicating that most tyrannosaurids would have died before this age (Erickson et al., 2004). The same also seems to be true of *Allosaurus* (Bybee et al., 2006). The aberrant alvarezsaurid *Shuvuuia* has also been studied histologically, and it probably attained full size in about 3 years (Erickson et al., 2001).

Major Questions about Dinosaur Growth

How did some dinosaurs become so large?

Gigantic dinosaurs like *Tyrannosaurus* and *Brachiosaurus* are some of the most awe-inspiring, yet mysterious, creatures to ever live. How exactly were these and other dinosaurs able to attain their enormous sizes? Did they grow faster than other dinosaurs? Did they simply live longer? Did they delay their sexual maturity until later in life, thereby extending the rapid juvenile phase of growth? These questions are only beginning to be answered, but an exceptional study has focused on the evolution of gigantism in tyrannosaurids (Fig. 8.8A). Erickson et al. (2004) constructed growth curves for four genera of Late Cretaceous tyrannosaurids: *Albertosaurus*, *Daspletosaurus*,

Gorgosaurus, and *Tyrannosaurus*. All four tyrannosaurids attained large size as adults, and all four grew at fast speeds, but the 6-tonne *Tyrannosaurus* itself demonstrated an unusual growth trajectory relative to its smaller cousins (which probably grew to adult sizes of about 2 tonnes). Between about 10 and 20 years of age *Tyrannosaurus* experienced an exponential growth phase, in which growth rates were elevated fourfold compared with those in other tyrannosaurids. During this time, *T. rex* grew at the astounding rate of over 2 kg per day. *Tyrannosaurus* therefore achieved its great stature by growing faster than its closest relatives, which is quite unlike the growth strategy employed by super-sized crocodilians, which simply grow for longer than other species (Erickson and Brochu, 1999; see also Horner and Padian, 2004).

Gigantism evolved several times independently within Dinosauria, and the mechanisms by which other theropods, as well as sauropods and ornithopods, became large are not as well studied. Sander et al. (2004) showed that sauropods as a whole had an elevated growth rate compared with "prosauropods" (Fig. 8.10), but exactly how individual sauropod clades developed species-specific instances of colossal size is unknown. This is largely due to the difficulty of constructing growth curves for sauropods, due to the rarity of growth lines in most specimens. Lehman and Woodward (2008) found suggestive evidence that titanosaurs may have grown faster than other sauropods, and perhaps it is no coincidence that several titanosaurs are among the largest land animals to have ever lived. More work is clearly needed on the fine-scale histology and growth dynamics of sauropods in order to assess whether rate acceleration or another strategy (such as delayed onset of maturity) was responsible for the most colossal species.

How did some dinosaurs become so small?

Not all dinosaurs were gigantic. Forgetting for a moment about birds, which are the smallest dinosaurs of all, it has been suggested that several island-dwelling dinosaurs were dwarfed relative to their closest cousins and mainland contemporaries (Fig. 8.11). Such dwarfing is thought to reflect the so-called "island rule," in which

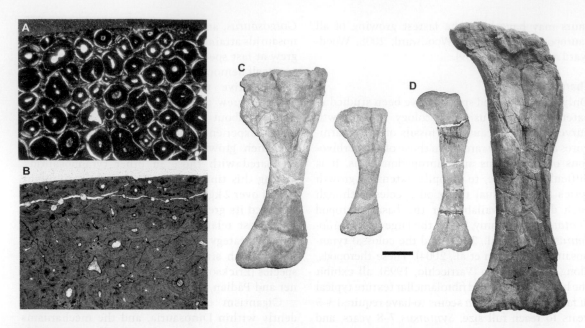

Figure 8.11 Dwarf dinosaurs. (A, B) Histological thin sections of the tibia (450 mm long) of the dwarf sauropod *Magyarosaurus* from the Late Cretaceous of Romania. Note the intense remodeling (numerous cross-cutting secondary osteons), which is usually seen in very mature sauropod individuals that are much larger than this individual of *Magyarosaurus*. This suggests that *Magyarosaurus* stopped growing at a much smaller size than its more normal-sized close relatives. (C, D) Comparison of the humerus (C) and femur (D) of *Magyarosaurus* with a more traditionally sized large sauropod (*Ampelosaurus*). Scale bar equals 10 cm. Images (A) and (B) courtesy of Koen Stein; images (C) and (D) courtesy of Dr Zoltán Csiki.

island taxa are morphologically modified because of the limited resources of their reduced habitat (Benton et al., 2010). Histological studies have firmly demonstrated that many supposed island dwarves really are small adults and not simply juveniles, because these specimens possess external fundamental systems and/or extensive remodeling indicative of advanced age (Sander et al., 2006; Stein et al., 2010; Company, 2011) (Fig. 8.11A,B). *Europasaurus*, a sauropod from the Late Jurassic of Germany that does not seem to have grown to more than 6 m in length, is known from 9-year-old specimens with an external fundamental system and heavy remodeling, both of which occur at age 15 or greater in close relatives (Sander et al., 2006). The presence of growth lines sets this taxon apart from most other sauropods, and is additional evidence that it grew slowly. Another dwarf sauropod, *Magyarosaurus* from the Late Cretaceous of Romania, exhibits the exten-

sively remodeled bones characteristic of Klein and Sander's (2008) oldest histological ontogenetic stages, yet these bones are less than half the size of those of close relatives (Stein et al., 2010) (Fig. 8.11A,B). Furthermore, the internal bone texture is dominated by slow-growing laminar tissue, not the faster-growing fibrolamellar bone characteristic of most dinosaurs, indicating that *Magyarosaurus* grew slower than its close relatives. A similarly decreased growth rate seems to have been present in the small titanosaur *Lirainosaurus* from the Late Cretaceous of Spain: it exhibits highly organized bone tissue and numerous lines of arrested growth, both of which are rare in fast-growing sauropods (Company, 2011). Therefore, it seems as if dwarfing was common among sauropod dinosaurs, and perhaps other dinosaurs, that lived on islands, and was primarily achieved by decreasing growth rates relative to closely related species.

At what age did dinosaurs become sexually mature?

Living crocodiles and lizards become sexually mature (i.e., capable of reproducing) long before they reach full adult size, which is understandable because their growth strategies permit relatively slow continuous growth for many years. Birds, on the other hand, grow so rapidly during their first year of life that they become sexually mature only after full size is reached. Which condition characterized non-avian dinosaurs? Did the avian condition originate among non-avian dinosaurs, or only after the origin of birds? It has sometimes been suggested that the age at which a dinosaur became sexually mature can be estimated by identifying the point on a growth curve where fast juvenile growth rates transition to markedly slower adult rates, because this is thought to reflect when primary energy allocation shifted from growth to reproduction (Sander, 2000; Erickson, 2005) (Fig. 8.8B). However, this indirect method is not always convincing, and it would be far more compelling if some direct indicators of sexual maturity could be identified in the dinosaur fossil record.

There are two unequivocal signs that an individual dinosaur specimen belonged to a sexually mature animal: either it possesses medullary bone, the ephemeral tissue used in living birds to shell eggs, or it is found brooding a nest of eggs. Erickson et al. (2007) studied the histology of seven theropod specimens found associated with nests, whereas Lee and Werning (2008) examined the histology of three specimens with medullary bone (one each of the theropods *Allosaurus* and *Tyrannosaurus* and the ornithopod *Tenontosaurus*). All these specimens could be aged using growth lines, and because each of these species is represented by growth series, these unequivocally sexually mature individuals could be placed on a growth curve to assess whether they were fully grown or not. Some of these reproductively capable dinosaur specimens, including all three specimens with medullary bone and one oviraptorid and one deinonychosaur brooding a nest, were found to be much younger full-grown adults. This is prime evidence that sexual maturity occurred prior to the attainment of anatomical maturity, similar to the condition in living reptiles (indeterminate growth) but different from that in birds (determinate growth). Even the most derived

bird-like theropods do not possess the avian condition, which must have evolved at some point after the origin of birds.

Were some dinosaurs able to alter their growth rates?

There are many dinosaurs whose growth rates clearly slowed during ontogeny, such that adults grew slower than juveniles (e.g., Varricchio, 1993; Erickson et al., 2004; Erickson, 2005). However, these dinosaurs seem to follow a constrained growth trajectory, where there was little variation in growth rate among individuals of a similar age. This is true of most living endothermic vertebrates (birds and mammals), but not necessarily true of reptiles, which can alter their growth rates depending on environmental factors such as changes in climate or the availability of food. This growth strategy is known as "developmental plasticity" (Smith-Gill, 1983). It seems as if at least one dinosaur, the "prosauropod" *Plateosaurus*, was capable of this reptilian tactic. Sander and Klein (2005) found that an external fundamental system, which is an unequivocal indicator of adult size and growth cessation, was present in a number of *Plateosaurus* specimens of varying sizes ranging from 5 to 10 m in body length. Furthermore, growth line counts indicated that these fully grown animals ranged from 12 to 26 years old. This is intriguing evidence that *Plateosaurus* was able to vary its growth rate, and the age at which it attained adult size, and therefore was developmentally plastic. Intriguingly, a close "prosauropod" relative, *Massospondylus*, was evidently not capable of varying its growth rate, and followed the narrow growth trajectory common to dinosaurs (Chinsamy, 1993; Erickson et al., 2001). Alternatively, however, it may be that *Massospondylus* was capable of developmental plasticity but did not utilize this strategy because it was living in more stable environmental conditions than *Plateosaurus*.

Sander and Klein (2005) suggest that *Plateosaurus* may have evolved developmental plasticity in order to deal with fluctuating environmental conditions, but this hypothesis remains to be critically tested. They also note that developmental plasticity in living animals is correlated with low metabolic rates and behavioral (rather than internal) regulation of body temperature, suggesting that

Plateosaurus may have been ectothermic. However, *Plateosaurus* does possess fast-growing fibrolamellar bone and no other dinosaurs are known to have exhibited developmental plasticity. This suggested to Sander and Klein (2005) that *Plateosaurus* may represent a transitional or initial stage in early dinosaur evolution, in which faster growth and higher metabolism were becoming decoupled from more primitive, reptile-like ectothermic physiology.

How did high-latitude dinosaurs grow?

As discussed in more detail later in this chapter, some dinosaurs lived in high-latitude polar regions during the Mesozoic. These regions were not as cold as today, but they would still have had a substantially cooler and more seasonal climate than the rest of the Mesozoic world. Many dinosaurs inhabited these polar regions, including various species of theropods and ornithopods, whose fossils are commonly found in Australia, Alaska, and other areas that were positioned in or near the polar belt in the Cretaceous. Woodward et al. (2011) recently presented a histological examination of ornithopod and theropod fossils from Cretaceous polar Australia, and showed that their bone texture (fibrolamellar bone early in ontogeny, more ordered parallel-fibered bone later in ontogeny, LAGs throughout) was similar to the textures seen in low-latitude relatives. This indicates that there may have been nothing inherently special about the physiology and growth dynamics of polar dinosaurs, but rather that the generalized dinosaurian condition of rapid growth (especially early in ontogeny) may have enabled them to successfully inhabit even extreme climates.

The evolution of "avian-style" growth strategies

To any birdwatcher, one of the rarest and most desired sightings is that of a baby bird. On the face of it, observing a still-growing juvenile in nature doesn't seem like anything too spectacular. After all, it's not that uncommon to see young mammals or lizards. Birds, however, are another story: even those of us that live in large cities rarely, if ever, see still-growing babies among the throngs of sidewalk pigeons. Why is this so? Living birds are characterized by a remarkable growth strategy unique among vertebrates: their growth rates are so explosive that the transition from hatchling to adult usually takes only a few weeks, or in some cases, even a handful of days. In other words, baby birds grow into adults so quickly that the window of opportunity to observe a still-growing juvenile is incredibly small.

Because birds evolved from dinosaurs, it is tempting to ask whether the unusual "avian-style" growth strategy originated in dinosaurs or evolved long after the dinosaur–bird evolutionary transition. A growing body of evidence firmly indicates that the latter is the case: the most bird-like theropod dinosaurs do not exhibit remarkably fast growth rates (Erickson et al., 2001, 2009a) and numerous species of bona fide Mesozoic birds, which are among the most primitive members of the avian clade, seem to have grown very slowly (Chinsamy et al., 1994, 1995, 1998; Chinsamy and Elzanowski, 2001; Chinsamy, 2002; Starck and Chinsamy, 2002; de Ricqlès et al., 2003b; Chinsamy-Turan, 2005; Erickson et al., 2009a). Histological studies show that several Mesozoic birds possess more organized and less vascular bone textures indicative of slow growth (compared with living birds) and exhibit growth lines, which are not deposited during the rapid growth cycle of living birds (Chinsamy et al., 1994; de Ricqlès et al., 2003b; Chinsamy-Turan, 2005). Perhaps most unexpected of all, histological analysis of the oldest known bird, *Archaeopteryx*, found this iconic taxon to possess highly organized, parallel-fibered bone with very little vascularity, a texture commonly seen in living reptiles but completely unknown among similarly sized modern birds (Erickson et al., 2009a) (Fig. 8.12A–C). However, similar tissues are present in some of the closest dinosaurian relatives to birds, which happen to be among the smallest known non-avian dinosaurs. It seems, therefore, that slow-growing bone tissue in early birds and their closest relatives was associated with small size, and probably indicates that these animals attained their miniature stature by growing more slowly than larger relatives (Erickson et al., 2009a). *Archaeopteryx* would not have exploded to full adult size in a few weeks, but rather probably took 2–3 years to fully mature (Fig. 8.12D). Similar slow-growing bone tissues, and similar inferred growth rates, are also present in the basal birds *Jeholornis* and *Sapeornis*, persuasively demonstrating that the growth strategy of *Archaeopteryx* was not aberrant but rather characteristic of primitive birds (Erickson et al., 2009a).

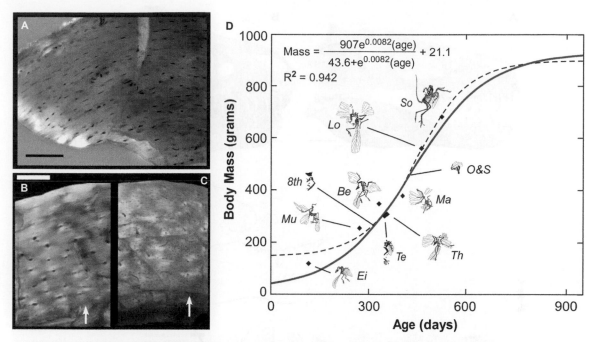

Figure 8.12 Bone histology and growth rates of primitive birds. (A) Histological thin section of the femur of *Archaeopteryx* showing parallel-fibered bone with poor vascularization. (B, C) Thin section of the femur of the primitive birds *Jeholornis* (B) and *Sapeornis* (C), both also exhibiting parallel-fibered bone, low vascularization, and lines of arrested growth (best example denoted with arrow). (D) Growth curve of *Archaeopteryx*, based on an equation (derived from the bone tissue formation rates of living birds) that predicts age from body size, based on the noted bone histology of *Archaeopteryx*. *Archaeopteryx* probably took more than 2 years to mature to adult size, and did not achieve full size within a few days or weeks as in most living birds. Abbreviations next to each specimen illustration denote the name (usually the locality) of each specimen and the equation at top describes the shape of the growth curve (see Erickson et al., 2009a for full details). All images courtesy of Dr Gregory Erickson and modified from Erickson et al. (2009a).

In sum, the oldest and most primitive birds grew in the same style as their closest dinosaurian relatives, but quite differently from living birds. *Archaeopteryx* and other primitive birds grew faster than living reptiles and had indeterminate growth, as is characteristic of dinosaurs as a whole, but grew approximately three times slower than living precocial birds (Erickson et al., 2009a). It is unclear exactly when the hyperelevated growth rates of living birds originated, but more derived Mesozoic birds such as *Hesperornis* and *Ichthyornis* have more randomly arrayed and highly vascularized bone textures indicative of faster growth, and seem to have grown continuously until adult size due to their lack of growth lines (Chinsamy et al., 1998). Although living birds employ their own distinctive growth strategy, this was some-

thing that evolved long after the origin of the avian clade itself.

Anatomical and functional changes during growth

The body sizes and skeletal proportions of dinosaurs would have changed, in some cases dramatically, as individuals matured from embryos to adults. Some dinosaurs are represented by sufficiently large fossil samples that ontogenetic growth series – an ordered sequence of specimens representing increasing age and growth stage – can be compiled (Dodson, 1975; Sampson et al., 1997; Carr, 1999; Carr and Williamson, 2004; Horner and Goodwin, 2006, 2009; Evans, 2010; Scannella and Horner, 2010) (Figs 8.13–8.17; see Plate 3). By looking at how the size, shape, and proportions of dinosaur skeletons changed across a growth series, paleontologists can gain a firm

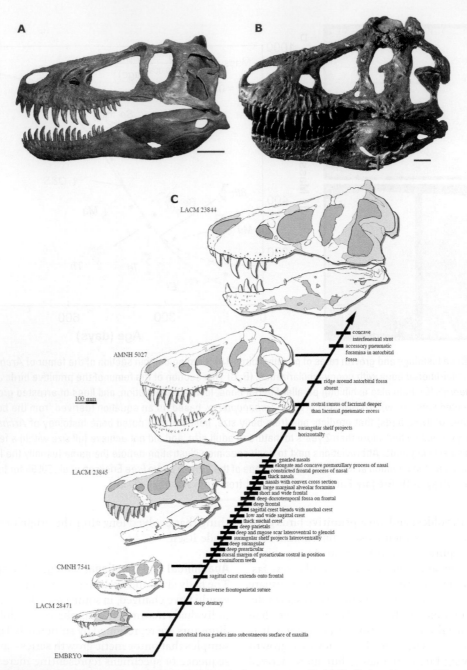

Figure 8.13 Changes in tyrannosaurid skull size and shape during ontogeny. (A, B) Skull of an 11-year-old juvenile (A) and adult (B) *Tyrannosaurus rex*, scaled to the same length, showing how the skull becomes proportionally shorter anteroposteriorly and deeper dorsoventrally during ontogeny. Scale bars for both equal 10 cm. (C) Ontogenetic growth trajectory of *Tyrannosaurus*, depicting how skull shape and size change during ontogeny, with a list of discrete changes in skull morphology that occur between different growth stages. (A) Courtesy of Scott Williams, Burpee Museum of Natural History; (B) courtesy of American Museum of Natural History Photo Archives (#2752), *Tyrannosaurus* skull as mounted in the old hall; (C) modified from Carr and Williamson (2004), and reproduced with permission.

appreciation of how form and function changed as an individual dinosaur matured.

Growth series clearly show that some species experienced profound anatomical changes during ontogeny, which in many cases probably related to changes in locomotory, feeding, and reproductive behaviors (Figs 8.13–8.17). In tyrannosaurids like *Tyrannosaurus* and *Albertosaurus*, the elongate and gracile skulls of juveniles became deeper, more robust, and more heavily ornamented by various cranial bosses in adults (Carr, 1999; Carr and Williamson, 2004) (Fig. 8.13). Furthermore, individual sutures became deeper, more rugose, and more interlocking during ontogeny, and the teeth became larger and more robust. Many of these adult features are integral components of the novel tyrannosaurid "bone-crunching" puncture–pull feeding abilities, suggesting that this technique was only employed by larger and older individuals. Not only that, but the weaker-skulled juveniles have much more elongate and gracile hindlimbs, which would have imparted greater speed than was possible for adults, and longer forelimbs, which unlike the stubby arms of adults may have been useful in grasping prey (Currie, 2003). Together, these major ontogenetic changes indicate a behavioral and dietary shift during ontogeny: juveniles were better suited for capturing smaller and faster prey, whereas adults could crunch through the bones of their prey but probably could not run down their dinner with such ease. A similar ontogenetic shift may have occurred in some hadrosaurids: in juveniles the hindlimbs are more robust and the forelimbs more gracile than those of adults, suggesting that juveniles may have walked primarily bipedally and adults quadrupedally (Dilkes, 2001; Kilbourne and Makovicky, 2010). Finally, embryos of the "prosauropod" *Massospondylus* indicate that juveniles walked quadrupedally, due to their equally sized forelimbs and hindlimbs, which differs from the adult condition in which proportionally longer hindlimbs is thought to reflect at least facultative bipedal posture (Reisz et al., 2005, 2010).

In some of the most dramatic ontogenetic changes of any dinosaurs, the ornately crested skulls of hadrosaurs and ceratopsians were progressively elaborated during growth, which is suggestive evidence that the ostentatious adult skulls were primarily used for sexual display (which

would have been unnecessary for juveniles) (Dodson, 1975; Sampson et al., 1997; Horner and Goodwin, 2006; Evans, 2010) (Figs 8.14, 8.16 and 8.17). In particular, the characteristic ceratopsian *Triceratops* underwent a typical series of cranial changes, as outlined in an excellent series of papers by Horner and Goodwin (2006) and Goodwin and Horner (2008) (Fig. 8.16). The postorbital horn core is straight in the youngest individuals, curves posteriorly in older individuals, and finally reorients itself to a more erect, but slightly anteriorly curving, position in adults. The juvenile frill is encircled by a series of triangular ossifications called epoccipitals, which fuse onto the frill later in ontogeny. Finally, sutures between individual bones, such as the opposing nasals, become fused in adults, and the postorbital horn core becomes progressively hollower during growth.

One of the great revelations emerging from the study of ontogenetic growth series is that many unusual dinosaur specimens once thought to represent distinct species are merely juveniles of well-established species. Carr (1999) and Carr and Williamson (2004) persuasively showed how a small and elongate tyrannosaurid skull, once thought to represent a distinct "pygmy tyrannosaur" called *Nanotyrannus*, is a juvenile. Because it shares so many features with the contemporary *Tyrannosaurus rex*, and because *T. rex* is the only large theropod currently known from the well-sampled Late Cretaceous of North America, it is parsimonious to conclude that "*Nanotyrannus*" is merely a juvenile *T. rex* (Fig. 8.13). More recently, Horner and Goodwin (2009) showed that two supposedly distinct genera of Late Cretaceous pachycephalosaurids, *Dracorex* and *Stygimoloch*, are only known from specimens that preserve characteristic juvenile features such as spongy fast-growing bone texture, open sutures in the skulls, and relatively flat cranial domes. Another genus from the same Late Cretaceous rock units, *Pachycephalosaurus*, is only known from mature specimens in which the skull is highly domed, individual cranial bones are indistinguishably fused together, and bone texture is denser and less vascularized. They concluded that all three genera actually represent different growth stages of one genus, which is referred to as *Pachycephalosaurus* because it was the first of the three given a scientific name (Fig. 8.15). However, sample sizes for Late

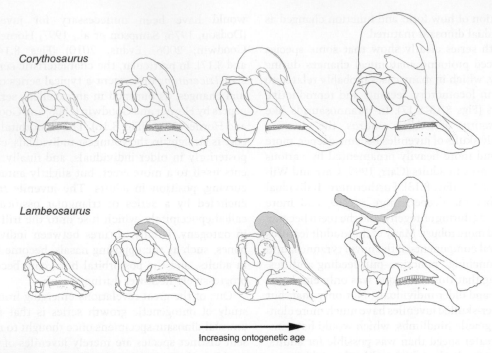

Corythosaurus

Lambeosaurus

Increasing ontogenetic age

Figure 8.14 Changes in hadrosaurid skull size and shape during ontogeny. Growth series of two common hadrosaurid dinosaurs from the Late Cretaceous of North America, *Corythosaurus* and *Lambeosaurus*, showing how skull morphology changes as an individual matures. Images courtesy of Dr David Evans (see Evans, 2010 for more details).

Cretaceous pachycephalosaurs are so low that it is possible that the known examples of *Dracorex* and *Stygimoloch* are juveniles of another taxon, not specifically *Pachycephalosaurus*. Only new discoveries can test this possibility.

The biggest bombshell of all, however, is certainly Scannella and Horner's (2010) provocative (and some would say controversial) hypothesis that ceratopsian specimens referred to as *Torosaurus* are actually the oldest and most mature adults of *Triceratops* (Fig. 8.17). *Torosaurus*, which is one of the rarest discoveries in the well-sampled Late Cretaceous rocks of North America, has long been differentiated from *Triceratops* by its proportionally longer squamosals and large fenestrae that pierce the frill. Scannella and Horner (2010) identified unequivocal *Triceratops* specimens whose frills appear to be thinning into the distinctive fenestrae of *Torosaurus*, and used a growth series to argue that the squamosals became progressively larger as *Triceratops* matured (Fig. 8.17). Circumstantial evidence also supports their claims: *Triceratops*, which is one

of the most common fossil discoveries in the Late Cretaceous of North America, is not represented by any fossil specimens pertaining to unequivocally old and fully mature individuals, whereas juvenile individuals of the rare *Torosaurus* are also unknown. That being said, their claim that these two characteristic genera are synonymous is bold, needs to be corroborated with additional evidence (including that from bone histology and growth rings), and is still generating debate among specialists (see Farke, 2011 for a dissenting view). Compiling ontogenetic series, and testing whether supposed distinct species are real or merely different growth stages of a single taxon, is currently one of the most exciting areas of research in taxonomic dinosaur paleontology, and future work will likely reveal many surprises.

Dinosaur Physiology

Were dinosaurs active and energetic creatures, with elevated activity levels and internally con-

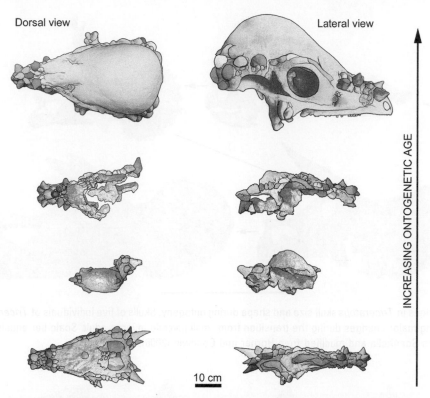

Dorsal view

Lateral view

INCREASING ONTOGENETIC AGE

10 cm

Figure 8.15 Changes in pachycephalosaurid skull size and shape during ontogeny. Hypothesized growth series for the Late Cretaceous genus *Pachycephalosaurus* (from Horner and Goodwin, 2009). The series illustrates how a small flat-headed form (previously given its own genus name, *Dracorex*) grows into a larger, more heavily domed form (previously thought to be the only form of *Pachycephalosaurus*). Images courtesy of Dr John Horner.

trolled body temperatures like living "warm-blooded" birds and mammals? On the contrary, were dinosaurs more akin to "cold-blooded" reptiles, which regulate their body temperatures behaviorally (i.e., by basking in the sun) and have much lower metabolic rates? Or, may dinosaurs have been wholly different in their metabolism, physiology, and temperature control? This question is one of the fundamental riddles of contemporary dinosaur research and has generated seemingly endless debate and speculation over the past several decades.

It was long thought that dinosaurs were nothing more than lethargic overgrown lizards, whose sluggish metabolism required behavioral control of body temperature. This conception of dinosaurs changed dramatically during the "Dinosaur Renaissance" of the 1970s, catalyzed by Robert

Bakker's (1972) startling arguments that dinosaurs behaved and grew much more like living warm-blooded animals than cold-blooded lizards. During the ensuing four decades, debate has oscillated between these two positions, and many workers have even suggested that dinosaurs had a physiology unlike that of any living animal. Currently, this debate has grown somewhat tedious, like a television program that has far outlived the originality and excitement that made it interesting in the first place. This is certainly not to say, however, that understanding dinosaur metabolism and physiology is not important; in fact, it is essential for understanding how dinosaurs grew, moved, and functioned as living animals. But, by now, arguments and evidence have been presented ad nauseam for cold-blood metabolisms, warm-blooded physiologies, and everything in between. Frustrat-

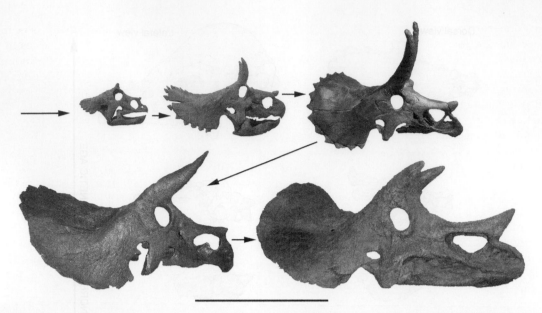

Figure 8.16 Changes in *Triceratops* skull size and shape during ontogeny. Skulls of five individuals of *Triceratops*, figured to scale, showing major changes during the transition from small juvenile to large adult. Scale bar equals 1 m. Images courtesy of John Scannella and modified from Horner and Goodwin (2006).

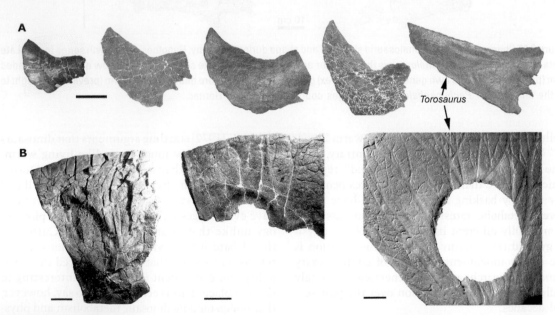

Figure 8.17 Osteological evidence that *Triceratops* and *Torosaurus* are different ontogenetic stages of the same species: (A) squamosals of *Triceratops* (first four specimens) and *Torosaurus* (final specimen, denoted with label) in right lateral view showing the general elongation of the squamosal during ontogeny; (B) parietals of *Triceratops* (first two specimens) and *Torosaurus* (final specimen, denoted with label) showing the sequential development of a fossa and later a perforated fenestra. Scale bar in (A) equals 30 cm; scale bars in (B) equal 10 cm. Modified from Scannella and Horner (2010), and reproduced with permission.

ingly, different scientists often examine the exact same evidence and come to different conclusions (for example, Chinsamy and Hillenius, 2004 versus Padian and Horner, 2004).

Throughout this debate one fact is inescapable: studying dinosaur physiology is not easy. Dinosaurs cannot be observed directly, and their body temperatures, activity levels, and energy expenditures cannot be measured. Not only that, but dinosaur fossils rarely, if ever, preserve unequivocal indicators of a specific metabolic strategy. All arguments are therefore based on interpretation of bone textures, growth rates, body posture, chemical composition of bones, theoretical mathematical arguments, and other indirect sources. And finally, it is not certain, or even likely that all dinosaurs had similar physiologies. Mesozoic dinosaurs ranged in mass from less than a kilogram to several tonnes, lived in a spectrum of different environments, and behaved in different ways. Some may have had fast metabolisms powered by internal temperature control, whereas others may have been more sluggish and varied their activity levels depending on the temperature of their environment.

Perhaps the greatest problem of all, however, is that links between body temperature control and metabolic activity are not always straightforward, even in living animals. At this point, some terminology is useful. Endotherms regulate their body temperature internally, meaning that they can maintain a constant temperature regardless of environmental conditions, whereas the temperature of an ectotherm relies on the temperature of the external environment. Homeotherms have constant internal body temperatures, whereas heterotherms (also known as poikilotherms) have varying body temperatures. Finally, tachymetabolic animals have high metabolic rates, whereas bradymetabolic organisms have slower metabolisms. Oftentimes endothermy, homeothermy, and tachymetaboly are associated with each other: an organism with internal temperature control often has a constant body temperature and a fast metabolism. However, this is not always the case: ectothermic animals can attain high metabolic rates if they live in a particularly warm environment, or can attain homeothermy if the temperature of their environment is constant or their size

is so large that they lose heat slowly (so-called "gigantothermy"). These subtle differences must be remembered when studying the physiology of extinct animals. Evidence for high metabolic rates may not imply endothermic temperature control. Similarly, evidence that an organism was ectothermic may not imply that it was sluggish and incapable of fast speeds and high activity. What, then, should researchers actually look for when trying to understand dinosaur metabolism? It is exceedingly difficult to find fossil evidence that an organism employed internal or external temperature control, so the focus is usually on determining whether dinosaurs had constant or fluctuating body temperatures and whether they were capable or incapable of high metabolic rates.

Numerous reviews have been written on the subject of dinosaur physiology (Farlow, 1990; Farlow et al., 1995b; Chinsamy and Hillenius, 2004; Hillenius and Ruben, 2004; Padian and Horner, 2004; Chinsamy-Turan, 2005). Together, these papers provide a nice summary of the history of the debate, specific studies that have addressed the physiological strategies of certain dinosaurs, and problematic questions that still need resolution. This information will not be rehashed in tiresome detail here; rather the following section will focus on the general evidence used to infer dinosaur physiology and emerging data that has come to light in recent years. Careful consideration of the available evidence does not yield any firm conclusions that hold for all dinosaurs – it is not possible, for instance, to say that all dinosaurs were tachymetabolic homeothermic endotherms – but several provocative lines of evidence suggest some general conclusions. First, dinosaurs clearly had heightened metabolisms compared with living reptiles, but whether these were on the same level as living birds and mammals, and whether they were powered by fully endothermic temperature control, is difficult to assess. Second, living birds have an active endothermic metabolism, so this specific physiological strategy must have evolved somewhere along the theropod–bird evolution continuum, perhaps in Mesozoic dinosaurs. It is currently unclear exactly when avian-style endothermy originated, but several features associated with high metabolism in living birds (feathers, unidirectional lungs, etc.) evolved in a piecemeal fashion in extinct dinosaurs.

Bone texture and microstructure

When examined in thin section, dinosaur bone is usually composed of highly vascular fibrolamellar tissue, which in living animals is produced during periods of rapid growth (Chinsamy-Turan, 2005) (Fig. 8.5). Furthermore, dinosaur bones often show evidence for extensive remodeling, in the form of densely packed secondary osteons characteristic of Haversian bone. Therefore, it is an inescapable conclusion that most, if not all, dinosaurs grew very fast and remodeled their bones at rapid rates throughout life. Is such rapid growth an unequivocal indicator of tachymetabolic metabolism or endothermic control of body temperature? The answer is not straightforward.

Because highly vascularized and extensively remodeled fibrolamellar bone is characteristic of living endothermic and tachymetabolic mammals, de Ricqlès (1980) hypothesized that these features are suggestive of high metabolic rates and internal temperature control in dinosaurs. However, Reid (1984, 1987) cautioned against this interpretation, noting that there is no distinctive "endothermic" or "exothermic" bone textures in living vertebrates. Indeed, high vascularization, fibrolamellar bone, and Haversian remodeling is not present only in living endotherms, but also in living ectotherms such as crocodiles. Clearly, these tissues are not restricted to endothermic or tachymetabolic animals and are therefore not an unequivocal indicator of high metabolic rates or internal temperature control (Chinsamy-Turan, 2005). Furthermore, unlike most living mammals and birds, dinosaur bones are usually punctuated by growth lines, which indicate a slowdown in growth during the year (Fig. 8.6). Living endothermic and tachymetabolic vertebrates are usually capable of growing throughout the year due to their high metabolism and body temperatures that are not depressed by seasonal (or other) environmental changes. Histological evidence may therefore not be a strong indicator that dinosaurs were endothermic or even had high metabolisms. On the contrary, some aspects of bone histology may suggest that dinosaurs grew more like living reptiles, whose bones are packed with growth lines.

That being said, although fibrolamellar bone is present in non-endotherms, and although dinosaurs exhibit growth lines, these arguments do not falsify the hypothesis that some dinosaurs were endothermic and/or had high metabolisms. As explained by Padian and Horner (2004), fibrolamellar bone is abnormal in living crocodiles, whereas it is present in almost all dinosaurs, regardless of body size, ecological habits, age, or environment. When fibrolamellar bone is produced by living reptiles, it is almost always in juveniles, which grow more rapidly than adults. Dinosaurs, on the other hand, continuously produce fibrolamellar bone until late in ontogeny, which is a persuasive indication that their physiology was different from that of living crocodiles and other fibrolamellar bone-producing ectotherms. Second, although growth lines do indicate a slowdown or cessation of growth, this in itself is not a telltale sign of slow metabolism. Some living mammals are known to produce growth lines: these do not reflect an inherent inability to grow year-round, but rather indicate that the animal simply did not grow continuously, probably because of seasonal shortages of food or extreme environmental fluctuations (Horner et al., 1999; Padian and Horner, 2004). Furthermore, not all dinosaurs possess growth lines: some species seem to lack growth lines altogether, whereas some species exhibit some bones with growth lines and others without (Chinsamy-Turan, 2005). The reason for this variability is not well understood but, at the very minimum, the ability of some dinosaurs to grow continuously falsifies the argument that dinosaurs were incapable of sustained growth as in living mammals and birds, and demonstrates that some dinosaurs were capable of a continuous growth strategy that is unknown in living ectotherms.

What, then, does histology conclusively reveal about dinosaur physiology? There is no doubt that fibrolamellar bone is produced by fast growth rates – everyone working on vertebrate bone histology agrees on this point (Chinsamy-Turan, 2005). The fact that this bone texture is so pervasive in dinosaurs, and was still deposited by older individuals that were no longer growing extremely fast, is unequivocal evidence that dinosaurs had rapid growth rates for long portions of their life. This is also borne out by the construction of growth curves, which often reveal periods of remarkably fast exponential growth as dinosaurs were maturing into adults (Erickson et al., 2001, 2004) (Fig. 8.8). In *Tyrannosaurus*, for example, individuals added an astounding 2 kg of mass per day during their exponential phase, which may have lasted up to 10 years. This type of growth differs from the more languid, prolonged growth strategy of living ecotherms and tachymetabolic organisms, and is more similar to the ways in which living birds and mammals grow. From a visceral perspective, it is inconceivable how an animal with a slow metabolism could grow at sustained rates of several kilograms per day and beef up to full size of 5 tonnes or more in less than 20 years. Whether *Tyrannosaurus* and other dinosaurs fueled their rapid metabolisms through endothermic temperature control, as opposed to another means, is open to debate. However, it is also difficult to imagine how a hatchling tyrannosaurid could power such a fast metabolism by anything but internal temperature control. There is simply no living animal that grows this fast, for such a long amount of time, which is not endothermic. In sum, if these dinosaurs did not control their temperature internally, they must have boasted truly spectacular and bizarre thermoregulatory strategies unlike anything alive today.

Posture and locomotion

Because of their higher metabolisms, living endotherms are able to sustain high levels of activity, such as longer runs, compared to ectotherms. In his landmark 1972 paper, Robert Bakker argued that dinosaurs were well suited for a sustained active lifestyle. One of his primary arguments was that the upright posture of dinosaurs, as well as their inferred ability to run at high speeds, made them more comparable to birds and mammals, which are capable of sustained activity levels, and quite unlike sprawling and slow-moving reptiles. This conclusion is borne out by the more recent study of Pontzer et al. (2009), which uses the latest in computer modeling techniques to estimate the metabolic rates required for various dinosaurs to walk or run at certain speeds (Fig. 8.18). These authors constructed digital models of several dinosaurs to measure two important parameters: the length of the hindlimb during locomotion and the estimated muscle volume needed to propel the body at a given speed. Importantly, like many modeling approaches, this study utilized a range of sensitivity analyses in order to control for uncertainties in fossil preservation and reconstruction, which result in varying measures of hindlimb length and muscle mass. Once limb length and muscle volume were calculated, they were inserted into equations that,

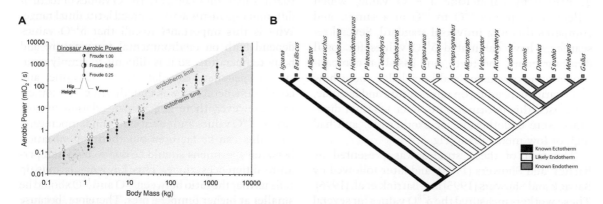

Figure 8.18 Dinosaur biomechanics and oxygen consumption. (A) Plot of oxygen consumption vs. body size showing that the estimated oxygen consumption rates of dinosaurs above 20 kg in mass fall into the range of oxygen consumption values known for living endotherms (and higher than those in living ectotherms). (B) Phylogeny showing the predicted endothermic physiology of dinosaurs, based on the oxygen consumption biomechanical analysis of Pontzer et al. (2009). Images courtesy of Dr John Hutchinson and modified from Pontzer et al. (2009).

in living animals, can confidently predict the amount of oxygen consumption required for an animal of a certain size moving at a certain speed. The results were fascinating: during both walking and running the oxygen consumption, and thus metabolic rates, required of the large-bodied dinosaurs (> 20 kg) were within the range of values for living endotherms, and far above the threshold for living ectotherms (Fig. 8.18A). In other words, these dinosaurs had such great energy demands that only an endothermic tachymetabolic lifestyle could fuel their metabolism, at least if living species are a good comparative guide. Smaller dinosaurs required less energy to move, even at higher speeds, but they also seem to have been more like living endotherms than ectotherms.

Oxygen isotopes

The chemical composition of dinosaur bones and teeth may give insight into the physiology of the living animal that produced them. Bone is largely composed of the mineral hydroxyapatite, which is a compound of calcium phosphate. Phosphate is further composed of phosphorus and oxygen, the latter of which can be in the form of several isotopes (atoms with different numbers of neutrons). The two most common oxygen isotopes are ^{16}O and ^{18}O, and the ratio of these in a bone depends on both the temperature of the environment and the body temperature of the organism. These ratios can be quantified by calculating a $\delta^{18}O$ value, which reflects the ratio of ^{18}O to ^{16}O in a sample and compares this to a universal standard. Several researchers have calculated $\delta^{18}O$ values for fossil dinosaur bones and teeth, and based on comparisons among organisms and knowledge about how $\delta^{18}O$ varies depending on temperature and latitude, have used this information to infer whether dinosaurs were endothermic or exothermic, and had high or low metabolism (Fig. 8.19).

The first of these studies was presented by Barrick and Showers (1994), and later followed by Barrick and Showers (1995) and Barrick et al. (1996). These workers measured the $\delta^{18}O$ values for several bones across a single skeleton and several different regions across a single bone. Most importantly, these studies found that $\delta^{18}O$ values within single skeletons and bones differed by only a small amount, which indicates that body temperature

only varied slightly between different parts of the skeleton. This, in turn, is suggestive evidence that these dinosaurs, which include various theropods (such as *Tyrannosaurus*) and ornithischians, were homeotherms with a constant body temperature. Because living homeothermic vertebrates are usually tachymetabolic endotherms, this physiological strategy can reasonably be implied for the studied Mesozoic dinosaurs. Although provocative, these studies have been criticized because it is known that the oxygen isotopes in fossil bone phosphate are especially prone to alteration during fossilization, burial, and weathering.

More recent studies have instead targeted teeth for sampling, because enamel is more resistant to alteration (Fricke and Rogers, 2000; Amiot et al., 2006). This has been demonstrated empirically: teeth of different species from the same localities have been studied, and because each species has a characteristic $\delta^{18}O$ value, this is firm evidence that the original isotopic composition of the living animal has been preserved (Fricke and Rogers, 2000). If the original material had been replaced, it would be expected that teeth of many species found together would show the same isotopic signal, because they had been subjected to the same fossilization and alteration processes.

Aside from using a better source of data in teeth, the recent studies of Fricke and Rogers (2000) and Amiot et al. (2006) have also made another conceptual advance: they compared $\delta^{18}O$ values of teeth of different organisms across a broad latitudinal range. Why is this important? Recall that $\delta^{18}O$ values depend both on environmental temperature and body temperature, so it is difficult to simply measure a specimen, calculate the $\delta^{18}O$ value, and conclude whether it was endothermic or exothermic. However, there is a known relationship between $\delta^{18}O$ values and environmental temperature, so if this can be factored out then any variation between specimens should be due to body temperature differences alone. This known relationship tells us that the ratio between ^{18}O and ^{16}O should be smaller at higher temperatures. Therefore, because of the differences in how homeotherms and heterotherms incorporate oxygen isotopes into their bones, it is expected that ectotherms should have higher $\delta^{18}O$ values than endotherms at high latitudes (above 50°), whereas endotherms should have

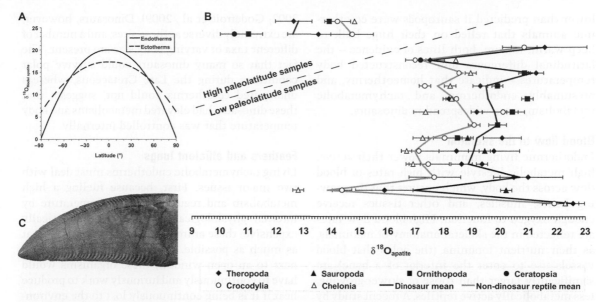

Figure 8.19 Isotopic evidence for dinosaur endothermy and high metabolisms. (A) Theoretical expected relationship between the $\delta^{18}O$ values of endotherms and ectotherms at different latitudes. (B) $\delta^{18}O$ values of dinosaurs and coexisting exothermic animals (crocodiles and turtles) at different latitudes. Note that $\delta^{18}O$ values are higher for dinosaurs (thick black line denotes mean value) than for ectotherms (thick gray line denotes mean value) at lower latitudes, but this relationship is reversed at higher latitudes, as would be expected if dinosaurs are endothermic. (C) Theropod dinosaur tooth, the type of specimen that provides $\delta^{18}O$ measurements for the analyses shown in the plots. Images (A) and (B) courtesy of Dr Romain Amiot and modified from images in Amiot et al. (2006); image (C) courtesy of Dr Roger Benson.

higher values than ectotherms at low latitudes (Fig. 8.19A). The details of this relationship, including exactly how it is derived using chemical arguments, are well summarized by Fricke and Rogers (2000) and will not be repeated here. What is important to remember, though, are the expected differences between $\delta^{18}O$ in homeotherms and heterotherms at different latitudes. These are not mere chemical predictions, but also borne out by study of living animals.

With this expectation in mind, the results of both studies are thought-provoking. Fricke and Rogers (2000) compared the $\delta^{18}O$ values of several theropod dinosaur teeth to those of the teeth of known ectotherms (crocodiles) at several different sites, which were at different latitudes during the Mesozoic. They found that, at lower latitudes, theropods consistently had higher $\delta^{18}O$ than crocodiles, regardless of which taxa were being analyzed, which hemisphere the fossil site was located in, or which environment the fossil site was formed in. Conversely, at higher latitudes, crocodiles consis-

tently had higher $\delta^{18}O$ values than theropods. This is exactly what would be predicted if theropods had homeothermic (and presumably ectothermic and tachymetabolic) metabolisms. As this study only focused on theropods, Amiot et al. (2006) later expanded sample sizes to include a range of theropods, sauropods, ornithopods, and ceratopsians represented by over 100 specimens from 11 localities (Fig. 8.19B). They reported the same result: at low latitudes dinosaurs had significantly higher $\delta^{18}O$ values than crocodiles and other exothermic organisms, whereas at high latitudes the dinosaurs had significantly lower $\delta^{18}O$ values (Fig. 8.19B). These authors also used $\delta^{18}O$ values to predict, perhaps somewhat coarsely, body temperatures for these dinosaurs, and found that all species had relatively high temperatures within the range of living endotherms, but higher than living ectotherms. The recent study of Eagle et al. (2011) reported a similar result: the carbon and isotope values of sauropod teeth indicated body temperatures similar to those of modern mammals, and lower than predicted if sauropods were ectothermic animals that relied on their huge bulk to keep warm. In sum, both lines of evidence – the latitudinal differences and reconstructed body temperatures – indicate that homeothermy, and presumably endothermy and tachymetabolic metabolism, were widespread in dinosaurs.

Blood flow to the skeleton

Endothermic living mammals power their active, high metabolic lifestyle with high rates of blood flow across the body, which ensures that the growing bones, muscles, and other tissues receive enough oxygen to function. High blood flow rates are reflected in the skeletal anatomy of mammals, as their nutrient foramina (the holes that blood vessels use to enter the interior of a bone) are significantly larger than those of slower-growing, less metabolically active reptiles. A recent study by Seymour et al. (2011) found that 10 analyzed species of dinosaur, including theropods, ornithopods, sauropodomorphs, ceratopsids, and stegosaurids, generally had even larger nutrient foramina relative to their body size than living mammals. This suggests that all these extinct dinosaurs experienced elevated rates of blood flow to their bones, which would be expected in highly active animals with fast metabolic rates.

Dinosaur distribution and paleoenvironments

Some dinosaurs lived in polar regions during the Mesozoic. Although there were probably no icecaps (or only small ones) during this time, and although global climates were on average warmer than today, these regions would still have been cold during the dark winter months. In the modern world, tachymetabolic endotherms are the most common, and in some cases the sole, organisms that can brave these environments. Organisms that have slower metabolisms and which cannot control their body temperature internally have a very difficult time surviving in extreme cold. What about polar ecosystems during the Mesozoic? As in today's world, unequivocally ecotothermic organisms such as crocodiles, turtles, and amphibians are usually completely missing from these localities. This is especially true of polar tetrapod faunas of the latest Cretaceous, which are extremely well sampled and well studied (Clemens and Nelms, 1993; Rich et al.,

2002; Godefroit et al., 2009). Dinosaurs, however, are extremely diverse at these sites, and a number of different taxa of varying body size are present. The fact that so many dinosaurs could survive polar conditions during the Late Cretaceous, whereas undisputed ectotherms could not, suggests that these dinosaurs had elevated metabolisms and body temperature that was controlled internally.

Feathers and efficient lungs

Living tachymetabolic endotherms must deal with two major issues. First, because fueling a high metabolism and regulating body temperature by internal physiological means is so energetically expensive, these animals must conserve body heat as much as possible. Much like a furnace placed next to an open window, these organisms would have to continuously and furiously work to produce heat if it is being continuously lost to the environment. Second, they must have an efficient respiratory and circulatory system to deliver oxygen and nutrients to their muscles and other body tissues, which power their rapid activity. This is analogous to a fast-moving car needing an efficient system for delivering gasoline to its engine; if this not present, the car will not be able to travel at high speeds and will eventually stop.

Many dinosaurs possess features that would have conserved body heat and efficiently delivered oxygen to the body tissues. Integumentary structures, ranging from simple filaments to fully pennaceous feathers, are known to be present on a number of dinosaurs, including several theropods and ornithischians (Xu and Guo, 2009). Such structures can only be fossilized under certain conditions, but as more exceptionally preserved fossil localities are discovered the roster of feathered dinosaurs continues to expand. This fact, and the known presence of integument in both saurischians and ornithischians, suggests that some form of feathers were probably very common among all dinosaurs. This is no trivial matter: living endotherms, such as mammals and birds, possess integumentary coverings, but living ectotherms such as reptiles and amphibians simply do not need them. Therefore, the presence of integument seems to be a prerequisite for tachymetabolic endotherms (especially small ones), whereas it is too metabolically expensive to maintain, or simply unnecessary, in ectotherms with slower metabolisms. With this in mind, the integumentary blankets of some dinosaurs are a strong indicator of high metabolism. Furthermore, the presence of hollowed bones and pneumatic foramina indicate that many dinosaurs had an avian-style unidirectional lung, which is a remarkably efficient respiratory structure (O'Connor and Claessens, 2005; Schachner et al., 2009, 2011b). This would also be expected in a tachymetabolic animal, and is consistent with high metabolisms and internal temperature control in dinosaurs.

Could dinosaur metabolism be related to body size?

Several workers have suggested that dinosaurs would not have needed endothermy to power constant body temperatures, because the large size of dinosaurs would have buffered these creatures from extreme temperature fluctuations ("gigantothermy") (Paladino et al., 1990; Spotila et al., 1991; O'Connor and Dodson, 1999; Seebacher, 2003; Gillooly et al., 2006; McNab, 2009). It is a basic physical principle that larger structures lose heat slower than smaller ones, because larger structures have a smaller ratio of surface area (which loses heat) to volume (which produces heat). Gigantothermy is an attractive hypothesis, and has led many scientists to label dinosaurs as having a "unique" physiology that is something of an "intermediate" between that of living ectothermic reptiles and endothermic birds and mammals. This general idea holds that giant dinosaurs did have nearly constant body temperatures, due solely to their large size, whereas smaller dinosaurs probably had fluctuating body temperatures, and probably lower metabolisms, similar to those of living reptiles.

However, some caveats immediately spring to mind. Dinosaurs spanned a range of body sizes, from less than 1 kg to several tonnes, and although it is easy to imagine a gigantothermic 75-tonne sauropod, this strategy would not have worked for a small theropod or ornithischian. O'Connor and Dodson (1999) hypothesized that dinosaurs greater than 500 kg were capable of gigantothermy (but see Eagle et al., 2011), but the majority of species were probably smaller than this. Many of these smaller species had feathery integument, an avian-style lung, rapid growth rates and extensive fibrolamellar bone, and oxygen isotope ratios indicative of high

metabolisms. Chinsamy and Hillenius (2004) speculated that the "mild" Mesozoic climates may have allowed these smaller species to enjoy constant body temperatures without high metabolisms or any internal temperature control, but this vague hypothesis would not explain why so many small dinosaurs lived in polar regions that were devoid of unequivocal ectothermic animals (Godefroit et al., 2009). In summary, gigantothermy may have been a legitimate thermoregulatory mechanism for the largest dinosaurs, but it cannot explain dinosaur physiology universally.

Dinosaur physiology: conclusions

So were dinosaurs "warm-blooded," "cold-blooded," or neither? Did they have fast or slow metabolisms, internal or external control of body temperature, constant or fluctuating body temperatures? As is evident, these questions are difficult to answer, but some general well-supported conclusions have emerged. Giant dinosaurs probably could maintain relatively constant body temperatures by their bulk alone, but there is other evidence (oxygen isotopes, bone histology, inferred avian-style lungs) which suggests that even large sauropods had fast metabolisms, and possibly endothermy. Smaller dinosaurs also probably had fast metabolisms and constant body temperatures powered by some type of endothermy. Although their giant bulk would not have buffered against temperature fluctuations, small dinosaurs could have controlled heat loss with feathery insulation. Their oxygen isotope compositions, bone histology, body posture, inferred locomotory speeds, and inferred avian-like lungs are all suggestive signs of fast metabolisms and endothermy, as is their presence in polar regions during the Late Cretaceous. Maybe dinosaurs did not have the exact same high-powered endothermic physiology of living birds, and maybe some dinosaurs (giant species) did largely control their body temperatures through external means (gigantothermy) rather than precise internal mechanisms. What seems clear, however, is that dinosaur physiology and metabolism was more similar to that of living birds and mammals than living reptiles. Dinosaurs clearly were not overgrown lizards, and most likely were biologically akin to modern endotherms.

Conclusions

Dinosaur reproduction, growth, and physiology are fascinating topics that have generated a large amount of research over the past several decades. All dinosaurs laid eggs, and some species appear to have built large nesting colonies that were used over successive generations. Dinosaurs exhibited varying levels of parental care: many species took great care in building their nests, some actively brooded their eggs, and some may have continued to feed and care for their young long after hatching. Many reproductive features typical of living birds, such as asymmetrical eggs and active incubation of nests, originated in non-avian dinosaurs, some of which appear to have exhibited paternal care as in many extant birds. Histological techniques now enable scientists to study dinosaur growth in unprecedented detail. By studying thin sections of bone, scientists can determine the age at which a dinosaur specimen died, identify bone textures characteristic of certain growth rates, and construct growth curves that show how body size changed during the transition from embryo to adult. These studies indicate that all dinosaurs grew fast due to the widespread possession of characteristic rapid-growing and well-vascularized fibrolamellar tissue. Dinosaurs became sexually mature before they reached adult size, exhibited indeterminate growth, and most species probably died by age 50–70, unlike living reptiles which grow slowly for long periods. However, it doesn't appear as if the hyperactive growth rates of living birds, which mature to adult size in a matter of weeks, originated until long after birds themselves evolved. Giant dinosaurs such as *Tyrannosaurus* developed their colossal size by amplifying their growth rate to an exponential pace, whereas dwarf island dinosaurs attained smaller size by slowing their growth rates. The body sizes, physical proportions, and functional and ecological habits of many species changed dramatically during ontogeny, so much so that many specimens once thought to represent distinct species simply pertain to juveniles or old adults of well-known species. Dinosaur physiology is the subject of endless debate and speculation, and it is uncertain whether some, or all, dinosaurs had high metabolisms, constant body temperatures, and the ability to control body temperature externally. It is clear, however, that dinosaurs had elevated metabolisms compared with living reptiles, and it is likely that their physiologies were quite similar (although perhaps not identical) to those of living "warm-blooded" mammals and birds.

9 Paleoecology and Dwelling

Like any living organism today, Mesozoic dinosaurs were components of larger ecosystems and would have interacted with each other and with other species (see Plate 16). Dinosaurs would have been part of complex food webs: certain dinosaur species would have eaten particular plants or animals, and other animals would have eaten dinosaurs. Some dinosaurs may have formed expansive, gregarious, and structured communities in which adults and juveniles lived side by side, but other species may have been more solitary. And of course, not all dinosaurs lived in the same regions, habitats, and environments: some species and groups were widely distributed across the globe, whereas others were restricted to a particular landmass, were specialized for a certain habitat, or could tolerate only a narrow range of environments.

Dinosaur paleoecology – how dinosaurs fit into their ecosystems, structured their communities, and were distributed across the globe – is one of the more fascinating topics in dinosaur paleontology. It is also one of the most difficult to study in a rigorous manner, because large samples of fossils are usually required to reach any convincing conclusions about how dinosaurs behaved and interacted with their environment. A single specimen of a dinosaur, no matter how complete or well preserved, will tell little about what habitats that species favored, whether it was gregarious or solitary, or what other species it would have frequently encountered. And even if many specimens of dinosaurs and other species are found together, there is no guarantee that they belonged to a community that would have interacted in life. Maybe, on the other hand, their remains were simply washed together after death, forming a random hodgepodge of fossils that masquerades as a prehistoric ecosystem preserved in stone. Many questions about dinosaur paleoecology and behavior are therefore currently intractable

Dinosaur Paleobiology, First Edition. Stephen L. Brusatte.
© 2012 John Wiley & Sons, Ltd. Published 2012 by John Wiley & Sons, Ltd.

based on our knowledge of the fossil record, and some questions may never be answered.

That being said, the study of dinosaur paleoecology is not entirely doomed. Some dinosaur-bearing rock formations, such as the Late Jurassic Morrison Formation and Late Cretaceous Hell Creek Formation of western North America, have been studied in such meticulous detail that fossil samples are large enough to make a reasonable reconstruction of the entire ecosystem (Dodson et al., 1980; White et al., 1998; Foster, 2003). When making these reconstructions, and drawing ecological hypotheses from even the most well-sampled formations, researchers must always be mindful that specimens found in the same rock unit, or even alongside each other, may not represent a true biological assemblage. Thankfully, the past few decades has brought renewed focus on the issue of taphonomy – the study of how fossils are preserved and how assemblages of fossils are created in the rock record (Behrensmeyer and Kidwell, 1985; Behrensmeyer et al., 2000; Fiorillo and Eberth, 2004). By carefully analyzing the rocks fossils were deposited in, the quality of preservation, and the relative abundance of different types of bones (compared with the entire skeleton), researchers can often differentiate between fossil assemblages representing animals that died (and presumably lived) together and those composed of specimens randomly gathered by geological processes (Voorhies, 1969; Rogers et al., 2008). Finally, as more and more dinosaur fossils are found across the globe, researchers are now able to use statistical techniques to assess whether certain types of dinosaurs are frequently found in association with other dinosaurs, are especially common in particular environments, and are distributed around the world in a recognizable pattern (Upchurch et al., 2002; Mannion and Upchurch, 2010a; Lyson and Longrich, 2011). As is evident, contemporary students of dinosaur paleoecology have many analytical tools at their disposal for studying what has long been one of the most elusive topics in dinosaur research.

The Dinosaur Fossil Record

Dinosaur fossils are known from all continents and have been found in almost every conceivable type of terrestrial sedimentary rock, ranging from arid wind-blown sandstones to mudstones formed on the floodplains of rivers and on the margins of lakes. Dinosaur specimens are most commonly encountered in those rocks formed by streams, rivers, and deltas in lowland environments, but their abundance in these sediments probably reflects sampling bias more than a genuine preference for these habitats, because lowland fluvial rocks are much more commonly preserved than those formed in mountains, deserts, or upland regions. Similarly, the vast majority of known dinosaur fossils come from North America, South America, Europe, and Asia, but their relative rarity in Africa, Antarctica, and Australia is also surely due to sampling bias. These continents have been explored in much less detail than the four well-sampled continents, preserve only a fraction of the terrestrial Mesozoic rock present on these other continents, and (except for Africa) are small landmasses that offer considerably less physical space for preserving and locating fossils. In sum, dinosaurs lived all over the globe and in many environments, including deserts and the cold polar regions. The dinosaur fossil record is constantly expanding as new areas are explored and more researchers specialize in paleontology. A good overview of the dinosaur record and the geographical and environmental distribution of dinosaur fossils was presented by Weishampel et al. (2004), but is becoming increasingly outdated as new discoveries are made. The best single source of information on the dinosaur fossil record, which is both authoritative and consistently updated, is the online Paleobiology Database (http://paleodb.org/).

Dinosaur Distribution and Biogeography

Dinosaurs were distributed across the globe, just as birds and mammals are today, but not every species ranged freely around the planet. Individual species and subgroups had their own characteristic distribution patterns, which were often intimately related to the physical movement of continents during the Mesozoic and the position of climate and temperature belts. Dinosaurs originated during a time when all the world's landmasses were joined into a single supercontinent, Pangaea, and the subsequent 160 million years of dinosaur evolution played out on a dynamic planet whose continents were fragmenting and drifting, and whose climatic zones were shifting (see Fig. 1.23). It is no surprise, therefore, that patterns in dinosaur distribution seem to correspond quite closely to patterns of physical geography and climate. Dinosaur distribution is now understood in such detail that some scientists are confident enough to predict the timing of certain continental fragmentation events by simply studying the dinosaur fossils on the continent in question, with little regard for what is often much more rigorous and convincing geophysical evidence in the rock record. Understandably, such exercises are prone to great error, but the fact that researchers frequently engage in this line of reasoning is a

strong sign that dinosaur distribution is an active and important area of research.

Understanding the distribution of dinosaurs – dinosaur biogeography – is a goal that has long excited paleontologists, largely because it is thought that by understanding the geographic spread of dinosaurs one may come to a more general appreciation of how major groups evolve and disperse over millions of years in relation to climate, continental drift, mass extinctions, and other factors (Cox, 1974). Dinosaur biogeography has emerged as one of the great case studies in biogeographic research more generally, and has engendered generations of questioning, analysis, and speculation. Most early workers relied on a literal reading of the fossil record, by noting where certain species were found and where they were absent. If similar species were found in, for example, North America and Asia, then it may be argued that these two continents were connected by dispersal corridors or land bridges during the time in question. This crude approach, however, is obviously hampered by sampling biases. In this example, similar dinosaurs may only be recorded in North America and Asia because they were genuinely absent from other landmasses, but perhaps they were present elsewhere but simply never preserved or have not yet been found. And there is also another problem: even if sampling is perfect, the distribution of one or a few species may simply be random. What is really needed is a method for uncovering statistically robust patterns of similar geographic distributions for many species.

These predicaments have motivated a newer generation of specialists to think more quantitatively and rigorously. Dinosaur distributions are commonly studied in the context of phylogeny: various cladistic biogeographic methods use parsimony or likelihood methods to search for congruent patterns of geographical distribution among many species of dinosaurs on a cladogram (Platnick and Nelson, 1981; Page, 1988; Humphries and Parenti, 1999; Crisci et al., 2003), and sensitivity analyses can ameliorate the effects of sampling biases (Turner et al., 2009). For instance, a cladistic study may show that many subgroups of dinosaurs contain Asian and North American species that are more closely related to each other than to European species. If this association of Asian and North

American taxa is statistically robust – that is, if it occurs more often than predicted by chance alone – then it can be taken as evidence for some sort of special relationship between these two continents, which can be further interpreted by reference to geological evidence. Perhaps, for instance, North America and Asia remained connected to each other after they separated from Europe or, conversely, perhaps North America and Asia collided with each other and their faunas intermingled. Cladistic biogeography can very rarely pinpoint exact reasons why faunas from certain regions are similar to each other (a process), but it can determine whether there is an overarching pattern in the geographic distribution of many species. This focus on pattern over process has molded biogeography into a more objective, rigorous analytical science that is the focus of much current research.

Cladistic biogeographic methods have been used in several studies of dinosaur biogeography, usually with the goal of constructing overarching "area cladograms" that take information from phylogenies to depict the degree to which areas were similar

to each other (Upchurch et al., 2002; Ezcurra, 2010b). In the above example, North America and Asia would be depicted as sister areas on an area cladogram because they share so many species with each other, akin to how two species may be recovered as sister taxa in a phylogenetic analysis because they share so many derived characters with each other. Other cladistic methods have been used to predict the ancestral areas of origin for major clades, using statistical modeling techniques that assign various costs to dispersal, extinction, and vicariance (the passive separation of organisms by a geographic barrier), in proportion to an assumed likelihood of their occurrence (Prieto-Márquez, 2010b). These explicit cladistic-based studies, along with other studies that interpret biogeographic patterns in light of a phylogeny but do not necessarily employ advanced statistics, have revealed several general patterns about the distribution of dinosaurs during the Mesozoic.

Although the earliest dinosaurs lived on a supercontinent, Late Triassic faunas were surprisingly provincial (Ezcurra, 2010b; Olsen et al., 2011;

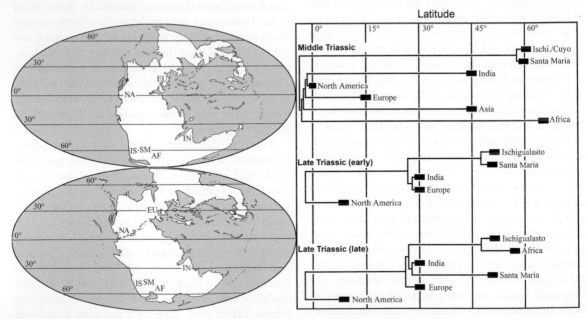

Figure 9.1 Middle to Late Triassic tetrapod biogeography. Plots at right are area cladograms depicting the degree of faunal similarity between different regions, which are shown on the paleogeographic maps at left. Note that regions at similar latitude, regardless of hemisphere, are often more similar to each other than physically closer regions at a different latitude. Image courtesy of Martin Ezcurra, modified from Ezcurra (2010b).

Whiteside et al., 2011) (Fig. 9.1). Major clades of dinosaurs and other terrestrial vertebrates were not evenly distributed across the globe, but rather faunas that occupied similar latitudinal belts (regardless of whether they were in the northern or southern hemisphere) were more similar to each other than to other regions, even those that were physically contiguous. This suggests that climate, which roughly correlates with latitude, was a more important influence on early dinosaur distribution than geography itself. Late Triassic endemicity is most strikingly demonstrated by the fact that basal sauropodomorphs, which are among the most common and diverse terrestrial vertebrates in Europe and South America, are completely unknown from the well-sampled North American fossil assemblages (Nesbitt et al., 2007, 2009b). Not all dinosaur taxa were endemic, however: basal theropods seem to have been more or less globally distributed (Nesbitt et al., 2009b).

Provincialism continued into the Early and Middle Jurassic, although this may have been more to do with the fragmentation of Pangaea than the influence of climate and latitude. Frustratingly, very few continents preserve a decent Early to Middle Jurassic fossil record, meaning that sampling biases are difficult to factor out. Regardless, it appears as if Asia, which separated early from Pangaea, had a largely endemic fauna of large theropods and sauropods, and such endemism may have persisted into the Late Jurassic (Upchurch et al., 2002). Counterintuitively, the Late Jurassic, and especially the Early Cretaceous, seem to have been intervals of widespread cosmopolitanism (Brusatte et al., 2009b; Barrett et al., 2011b). Although Pangaea was fragmenting, most major dinosaur subclades were globally distributed during the Early Cretaceous, including sauropods (Upchurch, 1995; Barrett et al., 2002), ornithopods (Norman, 1998), and theropods (Brusatte et al., 2009b, 2010c; Benson et al., 2010a, 2010b; Barrett et al., 2011b). Carcharodontosaurid theropods, which were once thought to be restricted to Gondwana, are now known from Africa, Asia, North America, and South America, and the closely related neovenatorids are known from Asia, Australia, Europe, and South America (Brusatte et al., 2009b; Benson et al., 2010a). Even the tyrannosauroids, a clade long considered a northern

hemisphere hallmark, have been reported from Gondwana (Benson et al., 2010b). Clearly, despite the slow drifting of continents, faunal connections between the now-dispersed Pangaean fragments must have persisted until at least the Early Cretaceous.

The Late Cretaceous, however, was undoubtedly an endemic world. Most continents had essentially reached their modern positions, in which northern and southern landmasses are largely separated by oceanic barriers and great distances (Smith et al., 1994) (see Fig. 1.23). Iconic groups such as the derived tyrannosaurids, pachycephalosaurs, and ceratopsids were restricted to the northern continents of Asia, North America, and Europe, which during much of the Cretaceous formed a coherent landmass called Laurasia (Sereno, 1997, 2000; Upchurch et al., 2002). Other major clades, such as the abelisaurid and noasaurid theropods, were mostly or entirely limited to the southern continents, which until their break-up during the Middle to Late Cretaceous comprised a united landmass called Gondwana (Bonaparte, 1991; Sampson et al., 1998; Sereno et al., 2004; Sereno and Brusatte, 2008). Different clades seemed to have filled the same niches on different continents: tyrannosaurids were the apex predators in the north, whereas abelisaurids and perhaps some late-surviving allosauroids were the largest predators in the south, and sauropods were abundantly common in the south but much rarer and less diverse in the hadrosaurid-dominated north.

The general pattern of distinct northern and southern faunas has been evident for quite some time, but two more nuanced questions about Late Cretaceous dinosaur biogeography have remained puzzling. First, there has been great debate about the sequence of Gondwanan fragmentation, and whether the distribution of vertebrate fossils can help elucidate when the various southern continents split from each other. It was once thought that the abelisaurids were present on all southern continents except for Africa, suggesting that they evolved after Africa split from Gondwana (Sampson et al., 1998). However, recent discoveries of African abelisaurids have falsified this hypothesis, and it is now unclear whether the dinosaur fossil record is complete enough to lend any robust evidence for understanding Gondwanan fragmentation (Sereno et al., 2004; Sereno and Brusatte, 2008). Instead, geological and geophysical evidence, which is generally subject to less extreme sampling biases than the vertebrate fossil record, will likely prove paramount in future studies.

Second, little is known about the biogeographic affinities of European dinosaurs during the Late Cretaceous. High sea levels reduced Europe to a series of small islands during this time, an environment poorly conducive to the preservation of fossils (Smith et al., 1994; Csontos and Voros, 2004). However, recent discoveries of spectacular new European fossils demonstrate that many European taxa (theropods, ceratopsians, ornithopods) are closely related to contemporary or near-contemporaneous Asian and North American species (Dalla Vecchia, 2009; Pereda-Suberbiola, 2009; Prieto-Márquez and Wagner, 2009; Csiki et al., 2010; Ősi et al., 2010). This indicates that Europe had a strong faunal connection with other Laurasian continents during the Late Cretaceous, and that the European islands were not isolated refugia for endemic species or primitive taxa stranded when sea levels rose. It is true, however, that some Late Cretaceous European islands supported dwarves and other morphologically bizarre taxa, as well as some primitive forms (Benton et al., 2010). These islands were therefore an unusual combination of aberrant taxa and those that were anatomically similar, and closely related, to mainland contemporaries (Csiki et al., 2010).

Dinosaur Ecosystems and Niche Partitioning

Any assemblage of dinosaur fossils represents part of an ecosystem, and if fossil sampling is extensive enough the entire ecosystem and a probable food web can be reconstructed with reasonable confidence. The literature is overflowing with paleoecological studies of individual dinosaur-bearing assemblages (see review in Fastovsky and Smith, 2004). Some of the more interesting studies have focused on ecosystem reconstructions for the Triassic–Jurassic Atlantic rift basins of North America (Olsen et al., 1989), the Late Jurassic Morrison Formation of western North America (Dodson et al., 1980; Foster, 2003) and Solnhofen island

archipelago of Europe (Barthel et al., 1990), the Early Cretaceous Jehol Biota (Zhou et al., 2003), and the Late Cretaceous Hell Creek Formation of North America (Lehman, 1987; White et al., 1998), European islands (Weishampel et al., 1991; Csiki and Benton, 2010), and arid Mongolian dune fields (Loope et al., 1998). The really intriguing question, however, is whether there are general rules for the composition and structure of dinosaur ecosystems across the board. Is there evidence that certain types of dinosaurs preferentially occupied a certain habitat? How did coexisting herbivores and carnivores partition niches? Are there any clear relationships between dinosaur ecosystem structure and climate? These questions are beginning to be answered by applying rigorous statistical analyses to ever-expanding datasets buoyed by new fossil discoveries.

Dinosaur habitats

Not all dinosaurs living at the same time lived in the same type of environment. Mesozoic environments were surely as heterogeneous as modern ones, and the fossil record clearly shows that dinosaurs were able to live along the coasts, on the floodplains of rivers, around the margins of lakes, within dune fields and deserts, and in forests and upland areas (Weishampel et al., 2004). This raises an important question: was the distribution of dinosaurs among different environments essentially random, or is there an orderly pattern in which certain species preferentially lived in certain habitats?

Two recent studies have attempted to answer this question using extensive databases that record individual fossil occurrences and the paleoenvironments they were found in (based on the lithology of the entombing rock). Butler and Barrett (2008) found that, during the Cretaceous, nodosaurid ankylosaurs and hadrosaurs are more commonly found in coastal and marginal marine rocks than would be expected by chance alone, whereas ceratopsians, pachycephalosaurs, theropods, sauropods, and ankylosaurid ankylosaurs are preferentially found in terrestrial sediments. Mannion and Upchurch (2010a) specifically examined the fossil record of sauropods from across the Mesozoic, including both skeletal fossils and footprints. They found that titanosaurs were significantly more common in inland environments, whereas non-titanosaurs were preferentially found in coastal and fluvial habitats. This study expanded on a previous analysis of the footprint record that recovered a strong association between sauropod trackways and coastal settings (Lockley et al., 1994). Exactly why titanosaurs may have preferred inland environments is debatable, but some of the clade's most conspicuous features (wide-gauge stance, highly flexible vertebral column, robust pelvic bones) may have allowed them to rear up to feed on the taller plants that may have been more common away from the coasts, or may have bestowed a more stable posture well suited for navigating the uneven terrain of the uplands.

Clearly, these types of studies, which subject expansive fossil databases to a litany of statistical analyses to look for robust patterns, can give provocative results. These studies are in their infancy and additional work is needed to examine the possible environmental preferences of theropod and ornithischian dinosaurs across the Mesozoic.

Niche partitioning

One of the most salient, yet puzzling, features of many dinosaur assemblages is the coexistence of numerous species that seem to share a similar feeding ecology. In the Late Jurassic Morrison Formation, for instance, there are more than 10 species of large-bodied, long-necked, barrel-chested sauropod herbivores, many of which obviously lived side by side because they are found together in individual quarries (Foster, 2003, 2007). How could so many similar herbivores persist in the same general environment? The answer is found in niche partitioning – the process by which similar, potentially competing species divide resources so that one does not drive the other to extinction. In general, there are three ways in which species can partition resources: they may live in slightly different habitats (spatial separation), feed on slightly different things (dietary separation), or simply avoid each other (temporal separation, such as noctural vs. diurnal foraging habits). How, then, did dinosaurs partition niches? It seems as if spatial and/or dietary separation were particularly important.

In many cases it is obvious, merely by observing morphology, that coexisting species fed on different foodstuffs or by using different techniques. Some Late Jurassic sauropods had long necks ideal for reaching high into the canopy, whereas others had shorter necks and may have been restricted to a low-browsing posture. Similarly, the teeth of these sauropods differed: some had pencil-like teeth for stripping vegetation, others more robust and spatulate teeth better suited for grinding (Upchurch and Barrett, 2000). Even among the primarily carnivorous theropods some differences are apparent: three large-bodied theropods were probably able to coexist in the lush Early Cretaceous ecosystems of northern Africa because one group (spinosaurids) may have primarily subsisted on fish, whereas the two hypercarnivorous groups (carcharodontosaurids and abelisaurids) differed in their body size and speed, which would have allowed them to target different prey (Brusatte and Sereno, 2007; Sereno and Brusatte, 2008). Many of these differences are discussed in Chapter 7 and will not be recounted in more detail here. The main point, though, is that reasonable hypotheses of niche partitioning can often follow from a careful consideration of morphology, especially if biomechanical information on feeding style, bite forces, and locomotory capabilities is also available.

In other cases, when morphology may not be a secure guide to understanding resource partitioning, there are other lines of evidence that some coexisting dinosaurs divided niches by dietary separation. Some of the most convincing evidence comes from the isotopic composition of dinosaur teeth (Fricke and Pearson, 2008) (Fig. 9.2C). The carbon and oxygen isotope composition of tooth enamel reflects the isotopic signature of the plants and water ingested by the animal during life. Recall from the preceding chapter that variations in isotopic composition can be quantified by calculating δ values, which denote the ratio of two common isotopes of a single element (^{16}O and ^{18}O in the case of oxygen, ^{12}C and ^{13}C in the case of carbon) to a universal standard. Fricke and Pearson (2008) calculated the $\delta^{18}O$ and $\delta^{13}C$ values of several hadrosaur and ceratopsian teeth from throughout the Late Cretaceous Hell Creek Formation. They found that these two herbivorous groups were characterized by significantly different isotope compositions whenever they coexisted, which indicates that they were feeding on different types of plants and perhaps occupying slightly different environments (with different types of groundwater) (Fig. 9.2C).

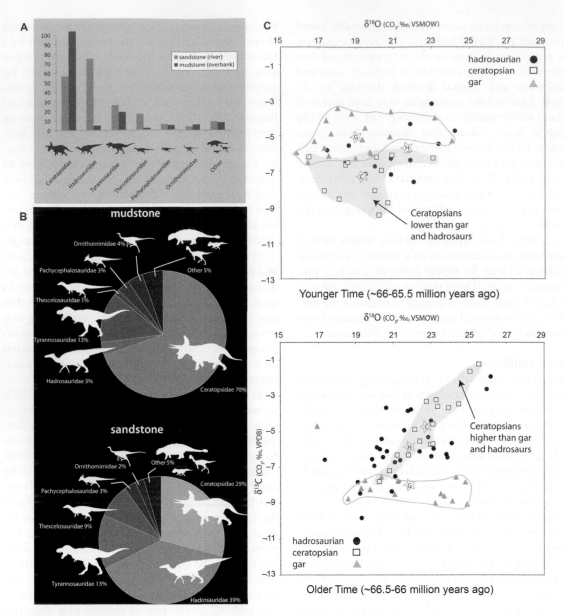

Figure 9.2 Niche partitioning in dinosaur ecosystems. (A) Abundance of different major groups of dinosaurs in sandstone and mudstone facies of the Hell Creek Formation. (B) Pie charts showing the relative dinosaur faunal composition of sandstone and mudstone facies in the Hell Creek Formation. Note from both images that ceratopsians preferentially occur in mudstones (floodplain environments) and hadrosaurs in sandstones (river systems), whereas tyrannosaurids are consistently found in both types of sediment. (C) Isotope plots showing the $\delta^{18}O$ (x-axis) and $\delta^{13}C$ (y-axis) isotope values for ceratopsian and hadrosaurid dinosaurs, with gar fish for references, from two general time intervals of the Hell Creek Formation (see Fricke and Pearson, 2008 for full details, including subtleties of the samples measured in each plot). Note that ceratopsians generally have higher isotope values than hadrosaurids and gars during the older time interval (below), but lower values in the younger time interval (above). Images (A) and (B) modified from Lyson and Longrich (2011) and courtesy of Tyler Lyson and Dr Nick Longrich; image (C) modified from Fricke and Pearson (2008) and used with permission from the Paleontological Society.

Furthermore, Fricke and Pearson (2008) found evidence for a shift in the isotopic values of ceratopsian teeth over the course of time (higher to lower $\delta^{18}O$ and $\delta^{13}C$ values), whereas those of hadrosaurs remained constant (Fig. 9.2C). Furthermore, older ceratopsians had higher isotope values than hadrosaurs, whereas younger ceratopsians had lower values. Based on comparisons with modern ecosystems, high values are thought to reflect open habitats such as swamps and tidal flats, because plants in these environments are more susceptible to evaporative water loss. Lower values, on the other hand, are more indicative of heavily vegetated forest with a closed canopy, which would have prevented heavy water loss. Hadrosaurs therefore seem to have preferred more densely vegetated habits throughout the Hell Creek Formation. Older Hell Creek ceratopsians probably preferred open habitats, but younger species shifted to a more closed canopy environment. Within this new environment, their very low isotopic values (now lower than even the hadrosaur values) suggest that ceratopsians were preferentially feeding on the lowest understory vegetation, which in modern ecosystems is usually characterized by extremely low isotope values.

A further study by Lyson and Longrich (2011) also found that spatial separation was a likely strategy for resource partitioning in the Hell Creek dinosaur assemblage (Fig. 9.2A,B). They compiled a database of nearly 350 individual fossil occurrences of Hell Creek dinosaurs and recorded the type of rock each specimen was found in. Differences in rock type are probably a reliable indicator of environmental differences, because mudstones are much more common in floodplain environments and sandstones in fluvial habitats. Because it is always possible that fossils have been transported far from their original environment, Lyson and Longrich (2011) focused only on well-preserved specimens with multiple bones found in association, making it more probable that any associations between dinosaur type and rock type reflect genuine environmental preferences. They found that ceratopsian fossils preferentially occur in mudstones (floodplain environments) and hadrosaurs in sandstones (river systems), whereas tyrannosaurids are consistently found in both types of sediment (Fig. 9.2A,B). This suggests that the major groups of

Hell Creek herbivores exploited different environments, in agreement with Fricke and Pearson's (2008) isotopic findings, but that tyrannosaurids (the sole apex predator in the Hell Creek Formation) roamed widely. What is especially interesting about their study is that such clear results emerged despite the fact that these fluvial and floodplain sediments were formed at the same time and in close proximity to each other. Clear ecological signals, it seems, may indeed be uncovered in dinosaur fossil assemblages.

Additional isotopic and spatial ecological analyses are sorely needed for other dinosaur assemblages and there is almost unlimited possibility for young researchers to make significant discoveries. There has been speculation that other dinosaur faunas may exhibit spatial and dietary niche separation: Makovicky et al. (2010) noted, for instance, that maniraptorans and ceratopsians are particularly common in the dry xeric Late Cretaceous units in the Gobi Desert, whereas ornithomimosaurs, therizinosaurs, and hadrosaurs are more abundant in the wetter mesic facies. These suggestions remain to be tested by additional studies, however, and this work promises to remain exciting for many years to come.

Dinosaurs and climate

An interesting question, especially in light of increasing concern about human-induced climate change, is how dinosaur communities may have been affected by climate. Did dinosaur diversity fluctuate in concert with climate changes? Did sudden climate shifts cause major extinctions? Did different climatic zones – wet, arid, warm, cold, temperate, etc. – have characteristic types of dinosaur faunas? Few of these questions have been addressed, but the recent study of Noto and Grossman (2010) examined whether there was a recognizable pattern of dinosaur community structure shared by regions of similar climate and precipitation levels during the Late Jurassic (Fig. 9.3). They assessed 12 well-sampled dinosaur assemblages and classified species into several guilds based on body size, locomotor type (bipedal vs. quadrupedal), and trophic mode (carnivore vs. herbivore, and for the latter, low, medium, and high browsers). The proportion of species in each guild was calculated for each assemblage, giving

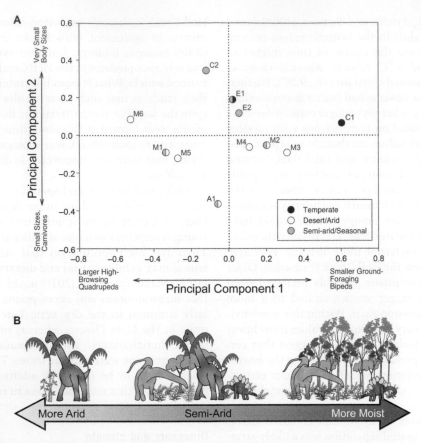

Figure 9.3 Dinosaur faunas and climate. (A) Multivariate statistical plot showing that dinosaur assemblages from similar climate zones (temperate, arid, and seasonal) share a similar characteristic set of dinosaur taxa. The position of each assemblage in multivariate space represents the proportion of taxa in that assemblage that belong to certain categories of body size, diet, and locomotory ability (as explained in the text). (B) Cartoon showing the major differences between the dinosaur faunas of the three main climate zones. Both images courtesy of Dr Christopher Noto and modified from Noto and Grossman (2010).

every assemblage an "ecological profile" score that could be investigated using multivariate statistics. The most important result of this exercise was that more arid assemblages were found to have a greater proportion of large-bodied species and high-browsing herbivores, probably because low-lying vegetation was scarce in these climates and because large animals are better able to cope with a lower density and quality of food resources. Wetter regions characteristically supported species from across the size spectrum, but with fewer high-browsing herbivores and more low-browsing species. Similar analyses that focus on other time periods, as well as additional studies addressing other aspects of dinosaur ecology in relation to climate, will likely be a major topic of research in the coming decades.

Dinosaur Populations

What, if anything, do we know about individual populations of dinosaurs: assemblages of interacting conspecifics, ranging from juveniles to adults, living in the same place at the same time? Occasionally, exceptionally preserved fossil deposits – bonebeds – record a mass death assemblage of a group of animals that lived, and then suddenly died, together. It is often difficult to determine whether a bonebed truly represents a biological population that had the bad fortune to die simultaneously, or whether it comprises a random array of fossils washed together by currents or other geological processes. Careful geological and taphonomic detective work can usually distinguish between these two possibilities (Voorhies, 1969; Ryan et al., 2001; Rogers et al., 2008). It would be a remarkable coincidence for currents to haphazardly gather together many individuals of the same species, so if a bonebed is dominated by one taxon then odds are it is a mass death assemblage representing a true population (Rogers et al., 2008; Rogers and Brady, 2010). Similarly, if all or most of the specimens have they same type of preservation – the same degree of articulation, completeness of bones, and external wear – then it is likely that they died and were buried together (Hunt, 1978; Ryan et al., 2001). Finally, the types of bones that are present can give some indication as to the degree of transport before deposition (Voorhies, 1969). Complete or near-complete skeletons are a good indication of little transport, but if larger and more plate-like bones like those of the pelvis and skull are preferentially abundant, whereas smaller elements such as vertebrae and phalanges are largely missing, then this is a sign that the specimens have been transported by currents, perhaps for long distances.

Many dinosaur bonebeds seem to represent true populations that lived, died, and were buried together, based on the fact that they are dominated by one taxon, are composed of hundreds or thousands of bones with a similar preservation style, and contain nearly complete and well-preserved skeletons that do not seem to have been transported long distances (Eberth and Getty, 2005; Rogers et al., 2008). The vast majority of these pertain to Late Cretaceous hadrosaurs and ceratopsians from western North America, and some contain more than 1000 individuals (Langston, 1975; Varricchio and Horner, 1993; Ryan et al., 2001; Dodson et al., 2004; Horner et al., 2004). This leaves little doubt that these dinosaurs were gregarious, and lived in large herds that may have migrated. Many of these bonebeds include the remains of juveniles, sub-adults, and adults, indicating that individuals of all ages were part of these herds. Gregarious behavior in some dinosaurs, including sauropods and ornithopods, is also recorded by footprint assemblages (Lockley et al., 1986; Meyer, 1993; Lockley and Hunt, 1995; Lockley and Matsukawa, 1999). Some footprint sites, called "megatracksites," include thousands of individual tracks facing in the same direction (Fig. 9.4). Although it is difficult to be certain that these tracks were deposited at the same

Figure 9.4 Gregarious behavior of dinosaurs recorded in tracksites: images of the Moab Megatracksite (Entrada Formation, Middle to Upper Jurassic, Utah). Courtesy of Dr Martin Lockley.

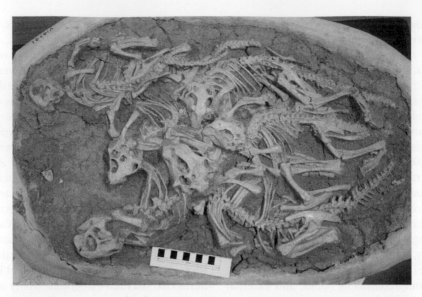

Figure 9.5 Gregarious behavior of juvenile dinosaurs (juvenile "herds" or aggregations) recorded in exceptional fossil deposits. The image depicts an aggregation of juveniles of the small ceratopsian *Psittacosaurus* from the Early Cretaceous of China (see Zhao et al., 2007). Image courtesy of Dr Paul Barrett.

time, and not over weeks or months, the dominance of a certain track type and the unidirectional nature of numerous coexisting trackways are good indications that megatracksites were formed simultaneously by a herd.

A remarkable pattern that has emerged from recent study of several bonebeds is that juvenile dinosaurs are often found together, in their own assemblages free of adults (see review in Varricchio, 2011) (Fig. 9.5). Examples are known from nearly every major dinosaur subclade, including ankylosaurs (Currie et al., 2011), sauropods (Myers and Fiorillo, 2009), theropods (Varricchio et al., 2008b), and both basal and derived ceratopsians (Zhao et al., 2007; Mathews et al., 2009) (Fig. 9.5). In some cases these bonebeds are unquestionably mass death assemblages in which juvenile herds were suddenly inundated by mudflows or floods and individuals preserved together in lifelike positions (Varricchio et al., 2008b; Currie et al., 2011). For some species, juvenile aggregations have been recorded but herds formed by adults are completely unknown. This is the case for the characteristic ceratopsid *Triceratops*, and researchers are confident that adult groups were not regular occurrences because hundreds of years of fossil collecting have yet to reveal

any examples (Mathews et al., 2009). It seems therefore that juvenile sociality was a regular occurrence in dinosaurs, and perhaps a characteristic feature of the group as a whole.

By studying fossil bonebeds that include individuals of a wide range of ages (as determined by bone histology and growth rings), paleontologists can gain an understanding of population dynamics – the balance between deaths and births that create a population's age structure (Fig. 9.6). Such studies have been carried out for the Late Cretaceous tyrannosaurid *Albertosaurus* (Erickson et al., 2006, 2010; Ricklefs, 2007a) and the Early Cretaceous basal ceratopsian *Psittacosaurus* (Erickson et al., 2009b). Erickson et al. (2006, 2010) found that *Albertosaurus* exhibited a survivorship profile like that in living mammals and birds, but quite unlike that of most reptiles (Fig. 9.6). Neonate mortality was probably very high, followed by few deaths after 2 years of age (presumably because such individuals were now large enough to protect themselves from most predators), and then increased mortality at mid-life (probably because of the rigors of reproduction, childbirth, and parental care). Few individuals therefore had a long reproductive lifespan (see also Ricklefs, 2007a). Erickson et al. (2009b) later found

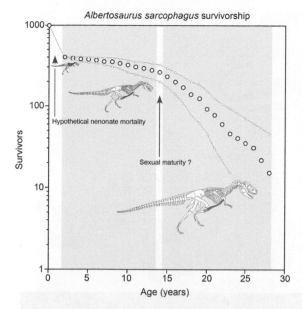

Figure 9.6 An ecological survivorship curve for a hypothetical population of 1000 individuals of the tyrannosaurid *Albertosaurus*. Neonate mortality was high, but once individuals reached a juvenile age (about 2 years old) they enjoyed a relatively stable period when the probability of death was low. Mortality then increased at mid-life, probably after the attainment of sexual maturity, most likely due to the rigors of reproduction. Few individuals reached old age, and no known fossil specimens are histologically aged at more than 30 years old. Image courtesy of Dr Gregory Erickson and modified from Brusatte et al. (2010d) and Erickson et al. (2010).

that *Psittacosaurus* had a similar life history. It must be warned, however, that sample sizes for *Albertosaurus* are quite low (26 individuals from the bonebed in question), and recent theoretical work has shown that larger samples may be needed to accurately capture population dynamics (Steinsaltz and Orzack, 2011).

Dinosaur Trophic Relationships

Food webs – the interwoven relationships between predators and prey in an ecosystem – can be reasonably hypothesized for many dinosaur communities. For instance, there is no doubt that

Tyrannosaurus was the apex predator in the Late Cretaceous of North America, and bite marks and coprolites prove that it fed on various herbivorous dinosaurs (ceratopsians and hadrosaurs), which in turn are known to have fed on certain plants due to the discovery of their coprolites and gut contents. Much of this information is summarized in Chapter 7 and will not be rehashed here. In closing this chapter, though, one provocative question about Mesozoic terrestrial food webs deserves comment: what animals, other than dinosaurs themselves, may have fed on dinosaurs? A few spectacularly preserved fossils have offered a clue. A specimen of the dog-sized Early Cretaceous mammal *Repenomamus* was found with the bones of a juvenile *Psittacosaurus* in its gut (Hu et al., 2005) (Fig. 9.7B), and a specimen of the 3.5-m Late Cretaceous snake *Sanajeh* was found coiled around a sauropod egg, with the bones of a hatchling nearby (Wilson et al., 2010) (Fig. 9.7A). It is also likely that enormous *Tyrannosaurus*-sized Cretaceous crocodylomorphs such as *Deinosuchus* and *Sarcosuchus* preyed on dinosaurs, although unequivocal trace fossils recording such a predator–prey interaction have yet to be found (Schwimmer, 2002). Dinosaurs may have been the dominant component of most Mesozoic terrestrial ecosystems, but they were still humble components of complex food webs and not immune from the predatory tastes of other animals.

Conclusions

Understanding dinosaur ecology is difficult because the fossil record rarely preserves complete unbiased snapshots of ancient ecosystems, and because most dinosaur species are only known from a small number of fossils. Increased fossil sampling, the construction of large datasets that record when and where dinosaur specimens are found, and the use of sophisticated statistical techniques are fueling something of a resurgence in the study of dinosaur ecology, and some major questions are tractable based on our current knowledge. Dinosaurs are known from across the globe and are found preserved in nearly every type of terrestrial sedimentary rock. Not every dinosaur lived on every continent and in every environment, however, and

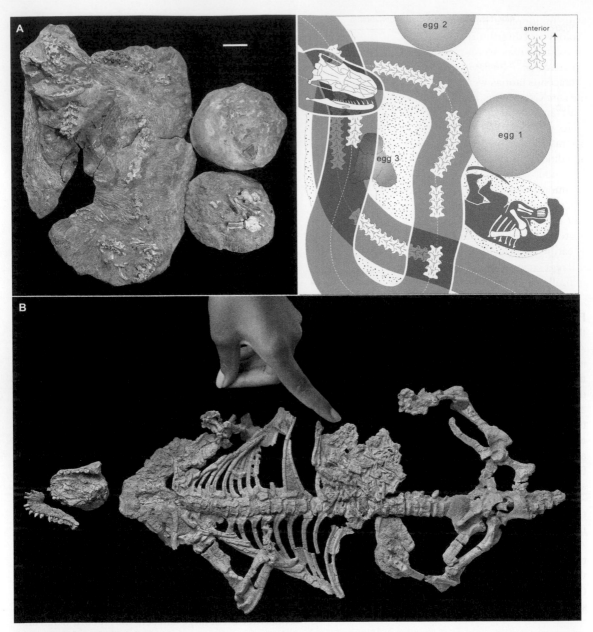

Figure 9.7 Animals that fed on dinosaurs: (A) photograph and interpretive drawing of a Late Cretaceous snake found within a nest of dinosaur eggs, prime evidence that some snakes preyed on dinosaurs; (B) photograph of a specimen of the cat-sized (or larger) Cretaceous mammal *Repenomamus* with a juvenile dinosaur preserved within its chest cavity (denoted by the pointing finger). Images in (A) courtesy of Dr Jeffrey Wilson and modified from Wilson et al. (2010); image (B) photograph © Mick Ellison.

some major patterns seem to characterize dinosaur ecology and distribution. Triassic and Early Jurassic dinosaur faunas were quite provincial, despite the existence of Pangaea, but Late Jurassic and Early Cretaceous faunas were more cosmopolitan. During the Late Cretaceous, as the continents drifted further apart and assumed a more modern configuration, there were marked differences between northern (Laurasian) and southern (Gondwanan) faunas. Certain dinosaur clades probably preferred certain habitats: nodosaurid ankylosaurs and hadrosaurs are commonly found in coastal settings, whereas most other dinosaurs are abundant in terrestrial sedimentary rocks. Isotopic and geological studies demonstrate that coexisting dinosaurs partitioned resources by eating slightly different types of food and living in different habitats. In particular, ceratopsians are commonly found in floodplain environments and

likely persisted on open and/or low-growing vegetation, whereas coexisting hadrosaurs are common in fluvial environments and fed on higher-growing canopy plants. The relationship between dinosaur distribution and climate is poorly understood, but during the Late Jurassic regions of similar climate and aridity supported similar dinosaur communities. Populations of conspecific dinosaurs that lived together are notoriously difficult to study, but spectacularly preserved bonebeds and expansive footprint sites indicate that some dinosaurs were gregarious, that juvenile dinosaurs often formed their own age-specific herds, and that (at least for some species) mortality was highest among juveniles and reproductively active adults. Additional work on dinosaur ecology will surely continue as fossil assemblages are better sampled and statistical and analytical techniques are further refined.

10 | Macroevolution and Extinction

The 160 million years or more of Mesozoic dinosaur evolution was a dramatic story that unfolded against a backdrop of drifting continents, changing climates, and the evolution of other major groups such as mammals and crocodiles. All research carried out by dinosaur paleontologists – every description of a new dinosaur specimen, every naming of a new species, every phylogenetic analysis, every functional study of locomotion or feeding – contributes primary data that is vital in piecing together the grand narrative of dinosaur evolution (dinosaur macroevolution). More than anything else, dinosaur researchers strive to understand this story, for it is *the* fundamental contribution of dinosaur research to the wider realm of science. Why study dinosaurs? In short, because they dominated terrestrial ecosystems for over 100 million years, included some of the largest and most spectacular animals to ever live, survived (but later succumbed to) mass extinctions and endured fluctuating climates and sea levels, and were the progenitors of one of today's most successful vertebrate groups (birds). By understanding the major patterns of dinosaur evolution across the Mesozoic, we hope to learn more about the general rules of evolution and how the earth and its inhabitants change over time.

What, then, do we know about the major patterns of dinosaur evolution during their 160 million year reign? Scientists have long speculated about what made dinosaurs successful, why they went extinct, and why their diversity waxed and waned during the Mesozoic. It has been said that dinosaurs rose to dominance by outcompeting other animals during the Triassic, or that dinosaurs simply took advantage of good fortune when other Triassic groups went extinct. Some have said that dinosaurs underwent a steady diversification throughout

Dinosaur Paleobiology, First Edition. Stephen L. Brusatte.
© 2012 John Wiley & Sons, Ltd. Published 2012 by John Wiley & Sons, Ltd.

the Mesozoic, whereas others have postulated that dinosaur history was punctuated by several mass extinctions. And there have been numerous hypotheses – too numerous to count – attempting to explain why non-avian dinosaurs ultimately went extinct. Did dinosaurs gradually wither away, perhaps because of disease or long-term climate change, or was their extinction triggered by an abrupt unpredictable event such as an asteroid impact? Clearly, the number of interesting questions about dinosaur macroevolution, as well as the number of possible answers, is almost limitless.

One of the great frustrations of macroevolutionary research is that it is often easy to speculate, to embellish grand narratives based on only a few scraps of evidence, and to simplify complex concepts into buzzwords. Many early scientists interested in dinosaur evolution offered hypotheses based on a literal reading of the fossil record combined with intuition, based on experience and assumptions about how macroevolution works over long intervals of time. Rigorous statistical analyses, or indeed numerical data of any kind, were rarely used, and sampling biases and other uncertainties were usually ignored. Over the past decade, however, paleontologists have become more objective when studying dinosaur evolution. Contemporary researchers have stepped away from flowery speculations and instead concentrate on quantifying trends and patterns. For instance, rather than grandstanding about how different waves of dinosaurs arose and went extinct over time, researchers now use large databases to quantitatively measure how the number of dinosaur species changed across the Mesozoic and utilize time-calibrated phylogenies to estimate when major groups arose. When objective patterns like these can be measured, they then may be marshaled as evidence in support of evolutionary processes. In other words, contemporary researchers aim to quantify patterns first and infer processes later.

The modern focus is therefore on quantifying evolutionary patterns in the most objective and rigorous manner possible. This usually involves compiling large databases that record something measurable about dinosaurs, such as all known species, total fossil occurrences, body size, or the possession of various anatomical characteristics. By applying statistical analyses to these datasets, and taking into account the possible effects of sampling biases, researchers can then outline major patterns. Some of the most salient macroevolutionary patterns, and those that have received the most attention from dinosaur paleontologists, are taxonomic diversity (number of species), morphological disparity (variety in anatomical features), absolute faunal abundance, and rates of morphological character change. These can be assessed temporally (i.e., measuring them over time to look for declines and peaks) and also comparatively (i.e., compared in sister taxa or other comparable clades). By quantifying and considering these patterns, dinosaur paleontologists have gained a rich understanding of how dinosaurs rose to dominance in the Triassic, why dinosaurs probably went extinct at the end of the Cretaceous, and how dinosaur evolution unfolded during the 160 million years in between.

Dinosaur Diversity

Dinosaurs are often described as a "diverse" group, but what does this mean? The term "diversity" has somewhat of an amorphous meaning, both in scientific and in popular lexicons, and is usually meant to describe variety, variability, or difference. Paleontologists, on the other hand, usually use the word "diversity" to refer to a specific macroevolutionary metric: the number of species, or other countable taxa (genera, families, major clades, etc.), within a certain time period, place, or group. Two questions about diversity have been the subject of particular attention among dinosaur workers. First, how did dinosaur diversity change over time? Second, are certain dinosaur subclades significantly more diverse than others? Quantifying these patterns can pinpoint major intervals of diversification during the Age of Dinosaurs, determine whether dinosaurs gradually or abruptly went extinct, reveal smaller mass extinction events that may have extirpated certain dinosaur subclades, and identify specific subclades or lineages that are exceptionally species rich (perhaps because of a morphological innovation that promoted speciation or success).

Temporal trends in dinosaur diversity

How did dinosaur diversity change across the Mesozoic? At first consideration, this seems like an easy question to answer. There are a finite number of known dinosaur species, whose names and geological ages are compiled in several print and online sources, and a solid day's work could easily produce a diversity curve showing the waxing and waning of dinosaur species counts across the Triassic, Jurassic, and Cretaceous (Fig. 10.1). Indeed, such compilations have been presented by many researchers over the past few decades (Weishampel and Norman, 1989; Dodson, 1990;

Sereno, 1999; Barrett and Willis, 2001; Taylor, 2006). Often, researchers will modify their diversity counts by adding in so-called "ghost taxa" that are not sampled, but whose existence is postulated by a phylogeny. The logic behind these corrections is simple: if a taxon is present in one interval then its sister lineage must be present by definition, so if half of the sister-taxon pair is sampled then the other must have also been present even if it is not sampled (Norell, 1992). Phylogenetically corrected dinosaur diversity curves, which include both sampled and ghost taxa, have been presented by Weishampel and Jianu (2000), Upchurch and Barrett (2005), Lloyd et al. (2008), and Brusatte et al. (2011b).

There are some problems with these approaches, however. The most vexing issue is that the fossil record is far from perfect, a reality that is grudgingly accepted by all paleontologists. For a dinosaur fossil to form, the animal must be killed and buried quickly, before all the bony tissues degrade into dust. It must also be buried in the right type of rock – that which has the proper range of pressure and pH so that a fossil is preserved and not destroyed – and survive within that rock for tens or hundreds of millions of years, without being pulverized by episodes of volcanism and mountain building. Then, if the fossil survives all these filters, it still remains to be discovered by a paleontologist. Using the modern world as an analogy, there are only a fleetingly small number of environments where conditions are ideal for fossilization to occur. Sewer rats and campus squirrels will probably never be preserved as fossils: they do not live and die in areas where sediment is being turned into rock. Even if one of these was miraculously fossilized, what would be the odds that its bones would survive millions of years of geological pressure and uplift, and then catch the glimpse of a wandering paleontologist? How, then, can paleontologists be at all confident that they are sampling

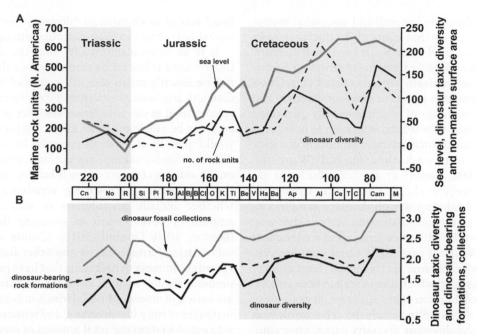

Figure 10.1 Dinosaur diversity and sampling. Plots of dinosaur diversity, various sampling proxies, and sea level (modified from Butler et al., 2011). (A) Dinosaur diversity (solid black line) compared with one estimate of sea level (solid gray line) and the number of Mesozoic marine rock units in North America (dotted line). Number of rock units on left y-axis and dinosaur diversity (number of species) and sea level (meters below or above present-day sea level) on right y-axis. (B) Dinosaur diversity (solid black line) compared with two sampling proxies, the number of dinosaur-bearing rock formations (dotted line), and number of discrete collections of dinosaurs compiled by paleontologists (gray line). Data plotted on a logarithmic scale. Images courtesy of Dr Richard Butler.

enough of the fossil record to capture a valid diversity signal? Perhaps times of high fossil diversity are simply those in which more environments preserved fossils, or more fossil-bearing rock was produced, or which have been the subject of more extensive exploration by paleontologists.

This concern has captured the rapt attention of paleontologists over the past decade, and much work has focused on quantifying the relationship between observed diversity and sampling (preservation of fossils and extent of exploration of fossil-bearing deposits) (Fig. 10.1). Most of this work has concerned the marine fossil record, and has persuasively demonstrated that, in many cases, sampling biases are real and may distort diversity patterns (Peters and Foote, 2001; Smith, 2001; Crampton et al., 2003; Peters, 2005; Smith and McGowan, 2007; McGowan and Smith, 2008). In some cases, trends in observed diversity over time closely match trends in fossil sampling: those intervals

that are relatively better sampled also record a relatively more diverse fossil record, and vice versa. The question, therefore, is whether the fossil record is good enough to reveal any genuine information about diversity over time. Everybody knows that the fossil record is biased, but may it still be adequate enough to answer certain fundamental questions? The answer is nuanced: it depends on the group in question, and the evolutionary questions being asked. In the case of dinosaurs, several recent studies have demonstrated that sampling biases do confound the observed dinosaur fossil record, but if these biases are carefully quantified and considered, meaningful understanding of Mesozoic diversity is attainable (Fastovsky et al., 2004; Wang and Dodson, 2006; Lloyd et al., 2008; Barrett et al., 2009; Mannion and Upchurch, 2010b; Vavrek and Larsson, 2010; Benson and Mannion, 2011; Butler et al., 2011; Lloyd, 2011; Mannion et al., 2011a, 2011b; Upchurch et al., 2011).

One of the more general, and somewhat worrying, revelations is that many of the fluctuations in observed (and phylogenetically corrected) dinosaur diversity are probably figments of sampling. Statistical analyses indicate a strong overall correlation between temporal trends in dinosaur diversity and sampling, the latter of which is usually quantified using a straightforward measure such as number of dinosaur-bearing formations or number of dinosaur specimens from each time interval (Wang and Dodson, 2006; Barrett et al., 2009; Butler et al., 2011; Upchurch et al., 2011) (Fig. 10.1). Lloyd et al. (2008) used information on dinosaur sampling to perform a statistical resampling analysis that simulated what the Mesozoic dinosaur diversity curve would look like if essentially equal samples were available for each time bin ("samples" were number of localities from which dinosaurs had been collected; see also Mannion et al., 2011a for an example of a similar subsampling analysis). The result was startling: the dinosaur diversity curve, once composed of peaks, valleys, and wiggles, was rendered into a nearly flat line (Fig. 10.2). This suggests that dinosaur diversity is heavily influenced by sampling, and indeed much of what is recorded in the

fossil record as changes in "diversity" over time may simply reflect variable sampling through time.

It is an unavoidable reality that the dinosaur fossil record is biased by sampling, but that does not necessarily mean that all studies of dinosaur diversity are doomed to failure. First, the sampling proxies used in the correlation analyses are coarse and may be problematic. Ideally, a sampling proxy would quantify the overall potential for sampling dinosaur fossils, taking into account information on the amount of potential fossil-bearing rock (rock that is capable of preserving vertebrate fossils, whether actually fossiliferous or not) and the amount of worker effort in sampling that rock (Benton, 2010; Dunhill, 2011). Counts of dinosaur-bearing formations, on the other hand, only record the number of rock units that have preserved dinosaur fossils, without counting rock units that are barren of dinosaur fossils (which may represent instances of truly low diversity) and without taking into consideration the total amount of rock that is being sampled (one dinosaur-bearing formation may be exposed over hundreds of square kilometers, another at only a few roadcuts). An additional sampling proxy used by some studies – the

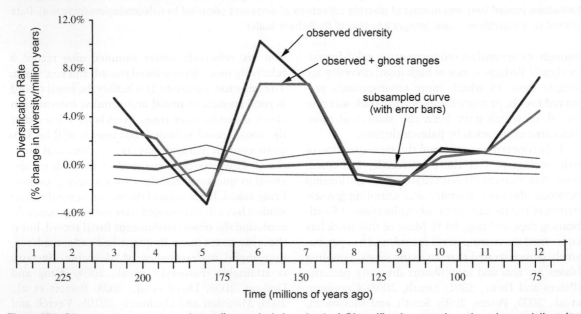

Figure 10.2 Dinosaur diversity after subsampling analysis (rarefaction). Diversification rates based on observed diversity and observed diversity with a ghost range correction, along with a subsampled curve. Image courtesy of Dr Graeme Lloyd and modified from Lloyd et al. (2008).

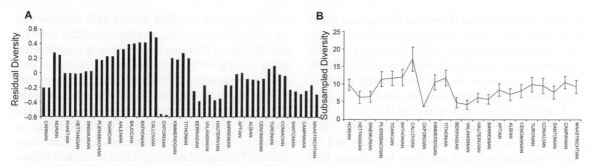

Figure 10.3 Two common methods for "correcting" dinosaur diversity curves to ameliorate the effects of sampling biases. (A) The residual approach depicts how much diversity is present in each interval after sampling is taken into account (by taking the observed diversity for each interval and subtracting how much diversity is predicted to be present if sampling is a perfect indicator of diversity, based on a correlation analysis of diversity and sampling proxies). (B) A subsampling approach in which a statistical computer program simulates what the diversity curve would look like if even sample sizes were available for each time interval. Both images modified from Mannion et al. (2011a), and reproduced with permission. They are modified versions of figures in Mannion et al. (2011a), which depict sauropod diversity across the Mesozoic.

number of discrete fossil collections made by dinosaur paleontologists – may be a better representation of both geological and anthropomorphic sampling (Lloyd et al., 2008; Butler et al., 2011; Upchurch et al., 2011). Importantly, this proxy is also significantly correlated with counts of dinosaur diversity over time, suggesting that correlation between diversity and sampling is not simply a figment of coarse sampling proxies such as dinosaur-bearing formations. That being said, further work is needed to determine if other, perhaps more sophisticated, sampling proxies also correlate with dinosaur diversity, but compiling detailed records of sampling is time-consuming and difficult.

Second, even if a strong correlation between sampling and diversity holds true, it is important to remember that this is a general relationship. The two measures will not always match perfectly, and it is possible to identify intervals in which diversity is higher or lower than expected after taking sampling into account (the "residual" approach: Barrett et al., 2009, Butler et al., 2011; Upchurch et al., 2011) (Fig. 10.3A). Similarly, it is possible to use statistical modeling techniques, such as the subsampling method employed by Lloyd et al. (2008), to try to ameliorate the effects of sampling biases (see also Lloyd, 2011 and Benson and Mannion, 2011 for similar types of analyses) (Figs 10.2 and 10.3B). These methods allow researchers to be honest and open about the limitations of the fossil record, and

determine what diversity patterns are most likely to be robust given inescapable (but quantifiable) sampling biases.

Despite the overall correlation between dinosaur diversity and sampling, several patterns do seem to be genuine. First, it is clear that dinosaurs did not explosively diversify immediately after they arose during the Late Triassic, but rather that sauropodomorphs underwent a major increase in diversification during the final part of the Late Triassic and theropods and ornithischians during the Early Jurassic (Upchurch et al., 2011). Second, there seems to be a genuine loss of dinosaur diversity across the Jurassic–Cretaceous boundary (Upchurch et al., 2011), and this is especially pronounced for sauropodomorphs (Upchurch and Barrett, 2005; Barrett et al., 2009; Benson and Mannion, 2011; Mannion et al., 2011a). Third, there is no overall trend of increasing dinosaur diversity throughout the Mesozoic, but rather that a supposed diversity peak during the final 20 million years of the Cretaceous is probably due to the fact that this interval boasts substantially more fossiliferous rock than most other intervals (Lloyd et al., 2008; Lloyd, 2011; Upchurch et al., 2011). Fourth, there is no clear evidence for a gradual decline in dinosaur diversity during the final several million years of the Cretaceous, as would be expected if dinosaurs wasted away to extinction, and any changes in diversity during this interval were on a similar scale to diversity increases and decreases throughout most of the remainder of the Mesozoic (Fastovsky et al., 2004; Wang and Dodson, 2006; Lloyd et al., 2008; Upchurch et al., 2011). Although Barrett et al. (2009) did report a decrease in diversity during the Late Cretaceous, this is probably an artifact of their methodology (Upchurch et al., 2011).

Additionally, fifth, there is no significant correlation between Mesozoic dinosaur diversity and sea levels, arguing against a common hypothesis that variations in sea levels drove dinosaur diversity (Butler et al., 2011). Sixth, focusing intensely on the dinosaur record of western North America during the latest Cretaceous, there seems to be little overall difference in the beta diversity (differences in the types of species found in an area) between different regions (Fig. 10.4), suggesting that a single major dinosaur community ranged across the entire western North American landscape (Vavrek and Larsson, 2010; although see alternative, but non-numerically supported, arguments by Sampson et al., 2010 and Longrich et al., 2010a). Seventh, and finally, global patterns of dinosaur diversity indicate that species were most diverse in the temperate middle latitudes, not the warmer equatorial latitudes as are present-day faunas (Mannion et al., 2011b).

Phylogenetic trends in dinosaur diversity

Are certain dinosaur subclades significantly more diverse than others? If so, are these clades concentrated in a certain portion of the dinosaur cladogram or in a specific time interval? In order to consider a clade as "exceptionally diverse," it must be demonstrated that the group in question is more species-rich than predicted by chance. The most straightforward way to test this is to compare an observed cladogram with a null expectation for how lineages should ideally split over time if such splitting is random. Random splitting can be approximated using an equal-rates "birth–death" model, which assumes that each lineage has an equal, but independent, probability of splitting at any given time (Chan and Moore, 2002; Nee, 2006; Ricklefs, 2007b; Purvis, 2008).

Using the birth–death model as a basis for comparison, the analytical program SymmeTree can determine significantly diverse clades on a phylogeny (Chan and Moore, 2005). First, SymmeTree calculates what is called a delta shift statistic (Δ_2) for each node on a given phylogeny, which assesses the likelihood that one sister taxon is significantly more lineage-rich than the other. However, this observed statistic itself does not indicate whether a clade is exceptionally diverse. In order to make this assessment, it must be determined how the observed statistic compares with a range of statistics calculated from random data. Therefore, SymmeTree calculates simulated statistics for hundreds of thousands of trees that include the same number of taxa as the real tree, but with different (random) phylogenetic relationships that are generated by the birth–death model. If the observed Δ_2 shift statistic falls within the upper 5% tail of the hundreds of thousands of randomized statistics, then the clade it describes is considered to be significantly more diverse than random. Oftentimes such clades are described as having a "significant diversification shift" at their bases.

This type of analysis has been used to study dinosaur phylogeny in general (Lloyd et al., 2008), as well as dinosaur phylogeny within the larger context of archosaur genealogy (Brusatte et al., 2011b). Lloyd et al. (2008) found that few dinosaur clades were exceptionally species-rich, and those that passed the statistical test were concentrated near the base of the dinosaur cladogram and mostly estimated to have diverged during the Early Jurassic (Fig. 10.5A). Significant diversification shifts therefore occurred early in dinosaur history, and dinosaur diversification rates during most of the Jurassic and Cretaceous were approximately steady. Brusatte et al. (2011b) addressed the early diversification of archosaurs more broadly, by subjecting a comprehensive phylogeny of most Triassic and Early Jurassic taxa to a SymmeTree analysis (Fig. 10.5B). Surprisingly, they found only a single archosaur clade to be more species-rich than predicted by chance: a group of basal saurischian dinosaurs. Two other clades, Dinosauria and Sauropodomorpha, exhibited delta shift statistics that were nearly significant, but barely missed the 95% probability cutoff. On the contrary, no clades on the crurotarsan side of archosaur phylogeny were found to have significant or near-significant diversification shifts at their bases. Therefore, when assessed within the context of archosaur phylogeny in general, most of early dinosaur

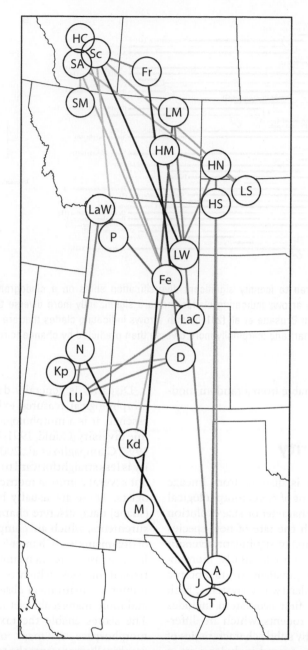

Figure 10.4 Regional dinosaur diversity during the Late Cretaceous in North America showing the similarities between individual fossil localities, with each locality indicated by a circle. The more similar two localities are to one another (based on dinosaur species present), the darker the line connecting them; the more dissimilar, the lighter the line. The important result is that there is no clear pattern: nearby localities are not more similar to each other than distant localities, and no clusters of similarity (indicative of regional faunas) are present. Many localities, on the other hand, are more similar to distant localities than nearby ones. This indicates that there is no strong evidence for regional endemicity of dinosaur faunas during the Late Cretaceous, but rather that a single major assemblage of dinosaurs roamed across western North America at this time. Modified from Vavrek, M.J., Larsson, H.C.E. (2010) Low beta diversity of Maastrichtian dinosaurs of North America. *Proceedings of the National Academy of Sciences (USA)* 107, 8265–8268, and reproduced with permission.

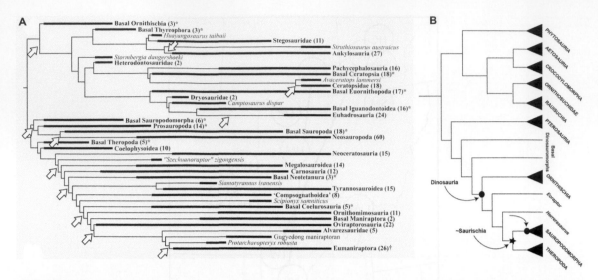

Figure 10.5 Using SymmeTree to identify significant diversification shifts on a cladogram: (A) dinosaur supertree from Lloyd et al. (2008) with arrows indicating clades that are significantly more diverse than predicted by chance; (B) archosaur supertree from Brusatte et al. (2011b) with arrows indicating clades that are significantly more diverse than predicted by chance (star) and marginally more diverse than predicted by chance (circle). Image (A) courtesy of Dr Graeme Lloyd.

evolution is indistinguishable from a random model of lineage splitting.

Dinosaur Disparity

Morphological evolution is distinct from lineage evolution: the development of novel morphological features or high rates of character or size evolution need not correspond with the rate of new species formation (lineage splitting) or significant diversification shifts on the cladogram. Similarly, morphological variety is not equivalent to taxonomic variety (diversity). Consider two ecosystems, both with three species. The first ecosystem includes three species of rat-sized rodents, which are differentiated from each other by subtle characteristics of the skeleton, whereas the second includes a rat, a hippo, and an elephant. Both ecosystems are equal in diversity (number of species), but the second includes a much greater amount of morphological variety, including a greater range of body size, dietary and ecological habits, and skeletal characteristics among the three species. This second ecosystem is therefore said to exhibit higher disparity.

Disparity is generally defined as the variety of morphological features exhibited by a group; in essence, it is a morphological analogue to taxonomic diversity (Gould, 1991; Wills et al., 1994; Foote, 1997; Ciampaglio et al., 2001; Erwin, 2007). Disparity is less straightforward to measure than diversity, but several common metrics are used by paleontologists. These are usually based on morphometric (shape) data, discrete character data, or size measurements, which are compiled for a large number of organisms (see Chapter 5 for more details). These large datasets are then subject to multivariate statistical analysis, which combines and distills the numerous anatomical observations into a smaller and more manageable set of scores for each taxon. The scores enable the taxa to be plotted into a morphospace, a "map" of morphologies which graphically represents the total range of anatomical features exhibited by the group (see Figs 5.6 and 10.6A). In this morphospace, taxa that are similar to each other are plotted close together, whereas those that are progressively more different are plotted in increasingly distant regions. The location of taxa in morphospace is used to calculate various statistics that quantify disparity. The two

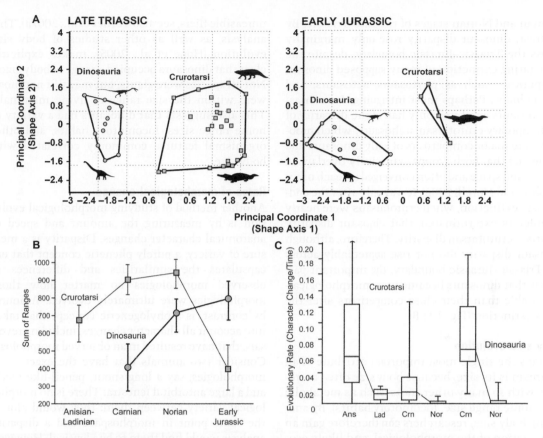

Figure 10.6 Morphological disparity and rates of change in early dinosaurs. (A) Morphospace plots for Late Triassic and Early Jurassic archosaurs depicting the relative morphospace size of dinosaurs and crurotarsans (modified from Brusatte et al., 2008b). (B) Morphological disparity plot for dinosaurs and crurotarsans across the Middle Triassic to Early Jurassic. Note that crurotarsans were significantly more disparate than dinosaurs during the Triassic, but dinosaurs became more disparate than crurotarsans after crurotarsan disparity crashed at the end-Triassic extinction (statistical significance denoted by non-overlap of error bars between two measurements) (modified from Brusatte et al., 2008a, 2008b, 2010b). (C) Morphological rates of evolution for crurotarsans and dinosaurs during the Late Triassic. Note that rates are highest during the early history of each clade and that Carnian dinosaurs have a statistically higher rate of change than coexisting crurotarsans (but that rates of change were indistinguishable in the two groups across the Late Triassic as a whole) (modified from Brusatte et al., 2008a).

most useful and intuitively understandable metrics are range and variance. Range denotes the entire size of morphospace occupied by the taxa in question (i.e., the overall spread of their morphological variation), whereas variance measures the spread of taxa in morphospace compared with the center (i.e., the average dissimilarity among members of the group).

Invertebrate groups such as trilobites and crinoids have been the frequent subject of disparity studies (Wills et al., 1994; Foote, 1997), but vertebrate groups have received comparatively little attention. One of the few vertebrate groups subjected to disparity analysis are Triassic and Early Jurassic dinosaurs and their archosaurian contemporaries, which were studied using a large discrete character dataset (Brusatte et al., 2008a, 2008b, 2011b, 2011c) (Fig. 10.6B). These studies found that dinosaur disparity generally increased during this time, with a significant increase between the

Carnian and Norian stages of the Late Triassic. By contrast, dinosaur disparity rose only marginally across the Triassic–Jurassic boundary, despite the devastating extinction of many supposed dinosaur competitors such as phytosaurs, aetosaurs, and rauisuchians. Perhaps most intriguing, Late Triassic dinosaurs exhibited only half of the disparity of contemporary crurotarsan archosaurs, which probably were major competitors of early dinosaurs due to the fact that both groups extensively overlapped in time and space and often converged on each other (Nesbitt and Norell, 2006). It was only after the end-Triassic extinction, when crurotarsans were nearly completely exterminated, that dinosaur disparity overtook crurotarsan disparity. Therefore, although dinosaur disparity did not rise appreciably across the Triassic–Jurassic boundary, the important pattern is that dinosaurs became more morphologically variable than their close competitors after the mass extinction (Fig. 10.6B).

Body size evolution

Perhaps the single most important attribute of an organism is its size, because body size often correlates with so many other traits (such as metabolic rate, home range size, locomotory habits). By analyzing body size, researchers can therefore gain an appreciation of the morphological, and likely ecological and behavioral, differences among species. In a comprehensive analysis of size evolution in dinosaurs, Carrano (2006) mapped out the estimated body size (using femur measurements as a proxy) of dinosaur species on an enormous cladogram and used a parsimony algorithm to predict the likely body sizes for hypothetical ancestors at each node on the tree. This allowed for the explicit identification of major size trends across the tree, including instances of size increases and decreases (see also Hone et al., 2005 and Irmis, 2011 for similar studies). Carrano (2006) found that major dinosaur clades consistently underwent trends in body size increase during the Mesozoic, and only two clades (coelurosaurian theropods and macronarian sauropods) exhibited significant size decreases. Exactly why these size changes occurred is difficult to understand, but at the very least the anomalous trend of decreasing body size in coelurosaurs was instrumental in the origin of avian flight (because large animals are aerodynamically unfeasible fliers; see also Turner et al., 2007a). This analysis, as well as other studies of body size evolution (Hone et al., 2005), makes explicitly clear that dinosaurs occupied a range of body sizes. Not all species were giants, but many (if not most) were within the size range of living mammals. This also suggests that dinosaurs had a variety of home range sizes, locomotory habits, and other organismal features commonly correlated with body size.

Rates of morphological change

Another method of studying morphological evolution is by measuring the amount and speed of anatomical character changes. Disparity is a measure of variety, a purely phenetic concept that encapsulates the similarities and differences of observed morphologies no matter how those morphologies were ultimately evolved. Amount, by contrast, is a phylogenetic concept that takes into account all character changes, including reversals, that have resulted in an observed morphology. Consider two animals that have the exact same morphologies, say a long snout, pencil-like teeth, and a large antorbital fenestra. There is no morphological variety between them: they would plot at the same point in morphospace and a disparity analysis would find them to be identical. However, perhaps one animal evolved from an ancestor with a short skull, stout teeth, and a small antorbital fenstra (three character changes), whereas the other evolved from an ancestor with a long snout, pencil-like teeth, and a small antorbital fenestra (one character change). In this case, the amount of evolution (one vs. three changes) differs between the two animals despite the fact that they look identical. The speed of evolution may also differ between the two animals: perhaps one evolved its morphological features over a shorter period of time than the other, and therefore underwent a higher rate of evolution.

Studies of amount and rate require a phylogeny for context, because it is necessary to know the number and sequence of character changes on the lines to observed morphologies and how much time has occurred between branching or speciation events (Wagner, 1997; Ruta et al., 2006; Brusatte, 2011). Put another way, it is necessary to measure how many characters change on each branch of the tree and over what length of time that branch existed (i.e., over what length of time those characters were changing). The number of characters changing on a branch is the amount, and the amount divided by the time duration is the rate. Amounts and rates can be measured across the phylogeny and binned according to time or clade, which gives an idea as to whether some time intervals or groups of organisms exhibited more or less, and faster or slower, evolution than others.

Morphological amounts and rates have been assessed for Triassic to Early Jurassic archosaurs (Brusatte et al., 2008a) (Fig. 10.6C) and amounts have been studied for theropod dinosaurs on the evolutionary line toward birds (Dececchi and Larsson, 2009). The most salient results of the Triassic to Early Jurassic study were that dinosaurs and their crurotarsan competitors had statistically indistinguishable amounts and rates of change during the Late Triassic, and that dinosaur rates were significantly highest early in their Triassic history (Carnian) compared with later time intervals (Fig. 10.6C). The latter result is consistent with a long-standing evolutionary hypothesis that morphological rates are highest during the early phases of a group's evolution (Gould, 2002). The most provocative finding of the theropod study is that amounts of forelimb evolution are essentially constant across theropod phylogeny, not concentrated in birds and their immediate ancestors (as may be expected because birds possess an aberrant forearm relative to other dinosaurs, used to power their volant lifestyle).

Dinosaur Molecular Evolution

The phylogeny and evolutionary rates of living species are often studied by reference to DNA sequences, proteins, and other molecular substances. Cladograms are often built by analyzing DNA and using computer algorithms to group species that share derived changes in the genetic sequence, much like organisms are grouped together as close relatives in a morphological phylogenetic analysis by their possession of shared derived features of the anatomy. When a molecular phylogeny

is available, the amount and speed of morphological evolution can be assessed by mapping out character changes onto the branches of the tree, much as how morphological evolution can be quantified using the rate methods discussed above. However, these types of studies are essentially impossible for dinosaurs and other extinct animals because microscopic soft tissues such as DNA are rarely, if ever, fossilized. Therefore, little is known about the molecular biology and evolution of dinosaurs, a problem that will probably always frustrate paleontologists.

Nevertheless, one major aspect of molecular evolution can be studied for dinosaurs. Because there is a tight correlation in living animals between the size of individual cells and the size of the genome (the total amount of DNA in an organism's genetic code), the genome sizes of dinosaurs can be estimated by measuring the sizes of the lacunae that hold osteocytes (bone cells) within dinosaur bones (Gregory, 2001). Genome size is an important measurement, because like body size it often correlates with many other organismal features such as metabolic rate and home range size. Furthermore, genome size in Mesozoic dinosaurs is an intriguing subject of study, because their descendants, living birds, have anomalously small genomes that have long been explained as an adaptation for saving energy (a necessary requirement for a metabolically expensive volant lifestyle).

Organ et al. (2007, 2009) compiled measurements of bone cell size for numerous dinosaurs, used this information to estimate genome size, and came to several important conclusions (Fig. 10.7). First, the small genome sizes of living birds are not unique, but are also seen in many of their closest non-flying theropod relatives. This is firm evidence that a small genome size was not an adaptation for flight in birds, but was simply inherited from distant ancestors. Second, saurischians have significantly smaller genomes than ornithischians, suggesting that there is something unique to the molecular biology of both major dinosaur subgroups. Third, there is no correlation between genome size and body size: *Apatosaurus*, the largest dinosaur measured in these studies, has an estimated genome size that is approximately the same as that of the primitive crow-sized bird *Confuciusornis*. These results are intriguing, but studies of

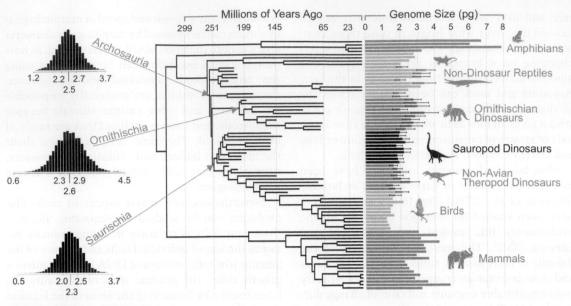

Figure 10.7 Genomic evolution in dinosaurs. Estimations of genome size (in picograms) of various dinosaurs, shown in context within a phylogeny of extant tetrapods with known genome sizes. The three histograms to the left depict the estimated ancestral genome sizes of three major clades (Archosauria, Ornithischia, Saurischia); the mean represents estimated genome sizes and the spread around the mean denotes uncertainty in these estimates. Note that saurischian dinosaurs have substantially smaller estimated genome sizes than ornithischians. Image courtesy of Dr Chris Organ and modified from Organ et al. (2009).

dinosaurian genomic and molecular evolution are still in their infancy and much work remains to be done (see Montanari et al., 2011).

The Grand Narrative of Dinosaur Evolution

An ever-expanding fossil record, studied carefully using rigorous phylogenetic and macroevolutionary methods, allows contemporary paleontologists to draft the grand story of dinosaur evolution in lucid detail. Like many historical tales, the narrative of dinosaur history must be pieced together using fragments – often incomplete fossil specimens from a biased geological record – but a commitment to quantitative rigor and a healthy realization of uncertainties allows these snippets to be interpreted in a meaningful way. Also like many historical tales, our understanding of dino-

saur evolution is consistently being updated in light of new discoveries and studies. Such is the burden, but also the excitement, of the sort of historical detective work that paleontologists specialize in. What we hold as true today may not survive the discoveries and testing of the next generation of researchers. With that caveat in mind, this book closes with a summary – a story – of the 160 million year Mesozoic evolutionary history of dinosaurs as understood by contemporary workers.

The origin and early evolution of dinosaurs

Footprints and body fossils clearly indicate that the archosaur clade originated in the immediate aftermath of the Permian–Triassic mass extinction, the greatest interval of mass death in earth history (Brusatte et al., 2011a, 2011b; Nesbitt et al., 2011) (Fig. 10.8). Dinosauromorpha, the dinosaur "stem lineage," must have originated around this time as well. The oldest dinosauromorph footprints are recorded merely 1–2 million years after the extinc-

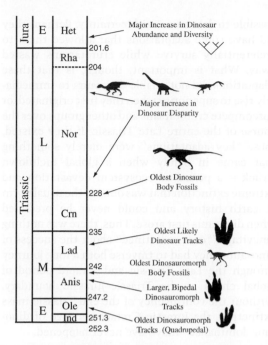

Figure 10.8 A general timeline of the first 50 million years of dinosauromorph history during the Triassic and Early Jurassic. Image modified from Brusatte et al. (2011a).

tion, although the oldest body fossils are not known until a few million years later (Nesbitt et al., 2010; Brusatte et al., 2011a). True dinosaurs (saurischians and ornithischians), which are distinguished from other dinosauromorphs by the fiat of a phylogenetic definition (see Chapter 1), probably originated in the Middle Triassic (c.240 million years ago), but their first unequivocal body fossils are found in Argentine rocks dating from near the Carnian–Norian boundary (c.228 million years ago) (Martinez et al., 2011). The dinosaurs of this assemblage, perhaps unexpectedly, are already quite morphologically and taxonomically diverse and somewhat abundant. Representatives of the three major subgroups (theropods, sauropodomorphs, and ornithischians) are present in this fauna, and dinosaur fossils comprise approximately 10% of the total vertebrate fossil record from these rocks (Rogers et al., 1993; Sereno et al., 1993; Brusatte et al., 2010b; Langer et al., 2010; Martinez et al., 2011).

These earliest dinosaurs, however, were a far cry from the iconic creatures that dominated Jurassic

and Cretaceous terrestrial ecosystems. They were still only a minor component of their faunas, and exhibited only a fraction of the taxonomic and morphological diversity, as well as the geographic range, that would characterize later dinosaur groups. The rise of dinosaurs – their journey from an ecologically and taxonomically marginal group to the preeminent terrestrial vertebrates of the later Mesozoic – has fascinated paleontologists for generations (Fig. 10.8). It was once accepted wisdom that dinosaurs rapidly rose to dominance during the Triassic, marshaling superior locomotory and physiological adaptations to outcompete other terrestrial vertebrates such as crurotarsan archosaurs (Bakker, 1971, 1972; Charig, 1984). Over the past few decades, however, an overwhelming weight of new fossil evidence, phylogenetic studies, and macroevolutionary analyses have dismissed this scenario as simplistic and incorrect (Benton, 1983, 2004; Irmis et al., 2007a; Brusatte et al., 2008a, 2008b, 2010b, 2011b; Langer et al., 2010; Irmis, 2011).

It is now understood that the dinosaur radiation was a complicated event that unfolded gradually, over the course of tens of millions of years (Fig. 10.8). In fact, about 50 million years elapsed between the origin of the dinosauromorph clade and the time, in the aftermath of the end-Triassic mass extinction, that dinosaurs could truly stake a claim as the most diverse, disparate, abundant, and widespread mid- to large-sized vertebrates in terrestrial ecosystems globally. Dinosaurs originated long before they became taxonomically diverse, morphologically disparate, and numerically abundant in their ecosystems (Brusatte et al., 2008a, 2008b). Morphological disparity and high rates of character change peaked early in dinosaur history (Carnian–Norian), millions of years before, and out of step with, taxonomic diversity (which underwent a significant increase in the Early Jurassic: Lloyd et al., 2008) (Fig. 10.6). The earliest dinosaurs were rare constituents in their ecosystems, but became progressively more abundant in the latest Triassic and finally, in the Early Jurassic, were consistently the most abundant large terrestrial vertebrates all across the globe (Benton, 1983).

Dinosaurs lived alongside many other major clades of terrestrial vertebrates during the first 30 million years of their history, but there is no sign that dinosaurs gradually outcompeted them. The crurotarsan archosaurs, which were likely competitors to early dinosaurs, were more morphologically disparate than dinosaurs throughout the Late Triassic, and the two groups exhibited indistinguishable rates of character evolution (Brusatte et al., 2008a, 2008b) (Fig. 10.6). Furthermore, crurotarsans were more numerically abundant than dinosaurs in many Late Triassic ecosystems (Benton, 1983). Dinosaurs eclipsed their close crurotarsan cousins only after the end-Triassic extinction, an unpredictable event of earth history that devastated global ecosystems and nearly wiped out the entire crurotarsan clade. Dinosaurs maintained their morphological disparity – their range of body plans, diets, and ecologies – across the extinction boundary, but most major crurotarsan body plans and major clades (phytosaurs, aetosaurs, rauisuchians) were exterminated. Now dinosaurs were more disparate and taxonomically diverse than crurotarsans, and also invariably more numerically abundant in global ecosystems (Benton, 1983; Brusatte et al., 2008a, 2008b, 2011b). There is also provocative evidence from footprints that dinosaurs substantially expanded in body size across the extinction interval as well (Olsen et al., 2002; although see alternative view in Irmis, 2011).

In sum, it seems as if the end-Triassic extinction was a critical turning point in dinosaur history. Before the extinction there was no overwhelming sign that dinosaurs would become so successful and dominant: they were merely one group of terrestrial vertebrates sharing ecosystems with other diverse, disparate, and abundant clades such as crurotarsans. Crurotarsans, it could be argued, were doing better than dinosaurs during the Late Triassic, based on their larger morphospaces, greater range of body types and diets, and numerical dominance in many ecosystems. However, the end-Triassic extinction decimated crurotarsans but apparently did not affect dinosaurs to any significant degree. Other more minor extinctions during the Late Triassic may have also wiped out or severely culled additional dinosaur competitor groups, such as the dicynodonts and rhynchosaurs, but spared dinosaurs (Benton, 1994, 2004; but see Irmis, 2011 for an alternative view).

Exactly why dinosaurs made it through these extinctions is currently unknown, and maybe im-possible to determine with certainty. Perhaps they did have some adaptations that allowed them to preferentially survive while crurotarsans wasted away. What is important, though, is that these adaptations did not enable dinosaurs to immediately rise to superiority when they first originated, or outcompete crurotarsans and other groups over the course of the entire Late Triassic. If they existed, these "key adaptations" were merely something that came in handy when a global meltdown struck – a period of ecosystem devastation and extreme extinction that was outside the usual norm of earth history and could never be predicted when dinosaurs first arose. Thus, there was nothing "inevitable" or "predestined" about the success of dinosaurs: they had to traverse both a long journey through the Late Triassic and a rapid period of global chaos at the Triassic–Jurassic boundary. Without the contingency of the end-Triassic mass extinction, the Jurassic–Cretaceous "age of dinosaur dominance" may have never happened.

The intervening years

After emerging from the end-Triassic devastation 200 million years ago, and diversifying taxonomically and morphologically during the Early Jurassic, dinosaurs remained the dominant vertebrates in most terrestrial ecosystems until their demise at the end of the Cretaceous, 65 million years ago. During this 130 million year interval, dinosaurs were the most diverse and abundant mid- to large-sized terrestrial vertebrates in most regions of the globe, and expanded into an astounding array of sizes, body types, and ecological niches. Some sauropod dinosaurs were the largest animals to ever live on land, nearly equaling the blue whale in size, whereas some theropods were the largest terrestrial predators to ever live. Some dinosaurs were fast runners, others burrowers, and one subgroup of small nimble predators evolved the ability to fly – one of the most remarkable evolutionary transformations in the history of life.

It seems as if birds, as well as many other major dinosaur subgroups, originated during the Early to Middle Jurassic, which was an explosive period of dinosaur diversification (Lloyd et al., 2008). Many dinosaur groups were restricted to their own corners of the globe at this time, likely due to climatic differences and the fragmentation of Pangaea

(Upchurch et al., 2002). It is difficult to understand how dinosaur diversity changed during most of the Jurassic and Cretaceous, because sampling biases so pervasively infect the dinosaur fossil record (Lloyd et al., 2008; Barrett et al., 2009; Butler et al., 2011; Upchurch et al., 2011). However, at the very least, it seems as if there was a substantial extinction event at the Jurassic–Cretaceous boundary 146 million years ago (Barrett et al., 2009; Upchurch et al., 2011), and it also seems like the major pulse of lineage splitting during the Early Jurassic was never again equaled in dinosaur history (Lloyd et al., 2008). Dinosaur faunas had a more cosmopolitan flavor in the Early Cretaceous, as many clades had a global distribution (Upchurch, 1995; Brusatte et al., 2009b; Benson et al., 2010a). During the latest Cretaceous, however, as the continents drifted further apart and high seas lapped onto the land, many dinosaur groups were isolated on their own endemic landmasses. Dinosaur diversity seems to have changed little during the final years of the Cretaceous: once uneven sampling is taken into account, there is no apparent significant increase or decrease in global diversity until all non-avian dinosaurs disappeared forever at the Cretaceous–Paleogene boundary (but see Lloyd, 2011).

Dinosaur Extinction

Why did the dinosaurs go extinct? Perhaps no other question in the history of paleontological research has generated such unbridled and fanciful speculation. Yet the riddle of dinosaur extinction has also motivated generations of careful detailed studies that creatively blend information from geology, biology, chemistry, physics, astronomy, and other disciplines. It truly is a puzzle: how was such a thoroughly dominant group as the dinosaurs exterminated from the face of the earth, after 160 million years of evolutionary success? Many opinions have been thrown around by scientists, but because the main focus of this book is on the biology and evolution of dinosaurs as living animals, there will be limited pontification here on dinosaur extinction. Readers are encouraged to follow up with some of the thousands of papers written on the subject, which are expertly summarized and discussed in books by Archibald (1996), Alvarez (1997),

Dingus and Rowe (1997), and Powell (1998), as well as the recent review paper of Schulte et al. (2010).

The meaning of the dinosaur extinction

Although the issue of dinosaur extinction may seem like a tired subject, some discussion is still warranted here. Clearly, understanding why non-avian dinosaurs went extinct is an important thread of evidence in understanding their biology and evolution, the overriding themes of this book. From a wider perspective, answering the riddle of the dinosaur extinction – which was but one aspect of a broader mass extinction at the end of the Cretaceous, 65 million years ago, that also wiped out or severely affected many other groups – has clear implications for understanding the effects of major environmental perturbations on organisms and ecosystems. The end-Cretaceous mass extinction is one of five such mass die-offs in the fossil record, and has undoubtedly been the most studied of the bunch (Raup and Sepkoski, 1982). While great extinctions and global catastrophes will always arouse the interest of the general public, the true importance of understanding the dinosaur extinction really relates to comprehending our place in nature, and coming to grips with the causes and effects of periods of mass global devastation. After all, dinosaurs were an immensely successful group that dominated ecosystems for a long stretch of time, but their reign eventually came to an end. There is mounting evidence that human-induced pollution and overpopulation is actively causing a so-called "sixth mass extinction," whose effects, and perhaps very reality, are obscured by humanity's inability to comprehend long-term trends (Barnosky et al., 2011). Only by studying previous extinction intervals in the fossil record can we hope to understand what actually happens during periods of worldwide extinction, and perhaps find a way to stop them.

Explanations for the dinosaur extinction

What do we currently know about the extinction of dinosaurs? Over the past century of research over 100 possible hypotheses have been presented to explain the extinction of dinosaurs and other organisms at the end of the Cretaceous, but most are so unlikely, or completely unsupported by evidence, that they cannot be taken seriously. The major competing theories hold that dinosaurs went extinct either suddenly or gradually at the end of the Cretaceous, either as the result of an asteroid or comet impact (suddenly) or due to long periods of volcanic eruptions, climate change, and sea-level fluctuations (gradually). It is also possible that a combination of multiple mechanisms may explain dinosaur extinction, with gradually changing climates or prolonged volcanic eruptions weakening dinosaur populations before a sudden impact delivered the final blow. And of course it must be remembered that not all dinosaurs went extinct: most did, but several lineages of birds made it through the end-Cretaceous extinction into the ensuing Paleogene period, and their descendants fill the skies today as approximately 10,000 species of extant living breathing dinosaurs.

What information, then, is most important when trying to understand the extinction of the non-avian dinosaurs? First of all, based on the pioneering work of the father-and-son team of Luis and Walter Alvarez and the ensuing research of hundreds of colleagues, there is unequivocal evidence that an enormous asteroid or comet hit the planet approximately 65 million years ago, at the same general time the dinosaurs went extinct. As first outlined by the electrifying study of Alvarez et al. (1980), a thin clay layer marks the Cretaceous–Paleogene boundary at many sites across the world, and it is invariably enriched in iridium, an element exceedingly rare on earth but common in extraterrestrial objects (Fig. 10.9). Not only that, but this layer is also bursting with other substances that can only result from asteroid or comet impacts, such as

Figure 10.9 The Cretaceous–Paleogene boundary, which marks the time that the non-avian dinosaurs went extinct, at the famous Bottaccione Gorge in Gubbio, Italy. It was at this site that Walter Alvarez and his colleagues first sampled an anomalous level of iridium in the thin clay layer marking the Cretaceous–Paleogene boundary, which was the first indication that an extraterrestrial impact may have caused the dinosaur extinction. Photo on the right shows the author (Steve Brusatte) examining the Cretaceous–Paleogene boundary clay layer, with Walter Alvarez in the foreground right. Photo on the left is a close-up of the clay layer, which is marked with an arrow. Photos by Nicole Lunning.

quartz grains whose mineral structure has been altered by intense pressure and spherules that are formed as liquefied material thrown up by an impact cools and falls back through the atmosphere (Alvarez et al., 1980; Alvarez, 1997; Smit, 1999; Schulte et al., 2010). Most convincing, however, is the presence of a huge crater, nearly 200 km wide, near Chicxulub, on the Yucatan Peninsula of Mexico, whose age exactly matches that of the clay layer (Hildebrand et al., 1991). There is no doubt, therefore, that a giant planetary body hit earth at the same time the dinosaurs went extinct.

That being said, there is also firm evidence that the dinosaur extinction happened in the aftermath of two other episodes of global perturbation. First, the latest Cretaceous was also a time of substantial sea-level regression (Haq et al., 1987; Peters, 2008). Sea levels apparently fell rapidly, and substantially, during the Maastrichtian (the final stage of the Cretaceous), and this quick fluctuation may have been among the most extreme sea-level changes during the entire Mesozoic. Second, there was a vast volcanic eruption in what is present-day India, which lasted for at least several tens of thousands of years and may have covered over $1,000,000\,km^2$ with lava (Wignall, 2001; Keller et al., 2008). This eruption, which produced the so-called Deccan Trap deposits that are more than 2000 m thick and spread over $500,000\,km^2$, seems to have occurred very close to the end of the Cretaceous, and may have even been occurring when the Chicxulub bolide hit (Keller et al., 2008).

Given that so many considerable global changes were occurring at, or near, the end of the Cretaceous, it may be exceedingly difficult to identify the primary causes of the mass extinction (Archibald et al., 2010). This is compounded by the fact that so many groups of organisms went extinct or were severely depleted at this time, and it is possible that different groups were individually affected by some, but not all, of the global catastrophes. Marine invertebrates, for instance, may have lost large swaths of habitable territory when sea levels dropped, but fluctuating oceans may not have affected land-living dinosaurs, which may have been more susceptible to acid rain and temperature drops caused by the bolide impact or an atmosphere poisoned by volcanic eruptions. As this book focuses on dinosaurs, I will not detail the myriad possible effects of end-Cretaceous global catastrophes on all ecosystems, but rather home in on what we currently know about the Late Cretaceous dinosaur fossil record and what this may tell us about the possible cause of their extinction.

Focusing on dinosaurs, it may be assumed that diversity patterns should hold a clue as to whether their extinction was sudden or gradual, which in turn would identify certain extinction agents as more or less likely. If, for instance, dinosaur diversity was decreasing throughout the Late Cretaceous then this would be a sign of a gradual extinction, whereas if it was steady (or increasing) but then rapidly fell at the end of the Cretaceous then this would be a hallmark of sudden demise. Sadly, the quality of the fossil record makes testing these scenarios difficult. Late Cretaceous dinosaur fossils are well known only from a few regions of the world, most notably western North America, and it is no guarantee that these samples are representative of any global reality. However, when these fossils are studied in detail, and statistical techniques are used to remedy sampling biases, there appears to be no major decrease in diversity near the end of the Age of Dinosaurs, either globally (Fastovsky et al., 2004; Wang and Dodson, 2006; Lloyd et al., 2008; Upchurch et al., 2011) or on the regional level within individual dinosaur-bearing rock deposits (Sheehan et al., 1991; Lillegraven and Eberle, 1999; Fastovsky and Sheehan, 2005). Again, this does not necessarily mean that there was no true reduction in diversity, but only that the current samples show no evidence for a precipitous drop prior to the Cretaceous–Paleogene boundary itself.

What the dinosaur fossil record clearly does not show is a major long-term decline in dinosaur diversity (both globally and locally) of the type that would be expected if dinosaurs underwent a prolonged (≥ 10 million year) regression that gradually resulted in their extinction (although see Lloyd, 2011). Therefore, out of all the uncertainty surrounding the dinosaur extinction, there is one conclusion that seems robust. The extinction of dinosaurs (and most likely other organisms at the end of the Cretaceous) was due to one or multiple remarkably unusual events that were outside the norm of usual geological happenstance and which were so extreme that they could have affected the entire planet. The extinction of the dinosaurs was therefore not the inevitable end product of millions of years of stagnating evolution, but was intimately tied to unpredictable global catastrophe.

That being said, is it possible to make a more general conclusion about the single most likely cause of the dinosaur extinction? Many researchers may disagree, but it is held here that the single best hypothesis for explaining the dinosaur extinction is that an overwhelming, unpredictable, and abrupt extraterrestrial impact was primarily responsible. It is known that an enormous asteroid or comet hit the planet at exactly the same time that the dinosaurs went extinct, and there is no solid evidence that dinosaurs were declining in diversity prior to this sudden impact. It is true that sea-level changes and volcanic eruptions were also occurring during the Late Cretaceous, and these may have indeed played a supporting role in the dinosaur extinction, but sea levels and climates were constantly changing during the 160 million year Age of Dinosaurs. Why, exactly, would changes in sea level or an increase in global temperature due to prolonged eruptions, on their own, completely wipe out non-avian dinosaurs when changes of similar magnitude had failed to do so for hundreds of millions of years? Similarly, if later studies do show a decrease in dinosaur diversity near the end of the Cretaceous, then why would this decrease result in total extinction when other diversity decreases during the Mesozoic were merely blips in the evolutionary history of dinosaurs? In the end, it would be a remarkable coincidence if the simultaneous (or near simultaneous) extinction of a long-lived and successful group and the impact of a giant extraterrestrial body were unrelated to each other. And we must not forget that not only dinosaurs went extinct: so did many other species that occupied a wide range of environments, ranging from planktonic microorganisms to ammonites. New evidence may always raise new questions, but for the time being, a sudden extraterrestrial impact is the best single unifying theory for explaining why so many species, including dinosaurs, went extinct 65 million years ago.

The post-impact world and the rise of mammals
Regardless of whether the Chicxulub bolide impact was the single most important cause of the dinosaur extinction, two things are undoubted. First, the impact – which we know definitely happened because of the reams of geological evidence reviewed above – would have had a profound effect on global ecosystems. Clouds of dust and debris kicked up by the impact would have entered the atmosphere and perhaps blocked out the sun for a long period of time, inhibiting photosynthesis and thereby causing a cascade of ecosystem collapse both in the oceans and on land (see reviews in Alvarez, 1997 and Pope et al., 1998). Intense infrared radiation would have occurred in the immediate wake of the impact, followed by toxic showers of acid rain, global wildfires, and mega-tsunamis. Amazingly, there is copious evidence in the geological record for many of these devastating consequences of impact (Wolbach et al., 1985; Prinn and Fegley, 1987; Bourgeois et al., 1988; MacDougall, 1988; Maurrasse and Sen, 1991; Alvarez, 1997; Pope et al., 1997, 1998; Schulte et al., 2010). It is no stretch to imagine how these various traumas may have affected dinosaurs and other organisms, although there is still considerable debate as to why all non-avian dinosaurs went extinct while other groups (such as turtles, lizards, birds, and mammals) were not completely wiped out (Robertson et al., 2004). This question, therefore, remains one of the fundamental riddles of the dinosaur extinction, and is sure to generate substantial research as our understanding of the Late Cretaceous fossil and geological record continues to improve.

Second, in the aftermath of both the impact itself and the extinction of the dinosaurs new groups of vertebrates had the opportunity to evolve and diversify in the Paleocene and Eocene. Chief among these were the birds, the evolutionary descendants of theropod dinosaurs (and, to be fair, true dinosaurs in their own right), and the mammals. The ensuing 65 million years after the end-Cretaceous extinction is commonly referred to as the Age of Mammals, and for good reason. Although mammals originated during the Triassic, around the same time that the first dinosaurs evolved, they remained mostly small and ecologically marginal during most of the Mesozoic, and their species-level diversity and range of body types and sizes paled in comparison to those of dinosaurs (Kielan-Jaworowska et al., 2004; Luo, 2007). In the Paleocene and Eocene, however, mammals explosively diversified: their range of body sizes increased to include much larger forms, their species-level diversity rose substantially, they began to exploit many niches and dietary strategies that were unavailable to them during the Mesozoic, and the earliest representatives of many major living clades appeared in the fossil record (Alroy, 1999; Rose, 2006; Archibald, 2011). The dinosaurs had enjoyed a 160 million year evolutionary run, but because of the contingencies of earth history mammals had now become the pre-eminent mid- to large-sized vertebrates in most terrestrial ecosytems. Dinosaurs, however, remained in the form of their own successful and diverse offshoot, the birds, which also enjoyed their own profound success during the past 65 million years.

Conclusions

Understanding the grand story of dinosaur evolution, from their Triassic origins to the end-Cretaceous extinction of all non-avian species, is one of the primary goals of many dinosaur researchers. It is also one of the great contributions of dinosaur paleontology to the wider knowledge of science in general, because we stand to learn many fundamental truths about evolution if we can understand how dinosaurs originated, diversified, evolved in concert with changing climates and drifting continents, and went extinct. Although sampling biases are always a source of concern, the ever-expanding dinosaur fossil record is proving to be a fruitful source of macroevolutionary studies. Researchers commonly use large databases of information from the dinosaur record to perform large-scale statistical analyses that examine major patterns in dinosaur diversity, disparity (anatomical variety), rates of morphological change, and faunal abundance across the Mesozoic. These studies reveal that dinosaurs did not explosively diversify after they originated in the Triassic, but rather the rise of dinosaurs was a gradual and complex event that unfolded over tens of millions of years. Dinosaurs split into their fundamental lineages (sauropodomorphs, theropods, ornithischians) and evolved a wide array of morphologies and body plans in the Late Triassic before undergoing their most significant period of diversification during the Early Jurassic, after the end-Triassic mass extinction wiped out several competitor groups. There is no sign that dinosaurs outcompeted other groups during the Late Triassic, but rather dinosaurs were able to endure one or more mass extinction events that preferentially decimated other clades. Many major dinosaur subgroups, including birds, probably originated in the Early to Middle Jurassic, there was a substantial extinction of dinosaur species across the Jurassic–Cretaceous boundary, and there is no sign that dinosaurs gradually withered away to extinction during the Late Cretaceous. Instead, it seems most likely that the impact of an enormous extraterrestrial body suddenly snuffed out the Age of Dinosaurs at the Cretaceous–Paleogene boundary, 65 million years ago. One small subgroup of dinosaurs, the birds, was able to survive this extinction, and today numbers approximately 10,000 living species.

References

Abler, W.L. (1992) The serrated teeth of tyrannosaurid dinosaurs, and biting structures in other animals. *Paleobiology* 18, 161–183.

Adams, D.C., Rohlf, F.J., Slice, D.E. (2004) Geometric morphometrics: ten years of progress following the "revolution." *Italian Journal of Zoology* 71, 5–16.

Ahlberg, P.E., Clack, J.A., Blom, H. (2005) The axial skeleton of the Devonian tetrapod *Ichthyostega*. *Nature* 437, 137–140.

Alexander, D.E., Gong, E., Martin, L.D., Burnham, D.A., Falk, A.R. (2010) Model tests of gliding with different hindwing configurations in the four-winged dromaeosaurid *Microraptor gui*. *Proceedings of the National Academy of Sciences USA* 107, 2972–2976.

Alexander, R.McN. (1976) Estimates of speeds of dinosaurs. *Nature* 261, 129–130.

Alexander, R.McN. (1985) Mechanics of gait and posture of some large dinosaurs. *Zoological Journal of the Linnean Society* 83, 1–25.

Alexander, R.McN. (1989) *Dynamics of Dinosaurs and Other Extinct Giants*. Columbia University Press, New York.

Alexander, R.McN. (2006) Dinosaur biomechanics. *Proceedings of the Royal Society of London, Series B* 273, 1849–1855.

Ali, F., Zelenitsky, D.K., Therrien, F., Weishampel, D.B. (2008) Homology of the "ethmoid complex" of tyrannosaurids and its implications for the reconstruction of the olfactory apparatus of non-avian theropods. *Journal of Vertebrate Paleontology* 28, 123–133.

Allain, R., Aquesbi, N. (2008) Anatomy and phylogenetic relationships of *Tazoudasaurus naimi* (Dinosauria, Sauropoda) from the late Early Jurassic of Morocco. *Geodiversitas* 30, 345–424.

Alroy, J. (1999) The fossil record of North American mammals: evidence for a Paleocene evolutionary radiation. *Systematic Biology* 48, 107–118.

Alvarez, L.W., Alvarez, W., Asaro, F., Michel, H.V. (1980) Extraterrestrial cause for the Cretaceous–Tertiary extinction. *Science* 208, 1095–1108.

Alvarez, W. (1997) *T. rex and the Crater of Doom*. Princeton University Press, Princeton, NJ.

Amiot, R., Lécuyer, C., Buffetaut, E., Escarguel, G., Fluteau, F., Martineau, F. (2006) Oxygen isotopes from biogenic apatites suggest widespread endothermy in Cretaceous dinosaurs. *Earth and Planetary Science Letters* 246, 41–54.

Amiot, R., Buffetaut, E., Lécuyer, C. et al. (2010) Oxygen isotope evidence for semi-aquatic habits among spinosaurid theropods. *Geology* 38, 139–142.

Anderson, J.F., Hall-Martin, A., Russell, D.A. (1985) Long-bone circumference and weight in mammals, birds and dinosaurs. *Journal of Zoology* 207, 53–61.

Andrade, M.B., Young, M.T., Desojo, J.B., Brusatte, S.L. (2010) The evolution of extreme hypercarnivory in Metriorhynchidae (Mesoeucrocodylia: Thalattosuchia) based on

Dinosaur Paleobiology, First Edition. Stephen L. Brusatte.
© 2012 John Wiley & Sons, Ltd. Published 2012 by John Wiley & Sons, Ltd.

evidence from microscopic denticle morphology. *Journal of Vertebrate Paleontology* 30, 1451–1465.

Apesteguía, S. (2004) *Bonitasaura salgadoi* gen. et sp. nov.: a beaked sauropod from the Late Cretaceous of Patagonia. *Naturwissenschaften* 91, 493–497.

Apesteguía, S. (2005) Evolution of the hyposphene–hypantrum complex within Sauropoda. In: Tidwell, V., Carpenter, K. (eds.), *Thunder Lizards: The Sauropodomorph Dinosaurs*. Indiana University Press, Bloomington, IN, pp. 248–267.

Arbour, V.M. (2009) Estimating impact forces of tail club strikes by ankylosaurid dinosaurs. *PLoS ONE* 4, e6738.

Arbour, V.M., Snively, E. (2009) Finite element analyses of ankylosaurid dinosaur tail club impacts. *Anatomical Record* 292, 1412–1426.

Archibald, J.D. (1996) *Dinosaur Extinction and the End of an Era*. Columbia University Press, New York.

Archibald, J.D. (2011) *Extinction and Radiation: How the Fall of Dinosaurs Led to the Rise of Mammals*. Johns Hopkins University Press, Baltimore.

Archibald, J.D., Clemens, W.A., Padian, K. et al. (2010) Cretaceous extinctions: multiple causes. *Science* 328, 973.

Archibald, J.K., Mort, M.E., Crawford, D.J. (2003) Bayesian inference of phylogeny: a non-technical primer. *Taxon* 52, 187–191.

Asara, J.M., Schweitzer, M.H., Freimark, L.M., Phillips, M., Cantley, L.C. (2007) Protein sequences from Mastodon and *Tyrannosaurus rex* revealed by mass spectrometry. *Science* 316, 280–285.

Bailey, J.B. (1997) Sailbacks or buffalo-backs? Neural spine elongation in dinosaurs. *Journal of Paleontology* 71, 1124–1146.

Baird, D. (1980) A prosauropod dinosaur trackway from the Navajo Sandstone (Lower Jurassic) of Arizona. In: Jacobs, L.L. (ed.), *Aspects of Vertebrate History*. Museum of Northern Arizona Press, Flagstaff, AZ, pp. 219–230.

Bakhurina, N.N., Unwin, D.M. (1995) A preliminary report on the evidence for "hair" in *Sordes pilosus*, an Upper Jurassic pterosaur from Middle Asia. In: Sun, A., Wang, Y. (eds.), *Sixth Symposium on Mesozoic Terrestrial Ecosystems and Biota. Short Papers*. China Ocean Press, Beijing, pp. 79–82.

Bakker, R.T. (1971) Dinosaur physiology and the origin of mammals. *Evolution* 25, 636–658.

Bakker, R.T. (1972) Anatomical and ecological evidence of endothermy in dinosaurs. *Nature* 238, 81–85.

Bakker, R.T. (1978) Dinosaur feeding behaviour and the origin of flowering plants. *Nature* 274, 661–663.

Bakker, R.T. (1986) *The Dinosaur Heresies*. William Morrow, New York.

Bakker, R.T., Galton, P.M. (1974) Dinosaur monophyly and a new class of vertebrates. *Nature* 248, 168–172.

Balanoff, A.M., Norell, M.A., Grellet-Tinner, G., Lewin, M.R. (2008) Digital preparation of a probable neoceratopsian preserved within an egg, with comments on microstructral anatomy of ornithischian eggshells. *Naturwissenschaften* 95, 493–500.

Balanoff, A.M., Xu, X., Kobayashi, Y., Matsufune, Y., Norell, M.A. (2009) Cranial osteology of the theropod dinosaur *Incisivosaurus gauthieri* (Theropoda: Oviraptorosauria). *American Museum Novitates* 3651, 1–35.

Balanoff, A.M., Bever, G.S., Ikejiri, T. (2010) The braincase of *Apatosaurus* (Dinosauria: Sauropoda) based on computed tomography of a new specimen with comments on variation and evolution in sauropod neuroanatomy. *American Museum Novitates* 3677, 1–29.

Barden, H.E., Maidment, S.C.R. (2011) Evidence for sexual dimorphism in the stegosaurian dinosaur *Kentrosaurus aethiopicus* from the Upper Jurassic of Tanzania. *Journal of Vertebrate Paleontology* 31, 641–651.

Barnosky, A.D., Matzke, N., Tomiya, S. et al. (2011) Has the Earth's sixth mass extinction already arrived? *Nature* 471, 51–57.

Barrett, P.M. (2000) Prosauropod dinosaurs and iguanas: speculations on the diets of extinct reptiles. In: Sues, H.-D. (ed.), *Evolution of Herbivory in Terrestrial Vertebrates*. Cambridge University Press, Cambridge, pp. 42–78.

Barrett, P.M. (2001) Tooth wear and possible jaw action of *Scelidosaurus harrisonii* Owen and a review of feeding mechanisms in other thyreophoran dinosaurs. In: Carpenter, K. (ed.), *The Armored Dinosaurs*. Indiana University Press, Bloomington, IN, pp. 25–52.

Barrett, P.M. (2005) The diet of ostrich dinosaurs (Theropoda: Ornithomimosauria). *Palaeontology* 48, 347–358.

Barrett, P.M., Rayfield, E.J. (2006) Ecological and evolutionary implications of dinosaur feeding behaviour. *Trends in Ecology and Evolution* 21, 217–224.

Barrett, P.M., Upchurch, P. (1994) Feeding mechanisms of *Diplodocus*. *Gaia* 10, 195–203.

Barrett, P.M., Upchurch, P. (2007) The evolution of feeding mechanisms in early sauropodomorph dinosaurs. *Special Papers in Palaeontology* 77, 91–112.

Barrett, P.M., Willis, K.J. (2001) Did dinosaurs invent flowers? Dinosaur–angiosperm coevolution revisited. *Biological Reviews* 76, 411–447.

Barrett, P.M., Hasegawa, Y., Manabe, M., Isaji, S., Matsuoka, H. (2002) Sauropod dinosaurs from the Lower Cretaceous of eastern Asia: taxonomic and biogeographical implications. *Palaeontology* 45, 1197–1217.

Barrett, P.M., McGowan, A.J., Page, V. (2009) Dinosaur diversity and the rock record. *Proceedings of the Royal Society of London, Series B* 276, 2667–2674.

Barrett, P.M., Butler, R.J., Nesbitt, S.J. (2011a) The roles of herbivory and omnivory in early dinosaur evolution. *Earth and Environmental Science Transactions of the Royal Society of Edinburgh* 101, 383–396.

Barrett, P.M., Benson, R.B.J., Rich, T.H., Vickers-Rich, P. (2011b) First spinosaurid dinosaur from Australia and the cosmopolitanism of Cretaceous dinosaur faunas. *Biology Letters* doi: 10.1098/rsbl.2011.0466

Barrick, R.E., Showers, W.J. (1994) Thermophysiology of *Tyrannosaurus rex*: evidence from oxygen isotopes. *Nature* 265, 222–224.

Barrick, R.E., Showers, W.J. (1995) Oxygen isotope variability in juvenile dinosaurs (*Hypacrosaurus*): evidence for thermoregulation. *Paleobiology* 21, 552–560.

Barrick, R.E., Showers, W.J., Fischer, A.G. (1996) Comparison of thermoregulation of four ornithischian dinosaurs and a varanid lizard from the Cretaceous Two Medicine Formation: evidence from oxygen isotopes. *Palaios* 11, 295–305.

Barrick, R.E., Stoskopf, M.K., Marcot, J.D., Russell, D.A., Showers, W.J. (1998) The thermoregulatory functions of the *Triceratops* frill and horns: heat flow measured with oxygen isotopes. *Journal of Vertebrate Paleontology* 18, 746–750.

Barthel, K.W., Swinburne, N.H.M., Conway Morris, S. (1990) *Solnhofen: A Study in Mesozoic Palaeontology*. Cambridge University Press, Cambridge.

Bates, K.T., Manning, P.L., Vila, B., Hodgetts, D. (2008a) Three dimensional modeling and analysis of dinosaur trackways. *Palaeontology* 51, 999–1010.

Bates, K.T., Rarity, F., Manning, P.L. et al. (2008b) High-resolution LiDAR and photogrammetric survey of the Fumanya dinosaur tracksites (Catalonia): implications for the conservation and interpretation of geological heritage sites. *Journal of the Geological Society London* 165, 115–127.

Bates, K.T., Manning, P.L., Hodgetts, D., Sellers, W.I. (2009a) Estimating mass properties of dinosaurs using laser imaging and 3D computer modeling. *PLoS ONE* 4, e4532.

Bates, K.T., Falkingham, P.L., Breithaupt, B.H., Hodgetts, D., Sellers, W.I., Manning, P.L. (2009b) How big was "Big Al"? Quantifying the effect of soft tissue and osteological unknowns on mass predictions for *Allosaurus*. *Palaeontologia Electronica* 12(3), Article 12.3.14A.

Baumel, J.J., Witmer, L.M. (1993) Osteologia. In: Baumel, J.J., King, A.S., Brazile, J.E., Evans, H.E., Vanden Berge, J.C. (eds.), *Handbook of Avian Anatomy: Nomina Anatomica Avium*, 2nd edn. Nuttall Ornithologial Society, Cambridge, pp. 45–132.

Behrensmeyer, A.K., Kidwell, S.M. (1985) Taphonomy's contribution to paleobiology. *Paleobiology* 11, 105–119.

Behrensmeyer, A.K., Kidwell, S.M., Gastaldo, R.A. (2000) Taphonomy and paleobiology. *Paleobiology* 26 (suppl.), 103–147.

Bell, P.R., Snively, E., Shychoski, L. (2009) A comparison of the jaw mechanics in hadrosaurid and ceratopsid dinosaurs using finite element analysis. *Anatomical Record* 292, 1338–1351.

Benson, R.B.J. (2010) A description of *Megalosaurus bucklandii* (Dinosauria: Theropoda) from the Bathonian of the United Kingdom and the relationships of Middle Jurassic theropods. *Zoological Journal of the Linnean Society* 158, 882–935.

Benson, R.B.J., Mannion, P.D. (2011) Multivariate models are essential for understanding vertebrate diversification in deep time. *Biology Letters* (in press).

Benson, R.B.J., Xu, X. (2008) The anatomy and systematic position of the theropod dinosaur *Chilantaisaurus tashuikouensis* Hu, 1964 from the Early Cretaceous of Alanshan, People's Republic of China. *Geological Magazine* 145, 778–789.

Benson, R.B.J., Carrano, M.T., Brusatte, S.L. (2010a) A new clade of archaic large-bodied predatory dinosaurs (Theropoda: Allosauroidea) that survived to the latest Mesozoic. *Naturwissenschaften* 97, 71–78.

Benson, R.B.J., Barrett, P.M., Rich, T.H., Vickers-Rich, P. (2010b) A southern tyrant reptile. *Science* 327, 1613.

Benson, R.B.J., Butler, R.J., Carrano, M.T., O'Connor, P.M. (2011) Air-filled postcranial bones in theropod dinosaurs: physiological implications and the "reptile"–bird transition. *Biological Reviews* doi: 10.1111/j.1469-185X.2011.00190.x

Benton, M.J. (1983) Dinosaur success in the Triassic: a noncompetitive ecological model. *Quarterly Review of Biology* 58, 29–55.

Benton, M.J. (1984) The relationships and early evolution of the Diapsida. *Symposium of the Zoological Society of London* 52, 575–596.

Benton, M.J. (1994) Late Triassic to Middle Jurassic extinctions among terrestrial tetrapods: testing the pattern. In: Fraser, N.C., Sues, H.-D. (eds), *In the Shadow of the Dinosaurs: Early Mesozoic Tetrapods*. Cambridge University Press, Cambridge, pp. 366–397.

Benton, M.J. (1999) *Scleromochlus taylori* and the origin of dinosaurs and pterosaurs. *Philosophical Transactions of the Royal Society of London, Series B* 354, 1423–1446.

Benton, M.J. (2003) *When Life Nearly Died: The Greatest Mass Extinction of All Time*. Thames and Hudson, London.

Benton, M.J. (2004) Origin and relationships of Dinosauria. In: Weishampel, D.B., Dodson, P., Osmólska, H. (eds.), *The Dinosauria*, 2nd edn. University of California Press, Berkeley, CA, pp. 7–24.

Benton, M.J. (2005) *Vertebrate Palaeontology*, 3rd edn. Blackwell Publishing, Oxford.

Benton, M.J. (2010) The origins of modern biodiversity on land. *Transactions of the Royal Society of London, Series B* 365, 3667–3679.

Benton, M.J., Clark, J.M. (1988) Archosaur phylogeny and the relationships of the Crocodylia. In: Benton, M.J. (ed.), *The Phylogeny and Classification of the Tetrapods*. Systematics Association Special Volume 35A, pp. 295–338.

Benton, M.J., Tverdokhlebov, V.P., Surkov, M.V. (2004) Ecosystem remodelling among vertebrates at the Permian Triassic boundary in Russia. *Nature* 432, 97–100.

Benton, M.J., Csiki, Z., Grigorescu, D. et al. (2010) Dinosaurs and the island rule: the dwarfed dinosaurs from Haţeg Island. *Palaeogeography, Palaeoclimatology, Palaeoecology* 293, 438–454.

Bergmann, U., Morton, R.W., Manning, P.L. et al. (2010) *Archaeopteryx* feathers and bone chemistry fully revealed via synchrotron imaging. *Proceedings of the National Academy of Sciences USA* 107, 9060–9065.

Berner, R.A. (2006) GEOCARBSULF: a combined model for Phanerozoic atmospheric O_2 and CO_2. *Geochimica et Cosmochimica Acta* 70, 5653–5664.

Bever, G.S., Brusatte, S.L., Balanoff, A.M., Norell, M.A. (2011) Variation, variability, and the origin of the avian endocranium and inner ear: insights from the anatomy of *Alioramus altai* (Theropoda: Tyrannosauroidea). *PLoS ONE* 6, e23393.

Bittencourt, J. de. S., Kellner, A.W.A. (2009) The anatomy and phylogenetic position of the Triassic dinosaur *Staurikosaurus pricei* Colbert, 1970. *Zootaxa* 2079, 1–56.

Bonaparte, J.F. (1976) *Pisanosaurus mertii* Casamiquela and the origin of the Ornithischia. *Journal of Paleontology* 50, 808–820.

Bonaparte, J.F. (1985) A horned Cretaceous carnosaur from Patagonia. *National Geographic Research* 1, 149–151.

Bonaparte, J.F. (1991) The Gondwanan theropod families Abelisauridae and Noasauridae. *Historical Biology* 5, 1–25.

Bonaparte, J.F., Novas, F.E. (1985) *Abelisaurus comahuensis*, n.g., n.sp., carnosauria del Cretacico tardio de Patagonia. *Ameghiniana* 21, 259–265.

Bonaparte, J.F., Powell, J.E. (1980) A continental assemblage of tetrapods from the Upper Cretaceous beds of El Brete, northwestern Argentina (Sauropoda-Coelurosauria-Carnosauria-Aves). *Mémoires de la Société Géologique de France* 139, 19–28.

Bonaparte, J.F., Vince, M. (1979) El hallazgo del primer nido de Dinosaurios Triásicos (Saurischia, Prosauropoda) Triásico superior de Patagonia, Argentina. *Ameghiniana* 16, 173–182.

Bonde, N., Christiansen, P. (2003) The detailed anatomy of *Rhamphorhynchus*: axial pneumaticity and its implications. In: Buffetaut, E., Mazin, J.-M. (eds.), *Evolution and Palaeobiology of Pterosaurs*. Geological Society Special Publication No. 217, pp. 217–232. The Geological Society, London.

Bonnan, M.F. (2003) The evolution of manus shape in sauropod dinosaurs: implications for functional morphology, forelimb orientation, and phylogeny. *Journal of Vertebrate Paleontology* 23, 595–613.

Bonnan, M.F. (2004) Morphometric analysis of humerus and femur shape in Morrison sauropods: implications for functional morphology and paleobiology. *Paleobiology* 30, 444–470.

Bonnan, M.F. (2007) Linear and geometric morphometric analysis of long bone scaling patterns in Jurassic neosauropod dinosaurs: their functional and paleobiological implications. *Anatomical Record* 290, 1089–1111.

Bonnan, M.F., Senter, P. (2007) Were the basal sauropodomorph dinosaurs *Plateosaurus* and *Massospondylus* habitual quadrupeds? *Special Papers in Palaeontology* 77, 139–155.

Bonnan, M.F., Yates, A.M. (2007) A new description of the forelimb of the basal sauropodomorph *Melanorosaurus*: implications for the evolution of pronation, manus shape and quadrupedalism in

sauropod dinosaurs. *Special Papers in Palaeontology* 77, 157–168.

Bookstein, F.L. (1991) *Morphometric Tools for Landmark Data: Geometry and Biology*. Cambridge University Press, Cambridge.

Bourgeois, J., Hansen, T.A., Wiberg, P.L., Kauffman, E.G. (1988) A tsunami deposit at the Cretaceous–Tertiary boundary in Texas. *Science* 241, 567–570.

Briggs, D.E.G. (2003) The role of decay and mineralization in the preservation of soft-bodied fossils. *Annual Review of Earth and Planetary Sciences* 32, 275–301.

Briggs, D.E.G., Wilby, P.R., Pérez-Moreno, B.P., Sanz, J.L., Fregenal-Martínez, M. (1997) The mineralization of dinosaur soft tissue in the Lower Cretaceous of Las Hoyas, Spain. *Journal of the Geological Society London* 154, 587–588.

Brinkman, D.B., Sues, H.-D. (1987) A staurikosaurid dinosaur from the Upper Triassic Ischigualasto Formation of Argentina and the relationships of the Staurikosauridae. *Palaeontology* 30, 493–503.

Britt, B.B. (1993) *Pneumatic postcranial bones in dinosaurs and other archosaurs*. PhD thesis, University of Calgary, Calgary.

Britt, B.B. (1997) Postcranial pneumaticity. In: Currie, P.J., Padian, K. (eds.), *Encyclopedia of Dinosaurs*. Academic Press, New York, pp. 590–593.

Brochu, C.A. (2000) A digitally-rendered endocast for *Tyrannosaurus rex*. *Journal of Vertebrate Paleontology* 20, 1–6.

Brochu, C.A. (2003) Osteology of *Tyrannosaurus rex*: insights from a nearly complete skeleton and high-resolution computed tomographic analysis of the skull. *Society of Vertebrate Paleontology Memoir* 7, 1–138.

Brougham, J., Brusatte, S.L. (2010) Distorted *Microraptor* specimen is not ideal for understanding the origin of avian flight. *Proceedings of the National Academy of Sciences USA* 107, E155.

Brown, B., Schlaikjer, E. (1940) The structure and relationships of *Protoceratops*. *Annals of the New York Academy of Sciences* 40, 133–166.

Brown, R.E., Brain, J.D., Wang, N. (1997) The avian respiratory system: a unique model for studies of respiratory toxicosis and monitoring air quality. *Environmental Health Perspectives* 105, 188–200.

Brusatte, S.L. (2011) Calculating the tempo of morphological evolution: rates of discrete character change in a phylogenetic context. In: Elewa, A.M.T. (ed.), *Computational Paleontology*. Springer-Verlag, Heidelberg, Germany, pp. 53–74.

Brusatte, S.L., Sereno, P.C. (2007) A new species of *Carcharodontosaurus* (Dinosauria: Theropoda) from the Cenomanian of Niger and a revision of the genus. *Journal of Vertebrate Paleontology* 24, 902–916.

Brusatte, S.L., Sereno, P.C. (2008) Phylogeny of Allosauroidea (Dinosauria: Theropoda): comparative analysis and resolution. *Journal of Systematic Palaeontology* 6, 155–182.

Brusatte, S.L., Benson, R.B.J., Carr, T.D., Williamson, T.E., Sereno, P.C. (2007) The systematic utility of theropod enamel wrinkles. *Journal of Vertebrate Paleontology* 27, 1052–1056.

Brusatte, S.L., Benton, M.J., Ruta, M., Lloyd, G.T. (2008a) Superiority, competition, and opportunism in the evolutionary radiation of dinosaurs. *Science* 321, 1485–1488.

Brusatte, S.L., Benton, M.J., Ruta, M., Lloyd, G.T. (2008b) The first 50 Myr of dinosaur evolution: macroevolutionary pattern and morphological disparity. *Biology Letters* 4, 733–736.

Brusatte, S.L., Carr, T.D., Erickson, G.M., Bever, G.S., Norell, M.A. (2009a) A long-snouted, multihorned tyrannosaurid from the Late Cretaceous of Mongolia. *Proceedings of the National Academy of Sciences USA* 106, 17261–17266.

Brusatte, S.L., Benson, R.B.J., Chure, D J., Xu, X., Sullivan, C., Hone, D. (2009b) The first definitive carcharodontosaurid (Dinosauria: Theropoda) from Asia and the delayed ascent of tyrannosaurids. *Naturwissenschaften* 96, 1051–1058.

Brusatte, S.L., Benton, M.J., Desojo, J.B., Langer, M.C. (2010a) The higher-level phylogeny of Archosauria (Tetrapoda: Diapsida). *Journal of Systematic Palaeontology* 8, 3–47.

Brusatte, S.L., Nesbitt, S.J., Irmis, R.B., Butler, R.J., Benton, M.J., Norell, M.A. (2010b) The origin and early radiation of dinosaurs. *Earth-Science Reviews* 101, 68–100.

Brusatte, S.L., Chure, D.J., Benson, R.B.J., Xu, X. (2010c) The osteology of *Shaochilong maortuensis*, a carcharodontosaurid (Dinosauria: Theropoda) from the Late Cretaceous of Asia. *Zootaxa* 2334, 1–46.

Brusatte, S.L., Norell, M.A., Carr, T.D. et al. (2010d) Tyrannosaur paleobiology: new research on ancient exemplar organisms. *Science* 329, 1481–1485.

Brusatte, S.L., Niedźwiedzki, G., Butler, R.J. (2011a) Footprints pull origin and diversification of dinosaur stem-lineage deep into Early Triassic. *Proceedings of the Royal Society of London, Series B* 278, 1107–1113.

Brusatte, S.L., Benton, M.J., Lloyd, G.T., Ruta, M., Wang, S.C. (2011b) Macroevolutionary patterns in the evolutionary radiation of archosaurs (Tetrapoda: Diapsida). *Earth and Environmental Science Transactions of the Royal Society of Edinburgh* 101, 367–382.

Brusatte, S.L., Montanari, S., Yi, H.-Y., Norell, M.A. (2011c) Phylogenetic corrections for morphological disparity analysis: new methodology and case studies. *Paleobiology* 37, 1–22.

Bryant, H.N., Russell, A.P. (1992) The role of phylogenetic analysis in the inference of unpreserved attributes of extinct taxa. *Philosophical Transactions of the Royal Society of London, Series B* 337, 405–418.

Bryant, H.N., Russell, A.P. (1993) The occurrence of clavicles within Dinosauria: implications for the homology of the avian furcula and the utility of negative evidence. *Journal of Vertebrate Paleontology* 13, 171–184.

Buchholtz, E.A. (1997) Paleoneurology. In: Currie, P.J., Padian, K. (eds.), *Encyclopedia of Dinosaurs*. Academic Press, New York, pp. 522–524.

Buckley, L.G., Larson, D.W., Reichel, M., Samman, T. (2010) Quantifying tooth variation within a single population of *Albertosaurus sarcophagus* (Theropoda: Tyrannosauridae) and implications for identifying isolated teeth of tyrannosaurids. *Canadian Journal of Earth Sciences* 47, 1227–1251.

Buffetaut, E., Martill, D.M., Escuillié, F. (2004) Pterosaurs as part of a spinosaur diet. *Nature* 430, 33.

Buffrénil, V. de, Farlow, J.O., de Ricqlès, A. (1986) Growth and function in *Stegosaurus* plates: evidence from bone histology. *Paleobiology* 12, 459–473.

Butler, R.J. (2005) The "fabrosaurid" ornithischian dinosaurs of the Upper Elliot Formation (Lower Jurassic) of South Africa and Lesotho. *Zoological Journal of the Linnean Society* 145, 175–218.

Butler, R.J. (2010) The anatomy of the basal ornithischian dinosaur *Eocursor parvus* from the lower Elliot Formation (Late Triassic) of South Africa. *Zoological Journal of the Linnean Society* 160, 648–684.

Butler, R.J., Barrett, P.M. (2008) Palaeoenvironmental controls on the distribution of Cretaceous herbivorous dinosaurs. *Naturwissenschaften* 95, 1027–1032.

Butler, R.J., Smith, R.M.H., Norman, D.B. (2007) A primitive ornithischian dinosaur from the Late Triassic of South Africa, and the early evolution and diversification of Ornithischia. *Proceedings of the Royal Society of London, Series B* 274, 2041–2046.

Butler, R.J., Upchurch, P., Norman, D.B. (2008a) The phylogeny of the ornithischian dinosaurs. *Journal of Systematic Palaeontology* 6, 1–40.

Butler, R.J., Porro, L.B., Norman, D.B. (2008b) A juvenile skull of the primitive ornithischian dinosaur *Heterodontosaurus tucki* from the "Stormberg" of Southern Africa. *Journal of Vertebrate Paleontology* 28, 702–711.

Butler, R.J., Barrett, P.M., Gower, D.J. (2009a) Postcranial skeletal pneumaticity and air-sacs in the earliest pterosaurs. *Biology Letters* 5, 557–560.

Butler, R.J., Barrett, P.M., Kenrick, P., Penn, M.G. (2009b) Diversity patterns amongst herbivorous dinosaurs and plants during the Cretaceous: implications for hypotheses of dinosaur/angiosperm co-evolution. *Journal of Evolutionary Biology* 22, 446–459.

Butler, R.J., Galton, P.M., Porro, L.B., Chiappe, L.M., Henderson, D.M., Erickson, G.M. (2010a) Lower limits of ornithischian dinosaur body size inferred from a new Upper Jurassic heterodontosaurid from North America. *Proceedings of the Royal Society of London, Series B* 277, 375–381.

Butler, R.J., Barrett, P.M., Penn, M.G., Kenrick, P. (2010b) Testing coevolutionary hypotheses over geological timescales: interactions between Cretaceous dinosaurs and plants. *Biological Journal of the Linnean Society* 100, 1–15.

Butler, R.J., Benson, R.B.J., Carrano, M.T., Mannion, P.D., Upchurch, P. (2011) Sea level, dinosaur diversity and sampling biases: investigating the "common cause" hypothesis in the terrestrial realm. *Proceedings of the Royal Society of London, Series B* 278, 1165–1170.

Bybee, P.J., Lee, A.H., Lamm, E.T. (2006) Sizing the Jurassic theropod dinosaur *Allosaurus*: assessing growth strategy and evolution of ontogenetic scaling of limbs. *Journal of Morphology* 267, 347–359.

Cadbury D. (2002) *Terrible Lizard: The First Dinosaur Hunters and the Birth of a New Science.* Henry Holt and Company, New York.

Carlson, W.D., Rowe, T., Ketcham, R.A., Colbert, M.W. (2003) Geological applications of high-resolution X-ray computed tomography in petrology, meteoritics and paleontology. In: Mees, F., Swennen, R., Van Geet, M., Jacobs, P. (eds.), *Applications of X-ray Computed Tomography in the Geosciences.* Geological Society Special Publication No. 215, pp. 7–22. The Geological Society, London.

Carpenter, K. (1990) Variation in *Tyrannosaurus rex.* In: Carpenter, K., Currie, P.J. (eds.), *Dinosaur Systematics: Approaches and Perspectives.* Cambridge University Press, Cambridge, pp. 141–145.

Carpenter, K. (1997) Agonistic behavior in pachycephalosaurs (Ornithischia: Dinosauria): a new look at head-butting behavior. *Contributions to Geology, University of Wyoming* 32, 19–25.

Carpenter, K. (1998) Armor of *Stegosaurus stenops*, and the taphonomic history of a new specimen from Garden Park, Colorado. *Modern Geology* 22, 127–144.

Carpenter, K. (1999) *Eggs, Nests, and Baby Dinosaurs.* Indiana University Press, Bloomington, IN.

Carpenter, K. (2000) Evidence of predatory behaviour by dinosaurs. *Gaia* 15, 135–144.

Carpenter, K. (2001) Phylogenetic analysis of the Ankylosauria. In: Carpenter, K. (ed.), *The Armored Dinosaurs.* Indiana University Press, Bloomington, IN, pp. 455–484.

Carpenter, K. (2002) Forelimb biomechanics of nonavian theropod dinosaurs in predation. *Senckenbergiana Lethaea* 82, 59–76.

Carpenter, K., Smith, M.B. (2001) Forelimb osteology and biomechanics of *Tyrannosaurus rex*. In: Tanke, D.H., Carpenter, K. (eds.), *Mesozoic Vertebrate Life*. Indiana University Press, Bloomington, IN, pp. 90–116.

Carpenter, K., Miles, C.A., Cloward, K. (2001) New primitive stegosaur from the Morrison Formation, Wyoming. In: Carpenter, K. (ed.), *The Armored Dinosaurs*. Indiana University Press, Bloomington, IN, pp. 55–75.

Carpenter, K., Sanders, F., McWhinney, L., Wood, L. (2005) Evidence for predator–prey relationships: examples for *Allosaurus* and *Stegosaurus*. In: Carpenter, K. (ed.), *The Carnivorous Dinosaurs*. Indiana University Press, Bloomington, IN, pp. 325–350.

Carr, T.D. (1999) Craniofacial ontogeny in Tyrannosauridae (Dinosauria, Coelurosauria). *Journal of Vertebrate Paleontology* 19, 497–520.

Carr, T.D., Williamson, T.E. (2004) Diversity of late Maastrichtian Tyrannosauridac (Dinosauria: Theropoda) from western North America. *Zoological Journal of the Linnean Society* 142, 479–523.

Carr, T.D., Williamson, T.E. (2010) *Bistahieversor sealeyi*, gen. et. sp. nov., a new tyrannosauroid from New Mexico and the origin of deep snouts in Tyrannosauroidea. *Journal of Vertebrate Paleontology* 30, 1–16.

Carr, T.D., Williamson, T.E., Schwimmer, D.R. (2005) A new genus and species of tyrannosauroid from the Late Cretaceous (Middle Campanian) Demopolis Formation of Alabama. *Journal of Vertebrate Paleontology* 25, 119–143.

Carrano, M.T. (1998) Locomotion in non-avian dinosaurs: integrating data from hindlimb kinematics, in vivo strains, and bone morphology. *Paleobiology* 24, 450–469.

Carrano, M.T. (1999) What, if anything, is a cursor? Categories versus continua for determining locomotor habits in mammals and dinosaurs. *Journal of Zoology* 247, 29–42.

Carrano, M.T. (2000) Homoplasy and the evolution of dinosaur locomotion. *Paleobiology* 26, 489–512.

Carrano, M.T. (2001) Implications of limb bone scaling, curvature and eccentricity in mammals and non-avian dinosaurs. *Journal of Zoology* 254, 41–55.

Carrano, M.T. (2006) Body-size evolution in the Dinosauria. In: Carrano, M.T., Blob, R.W., Gaudin, T.J., Wible, J.R. (eds.), *Amniote Paleobiology*. University of Chicago Press, Chicago, IL, pp. 225–268.

Carrano, M.T., Biewener, A.A. (1999) Experimental alteration of limb posture in the chicken (*Gallus gallus*) and its bearing on the use of birds as analogs for dinosaur locomotion. *Journal of Morphology* 240, 237–249.

Carrano, M.T., Hutchinson, J.R. (2002) Pelvic and hindlimb musculature of *Tyrannosaurus rex* (Dinosauria: Theropoda). *Journal of Morphology* 253, 207–228.

Carrano, M.T., Sampson, S.D. (2008) The phylogeny of Ceratosauria (Dinosauria: Theropoda). *Journal of Systematic Palaeontology* 6, 183–236.

Carrano, M.T., Wilson, J.A. (2001) Taxon distributions and the tetrapod track record. *Paleobiology* 27, 563–581.

Carrano, M.T., Janis, C.M., Sepkoski, J.J. (1999) Hadrosaurs as ungulate parallels: lost lifestyles and deficient data. *Acta Palaeontologica Polonica* 44, 237–261.

Carrano, M.T., Sampson, S.D., Forster, C.A. (2002) The osteology of *Masiakasaurus knopfleri*, a small abelisauroid (Dinosauria: Theropoda) from the Late Cretaceous of Madagascar. *Journal of Vertebrate Paleontology* 22, 510–534.

Carrier, D.R., Walter, R.M., Lee, D.V. (2001) Influence of rotational inertia on turning performance of theropod dinosaurs. *Journal of Experimental Biology* 204, 3917–3926.

Carroll, R.L. (1988) *Vertebrate Paleontology and Evolution*. W.H. Freeman and Company, New York.

Case, T.J. (1978) Speculations on the growth rate and reproduction of some dinosaurs. *Paleobiology* 3, 320–328.

Castanet, J., Rogers, K.R., Cubo, J., Boisard, J. (2000) Periosteal bone growth rates in extant ratites (ostrich and emu). Implications for assessing growth in dinosaurs. *Life Sciences* 323, 543–550.

Cerda, I.A., Powell, J.E. (2010) Dermal armor histology of *Saltasaurus loricatus*, an Upper Cretaceous sauropod dinosaur from northwest Argentina. *Acta Palaeontologica Polonica* 55, 389–398.

Chambers, P. (2002) *Bones of Contention: The* Archaeopteryx *Scandals*. John Murray Publishers, London.

Chan, K.M.A., Moore, B.R. (2002) Whole-tree methods for detecting differential diversification rates. *Systematic Biology* 51, 855–865.

Chan, K.M.A., Moore, B.R. (2005) SymmeTREE: whole-tree analysis of differential diversification rates. *Bioinformatics* 21, 1709–1710.

Chapman, R.E. (1990) Shape analysis in the study of dinosaur morphology. In: Carpenter, K., Currie, P.J. (eds.), *Dinosaur Systematics: Approaches and Perspectives*. Cambridge University Press, Cambridge, pp. 21–42.

Chapman, R.E., Brett-Surman, M.K. (1990) Morphometric observations on hadrosaurid ornithopods. In: Carpenter, K., Currie, P.J. (eds.), *Dinosaur Systematics: Approaches and Perspectives*. Cambridge University Press, Cambridge, pp. 163–177.

Chapman, R.E., Galton, P.M., Sepkoski, J.J., Wall, W.P. (1981) A morphometric study of the cranium of the pachycephalosaurid dinosaur *Stegoceras*. *Journal of Paleontology* 55, 608–618.

Chapman, R.E., Weishampel, D.B., Hunt, G., Rasskin-Gutman, D. (1997) Sexual dimorphism in dinosaurs. In: Wolberg, D.L., Stump, E., Rosenberg, G. (eds.), *Dinofest International: Proceedings of a Symposium Sponsored By Arizona State University*. Academy of Natural Sciences, Philadelphia, pp. 83–93.

Charig, A.J. (1984) Competition between therapsids and archosaurs during the Triassic Period: a review and synthesis of current theories. *Symposium of the Zoological Society of London* 52, 597–628.

Charig, A.J., Milner, A.C. (1997) *Baryonyx walkeri*, a fish-eating dinosaur from the Wealden of Surrey. *Bulletin of the Natural History Museum London (Geology)* 53, 11–70.

Chatterjee, S., Templin, R.J. (2007) Biplane wing planform and flight performance of the feathered dinosaur *Microraptor gui*. *Proceedings of the National Academy of Sciences USA* 104, 1576–1580.

Chen, P., Dong, Z., Zhen, S. (1998) An exceptionally well-preserved theropod dinosaur from the Yixian Formation of China. *Nature* 391, 147–152.

Chiappe, L.M. (2007) *Glorified Dinosaurs: The Origin and Early Evolution of Birds*. John Wiley & Sons, Inc., Hoboken, NJ.

Chiappe, L.M., Witmer, L.M. (eds.) (2002) *Mesozoic Birds: Above the Heads of Dinosaurs*. University of California Press, Berkeley, CA.

Chiappe, L.M., Coria, R.A., Dingus, L., Jackson, F., Chinsamy, A., Fox, M. (1998) Sauropod dinosaur embryos from the Late Cretaceous of Patagonia. *Nature* 396, 258–261.

Chiappe, L.M., Schmitt, J.G., Jackson, F.D., Garrido, A., Dingus, L., Grellet-Tinner, G. (2004) Nest structure for sauropods: sedimentary criteria for recognition of dinosaur nesting traces. *Palaios* 19, 89–95.

Chin, K. (2007) The paleobiological implications of herbivorous dinosaur coprolites from the Upper Cretaceous Two Medicine Formation of Montana: why eat wood? *Palaios* 22, 554–566.

Chin, K., Gill, B.D. (1996) Dinosaurs, dung beetles, and conifers: participants in a Cretaceous food web. *Palaios* 11, 280–285.

Chin, K., Tokaryk, T.T., Erickson, G.M., Calk, K.C. (1998) A king-sized theropod coprolite. *Nature* 393, 680–682.

Chinnery, B.J. (2004) Morphometric analysis of evolutionary trends in the ceratopsian postcranial skeleton. *Journal of Vertebrate Paleontology* 24 591–609.

Chinnery, B.J., Horner, J.R. (2007) A new neoceratopsian dinosaur linking North American and Asian taxa. *Journal of Vertebrate Paleontology* 27, 625–641.

Chinsamy, A. (1990) Physiological implications of the bone histology of *Syntarsus rhodesiensis* (Saurischia: Theropoda). *Palaeontologia Africana* 27, 77–82.

Chinsamy, A. (1993) Bone histology and growth trajectory of the prosauropod dinosaur *Massospondylus carinatus* Owen. *Modern Geology* 18, 319–329.

Chinsamy, A. (1995) Ontogenetic changes in the bone histology of the Late Jurassic ornithopod

Dryosaurus lettowvorbecki. Journal of Vertebrate Paleontology 15, 96–104.

Chinsamy, A. (2002) Bone microstructure of early birds. In: Chiappe, L.M., Witmer, L.M. (eds.), *Mesozoic Birds: Above the Heads of Dinosaurs*. University of California Press, Berkeley, CA, pp. 421–431.

Chinsamy, A., Elzanowski, A. (2001) Evolution of growth pattern in birds. *Nature* 412, 402–403.

Chinsamy, A., Hillenius, W.J. (2004) Physiology of non-avian dinosaurs. In: Weishampel, D.B., Dodson, P., Osmólska, H. (eds.), *The Dinosauria*, 2nd edn. University of California Press, Berkeley, CA, pp. 643–659.

Chinsamy, A., Tumarkin-Deratzian, A. (2009) Pathologic bone tissues in a turkey vulture and a nonavian dinosaur: implications for interpreting endosteal bone and radial fibrolamellar bone in fossil dinosaurs. *Anatomical Record* 292, 1478–1484.

Chinsamy, A., Chiappe, L.M., Dodson, P. (1994) Growth rings in Mesozoic avian bones: physiological implications for basal birds. *Nature* 368, 196–197.

Chinsamy, A., Chiappe, L.M., Dodson, P. (1995) Mesozoic avian bone microstructure: physiological implications. *Paleobiology* 21, 561–574.

Chinsamy, A., Martin L.D., Dodson, P. (1998) Bone microstructure of diving *Hesperornis* and the volant *Ichthyornis* from the Niobrara Chalk of western Kansas. *Cretaceous Research* 19, 225–235.

Chinsamy-Turan, A. (2005) *The Microstructure of Dinosaur Bone*. Johns Hopkins University Press, Baltimore, MD.

Choiniere, J.N., Xu, X., Clark, J.M., Forster, C.A., Guo, Y., Han, F. (2010) A basal alvarezsauroid theropod from the Early Late Jurassic of Xinjiang, China. *Science* 327, 571–574.

Christian, A. (2010) Some sauropods raised their necks: evidence for high browsing in *Euhelops zdanskyi*. *Biology Letters* 6, 823–825.

Christiansen, N.A., Tschopp, E. (2010) Exceptional stegosaur integument impressions from the Late Jurassic Morrison Formation of Wyoming. *Swiss Journal of Geosciences* 103, 163–171.

Christiansen, P. (1997a) Forelimbs and hands. In: Currie, P.J., Padian, K. (eds.), *Encyclopedia of Dinosaurs*. Academic Press, New York, pp. 245–253.

Christiansen, P. (1997b) Hindlimbs and feet. In: Currie, P.J., Padian, K. (eds.), *Encyclopedia of Dinosaurs*. Academic Press, New York, pp. 320–328.

Christiansen, P. (1999) Long bone scaling and limb posture in nonavian theropods: evidence for differential allometry. *Journal of Vertebrate Paleontology* 19, 666–680.

Christiansen, P. (2000a) Strength indicator values of theropod long bones, with comments on limb proportions and cursorial potential. *Gaia* 15, 241–255.

Christiansen, P. (2000b) Feeding mechanisms of the sauropod dinosaurs *Brachiosaurus*, *Camarasaurus*, *Diplodocus* and *Dicraeosaurus*. *Historical Biology* 14, 137–152.

Christiansen, P., Bonde, N. (2004) Body plumage in *Archaeopteryx*: a review, and new evidence from the Berlin specimen. *Comptes Rendus Palevol* 3, 99–118.

Christiansen, P., Fariña, R.A. (2004) Mass prediction in theropod dinosaurs. *Historical Biology* 16, 85–92.

Chure, D., Britt, B.B., Whitlock, J.A., Wilson, J.A. (2010) First complete sauropod dinosaur skull from the Cretaceous of the Americas and the evolution of sauropod dinosaurs. *Naturwissenschaften* 97, 379–391.

Ciampaglio, C.N., Kemp, M., McShea, D.W. (2001) Detecting changes in morphospace occupation patterns in the fossil record: characterization and analysis of measures of disparity. *Paleobiology* 27, 695–715.

Clack, J. (2002) *Gaining Ground: The Origin and Early Evolution of Tetrapods*. Indiana University Press, Bloomington, IN.

Claessens, L.P.A.M. (2004) Dinosaur gastralia; origin, morphology, and function. *Journal of Vertebrate Paleontology* 24, 89–106.

Claessens, L.P.A.M., O'Connor, P.M., Unwin, D.M. (2009) Respiratory evolution facilitated the origin of pterosaur flight and aerial gigantism. *PLoS ONE* 4, e4497.

Clapham, M.E., Shen, S., Bottjer, D.J. (2009) The double mass extinction revisited: reassessing the severity, selectivity, and causes of the end-Guadalupian biotic crisis (Late Permian). *Paleobiology* 35, 32–50.

Clark, J.M., Perle, A., Norell, M.A. (1994) The skull of *Erlicosaurus andrewsi*, a Late Cretaceous "segnosaur" (Theropoda: Therizinosauridae) from Mongolia. *American Museum Novitates* 3115, 1–39.

Clark, J.M., Norell, M.A., Chiappe, L.M. (1999) An oviraptorid skeleton from the Late Cretaceous of Ukhaa Tolgod, Mongolia, preserved in an avian-like brooding position over an oviraptorid nest. *American Museum Novitates* 3265, 1–36.

Clark, J.M., Norell, M.A., Makovicky, P.J. (2002) Cladistic approaches to the relationship of birds to other theropod dinosaurs. In: Chiappe, L.M., Witmer, L.M. (eds.), *Mesozoic Birds: Above the Heads of Dinosaurs*. University of California Press, Berkeley, CA, pp. 31–64.

Clark, J.M., Maryanska, T., Barsbold, R. (2004) Therizinosauroidea. In: Weishampel, D.B., Dodson, P., Osmólska, H. (eds.), *The Dinosauria*, 2nd edn. University of California Press, Berkeley, CA, pp. 151–164.

Cleland, T.P., Stoskopf, M.K., Schweitzer, M.H. (2011) Histological, chemical, and morphological reexamination of the "heart" of a small Late Cretaceous *Thescelosaurus*. *Naturwissenschaften* 98, 203–211.

Clemens, W.A., Nelms, L.G. (1993) Paleoecological implications of Alaskan terrestrial vertebrate fauna in latest Cretaceous time at high latitude. *Geology* 21, 503–506.

Codd, J.R. (2010) Uncinate processes in birds: morphology, physiology and function. *Comparative Biochemistry and Physiology, Part A* 156, 303–308.

Codd, J.R., Boggs, D.F., Perry, S.F., Carrier, D.R. (2005) Activity of three muscles associated with the uncinate processes in the giant Canada goose *Branta canadensis maximus. Journal of Experimental Biology* 208, 849–857.

Codd, J.R., Manning, P.L., Norell, M.A., Perry, S.F. (2008) Avian-like breathing mechanics in maniraptoran dinosaurs. *Proceedings of the Royal Society of London, Series B* 275, 157–161.

Colbert, E.H. (1955) *Evolution of the Vertebrates.* Wiley, New York.

Colbert, E.H. (1962) The weights of dinosaurs. *American Museum Novitates* 2181, 1–24.

Colbert, E.H. (1970) A saurischian dinosaur from the Triassic of Brazil. *American Museum Novitates* 2405, 1–39.

Colbert, E.H. (1989) The Triassic dinosaur *Coelophysis. Bulletin of the Museum of Northern Arizona* 57, 1–160.

Colbert, E.H. (1990) Variation in *Coelophysis bauri*. In: Carpenter, K., Currie, P.J. (eds.), *Dinosaur Systematics: Approaches and Perspectives*. Cambridge University Press, Cambridge, pp. 81–90.

Colbert, E.H. (1993) Feeding strategies and metabolism in elephants and sauropod dinosaurs. *American Journal of Science* 293A, 1–19.

Company, J. (2011) Bone histology of the titanosaur *Lirainosaurus astibiae* (Dinosauria: Sauropoda) from the latest Cretaceous of Spain. *Naturwissenschaften* 98, 67–78.

Coombs, W.P. (1975) Sauropod habits and habitats. *Palaeogeography, Palaeoclimatology, Palaeoecology* 17, 1–33.

Coombs, W.P. (1978a) Forelimb muscles of the Ankylosauria (Reptilia, Ornithischia). *Journal of Paleontology* 52, 642–657.

Coombs, W.P. (1978b) Theoretical aspects of cursorial adaptations in dinosaurs. *Quarterly Review of Biology* 3, 393–418.

Cooper, L.N., Lee, A.H., Taper, M.L., Horner, J.R. (2008) Relative growth rates of predator and prey dinosaurs reflect effects of predation. *Proceedings of the Royal Society of London, Series B* 275, 2609–2615.

Cope, E.D. (1869) Synopsis of the extinct Batrachia, Reptilia and Aves of North America. *Transactions of the American Philosophical Society* 14, 1–252.

Coria, R.A, Salgado, L. (1995) A new giant carnivorous dinosaur from the Cretaceous of Patagonia. *Nature* 377, 224–226.

Cox, C.B. (1974) Vertebrate palaeodistributional patterns and continental drift. *Journal of Biogeography* 1, 75–94.

Crampton, J.S., Beu, A.G., Cooper, R.A., Jones, C.M., Marshall, B., Maxwell, P.A. (2003) Estimating the rock volume bias in paleobiodiversity studies. *Science* 301, 358–360.

Crisci, J.V., Katinas, L., Posadas, P. (2003) *Historical Biogeography*. Harvard University Press, Cambridge, MA.

Crompton, A.W., Charig, A.J. (1962) A new ornithischian from the Upper Triassic of South Africa. *Nature* 196, 1074–1077.

Cruickshank, A.R.I., Benton, M.J. (1985) Archosaur ankles and the relationships of the thecodontian and dinosaurian reptiles. *Nature* 317, 715–717.

Csiki, Z., Benton, M.J. (eds.) (2010) European island faunas of the Late Cretaceous: The Haţeg Island. *Palaeogeography, Palaeoclimatology, Palaeoecology* 293, 265–454.

Csiki, Z., Vremir, M., Brusatte, S.L., Norell, M.A. (2010) An aberrant island-dwelling theropod dinosaur from the Late Cretaceous of Romania. *Proceedings of the National Academy of Sciences USA* 107, 15357–15361.

Csontos, L., Vörös, A. (2004) Mesozoic plate tectonic reconstruction of the Carpathian region. *Palaeogeography, Palaeoclimatology, Palaeoecology* 210, 1–56.

Currie, P.J. (1993) A new troodontid (Dinosauria, Theropoda) braincase from the Dinosaur Park Formation (Campanian) of Alberta. *Canadian Journal of Earth Sciences* 30, 2231–2247.

Currie, P.J. (1997) Braincase anatomy. In: Currie, P.J., Padian, K. (eds.), *Encyclopedia of Dinosaurs*. Academic Press, New York, pp. 81–85.

Currie, P.J. (2003) Allometric growth in tyrannosaurids (Dinosauria: Theropoda) from the Upper Cretaceous of North America and Asia. *Canadian Journal of Earth Sciences* 40, 651–665.

Currie, P.J., Chen, P.J. (2001) Anatomy of *Sinosauropteryx prima* from Liaoning, northeastern China. *Canadian Journal of Earth Sciences* 38, 1705–1727.

Currie, P.J., Rigby, J.K., Sloan, R.E. (1990) Theropod teeth from the Judith River Formation of southern Alberta, Canada. In: Currie, P.J., Carpenter, K. (eds.), *Dinosaur Systematics*. Cambridge University Press, Cambridge, pp. 107–125.

Currie, P.J., Nadon, G., Lockley, M.G. (1991) Dinosaur footprints with skin impressions from the Cretaceous of Alberta and Colorado. *Canadian Journal of Earth Sciences* 28, 102–115.

Currie, P.J., Badamgarav, D., Koppelhus, E.B., Sissons, R., Vickaryous, M.K. (2011) Hands, feet and behaviour in *Pinacosaurus* (Dinosauria: Ankylosauridae). *Acta Palaeontologica Polonica* 56, 489–504.

Curry, K.A. (1999) Ontogenetic histology of *Apatosaurus* (Dinosauria: Sauropoda): new insights on growth rate and longevity. *Journal of Vertebrate Paleontology* 19, 654–665.

Curry-Rogers, K. (2005) Titanosauria: a phylogenetic overview. In: Curry-Rogers, K., Wilson, J.A. (eds.), *The Sauropods: Evolution and Paleobiology*. University of California Press, Berkeley, CA, pp. 50–103.

Curry-Rogers, K.A., Wilson, J.A. (eds.) (2005) *The Sauropods: Evolution and Paleobiology*. University of California Press, Berkeley, CA.

Czerkas, S. (1992) Discovery of dermal spines reveals a new look for sauropod dinosaurs. *Geology* 20, 1068–1070.

Czerkas, S. (1997) Skin. In: Currie, P.J., Padian, K. (eds.), *Encyclopedia of Dinosaurs*. Academic Press, San Diego, CA, pp. 669–675.

Dalla Vecchia, F.M. (2009) *Tethyshadros insularis*, a new hadrosauroid dinosaur (Ornithischia) from the Upper Cretaceous of Italy. *Journal of Vertebrate Paleontology* 29, 1100–1116.

Dal Sasso, C., Signore, M. (1998) Exceptional soft-tissue preservation in a theropod dinosaur from Italy. *Nature* 392, 383–387.

Dal Sasso, C., Maganuco, S., Buffetaut, E., Mendez, M.A. (2005) New information on the skull of the enigmatic theropod *Spinosaurus*, with remarks on its size and affinities. *Journal of Vertebrate Paleontology* 25, 888–896.

D'Amore, D.C. (2009) A functional explanation for denticulation in theropod dinosaur teeth. *Anatomical Record* 292, 1297–1314.

Day, J.J., Norman, D.B., Upchurch, P., Powell, H.P. (2002) Dinosaur locomotion from a new trackway. *Nature* 415, 494–495.

Dececchi, T.A., Larsson, H.C.E. (2009) Patristic evolutionary rates suggest a punctuated pattern in forelimb evolution before and after the origin of birds. *Paleobiology* 35, 1–12.

Dececchi, T.A., Larsson, H.C.E. (2011) Assessing arboreal adaptations of bird antecedents: testing the ecological setting of the origin of the avian flight stroke. *PLoS ONE* 6, e22292.

Deeming, D.C. (2006) Ultrastructural and functional morphology of eggshells supports the idea that dinosaur eggs were incubated buried in a substrate. *Palaeontology* 49, 171–185.

de Queiroz, K., Gauthier, J. (1990) Phylogeny as a central principle in taxonomy: phylogenetic definitions of taxon names. *Systematic Zoology* 39, 307–322.

de Queiroz, K., Gauthier, J. (1992) Phylogenetic taxonomy. *Annual Review of Ecology and Systematics* 23, 449–480.

de Ricqlès, A. (1980) Tissue structure of dinosaur bone: functional significance and possible relation to dinosaur physiology. In: Olsen, E.C., Thomas, R.D.K. (eds.), *A Cold Look at Warm-Blooded Dinosaurs*. Westview Press, Boulder, CO, pp. 103–139.

de Ricqlès, A., Padian, K., Horner, J.R., Francillon-Vieillot, H. (2000) Palaeohistology of the bones of pterosaurs (Reptilia: Archosauria): anatomy, ontogeny, and biomechanical implications. *Zoological Journal of the Linnean Society* 129, 349–385.

de Ricqlès, A., Padian, K., Horner, J.R. (2003a) On the bone histology of some Triassic pseudosuchian archosaurs and related taxa. *Annales de Paléontologie* 89, 67–101.

de Ricqlès, A., Padian, K., Horner, J.R., Lamm, E.-T., Myhrvold, N. (2003b) Osteohistology of *Confuciusornis sanctus* (Theropoda: Aves). *Journal of Vertebrate Paleontology* 23, 373–386.

de Ricqlès, A., Padian, K., Knoll, F., Horner, J.R. (2008) On the origin of high growth rates in archosaurs and their ancient relatives: complementary histological studies on Triassic archosauriforms and the problem of a "phylogenetic signal" in bone histology. *Annales de Paléontologie* 94, 57–76.

de Margerie, E., Cubo, J., Castanet, J. (2002) Bone typology and growth rate: testing and quantifying "Amprino's rule" in the mallard (*Anas platyrhynchos*). *Comptes Rendu Biologie* 325, 221–230.

Dial, K.P. (2003a) Wing-assisted incline running and the origin of flight. *Science* 299, 402–404.

Dial, K.P. (2003b) Evolution of avian locomotion: correlates of flight style, locomotor modules, nesting biology, body size, development, and the origin of flapping flight. *Auk* 120, 941–952.

Dial, K.P., Jackson, B.E., Segre, P. (2008) A fundamental avian wingstroke provides a new perspective on the origin of flight. *Nature* 451, 985–989.

Dilkes, D.W. (1998) The Early Triassic rhynchosaur *Mesosuchus browni* and the interrelationships of basal archosauromorph reptiles. *Philosophical Transactions of the Royal Society of London, Series B* 353, 501–541.

Dilkes, D.W. (2000) Appendicular myology of the hadrosaurian dinosaur *Maiasaura peeblesorum* from the Late Cretaceous (Campanian) of Montana. *Transactions of the Royal Society of Edinburgh* 90, 87–125.

Dilkes, D.W. (2001) An ontogenetic perspective on locomotion in the Late Cretaceous dinosaur *Maiasaura peeblesorum* (Ornithischia: Hadrosauridae). *Canadian Journal of Earth Sciences* 38, 1205–1227.

Dingus, L., Rowe, T. (1997) *The Mistaken Extinction*. W.H. Freeman and Company, New York.

Dodson, P. (1975) Taxonomic implications of relative growth in lambeosaurine hadrosaurs. *Systematic Zoology* 24, 37–54.

Dodson, P. (1976) Quantitative aspects of relative growth and sexual dimorphism in *Protoceratops*. *Journal of Paleontology* 50, 929–940.

Dodson, P. (1990) Counting dinosaurs. *Proceedings of the National Academy of Sciences USA* 87, 7608–7612.

Dodson, P. (1993) Comparative craniology of the Ceratopsia. *American Journal of Science* 293(A), 200–234.

Dodson, P., Madsen, J.H. (1981) On the sternum of *Camptosaurus*. *Journal of Paleontology* 55, 109–112.

Dodson, P., Behrensmeyer, A.K., Bakker, R.T., McIntosh, J.S. (1980) Taphonomy and paleoecology of the Upper Jurassic Morrison Formation. *Paleobiology* 6, 208–232.

Dodson, P., Forster, C.A., Sampson, S.D. (2004) Ceratopsidae. In: Weishampel, D.B., Dodson, P., Osmólska, H. (eds.), *The Dinosauria*, 2nd edn. University of California Press, Berkeley, CA, pp. 494–513.

Domínguez Alonso, P., Milner, A.C., Ketcham, R.A., Cookson, M.J., Rowe, T.B. (2004) The avian nature of the brain and inner ear of *Archaeopteryx*. *Nature* 430, 666–669.

Dong, Z.M., Currie, P.J. (1996) On the discovery of an oviraptorid skeleton on a nest of eggs at Bayan Mandahu, Inner Mongolia, People's Republic of China. *Canadian Journal of Earth Sciences* 33, 631–636.

Donoghue, P.C.J., Sansom, I.J. (2002) Origin and early evolution of vertebrate skeletonization. *Microscopy Research and Technique* 59, 352–372.

Donoghue, P.C.J., Sansom, I.J., Downs, J.P. (2006) Early evolution of vertebrate skeletal tissues and cellular interactions, and the canalization of skeletal development. *Journal of Experimental Zoology Part B: Molecular and Developmental Evolution* 306B, 278–294.

Dumont, E.R., Grosse, I.R., Slater, G.J. (2009) Requirements for comparing the performance of finite element models of biological structures. *Journal of Theoretical Biology* 256, 96–103.

Duncker, H.R. (1971) The lung air sac system of birds. *Advances in Anatomy, Embryology, and Cell Biology* 45, 1–171.

Dunhill, A.M. (2011) Using remote sensing and a geographic information system to quantify rock exposure area in England and Wales: implications for paleodiversity studies. *Geology* 39, 111–114.

Dzemski, G., Christian, A. (2007) Flexibility along the neck of the ostrich (*Struthio camelus*) and consequences for the reconstruction of dinosaurs with extreme neck length. *Journal of Morphology* 268, 701–714.

Dzik, J. (2003) A beaked herbivorous archosaur with dinosaur affinities from the early Late Triassic of Poland. *Journal of Vertebrate Paleontology* 23, 556–574.

Dzik, J., Sulej, T. (2007) A review of the early Late Triassic Krasiejów biota from Silesia, Poland. *Palaeontologia Polonica* 64, 3–27.

Eagle, R.A., Tütken, T., Martin, T.S. et al. (2011) Dinosaur body temperatures determined from isotopic (^{13}C-^{18}O) ordering in fossil biominerals. *Science* 333, 443–445.

Eberth, D.A., Getty, M.A. (2005) Ceratopsian bonebeds: occurrence, origins, and significance.

In: Currie, P.J., Koppelhus, E.B. (eds.), *Dinosaur Provincial Park: A Spectacular Ancient Ecosystem Revealed*. Indiana University Press, Bloomington, IN, pp. 501–536.

Egi, N., Weishampel, D.B. (2002) Morphometric analyses of humeral shape in hadrosaurids (Ornithopoda, Dinosauria). *Senckenbergiana Lethaea* 82, 43–58.

Elewa, A. (ed.) (2004) *Morphometrics: Applications in Biology and Paleontology*. Springer-Verlag, Heidelberg, Germany.

Embery, G., Milner, A.C., Waddington, R.J., Hall, R.C., Langley, M.S., Milan, A.M. (2003) Identification of proteinaceous material in the bone of the dinosaur *Iguanodon*. *Connective Tissue Research* 44, 41–46.

Erickson, G.M. (2005) Assessing dinosaur growth patterns: a microscopic revolution. *Trends in Ecology and Evolution* 20, 677–684.

Erickson, G.M., Brochu, C.A. (1999) How the "terror crocodile" grew so big. *Nature* 398, 205–206.

Erickson, G.M., Olson, K.H. (1996) Bite marks attributable to *Tyrannosaurus rex*: a preliminary description and implications. *Journal of Vertebrate Paleontology* 16, 175–178.

Erickson, G.M., Tumanova, T.A. (2000) Growth curve of *Psittacosaurus mongoliensis* Osborn (Ceratopsia: Psittacosauridae) inferred from long bone histology. *Zoological Journal of the Linnean Society* 130, 551–566.

Erickson, G.M., Van Kirk, S.D., Su, J., Levenston, M.E., Caler, W.E., Carter, D.R. (1996) Bite-force estimation for *Tyrannosaurus rex* from tooth-marked bones. *Nature* 382, 706–708.

Erickson, G.M., Curry-Rogers, K., Yerby, S.A. (2001) Dinosaurian growth patterns and rapid avian growth rates. *Nature* 412, 429–432.

Erickson, G.M., Makovicky, P.J., Currie, P.J., Norell, M.A., Yerby, S.A., Brochu, C.A. (2004) Gigantism and comparative life-history parameters of tyrannosaurid dinosaurs. *Nature* 430, 772–775.

Erickson, G.M., Lappin, A.K., Larson, P. (2005) Androgynous rex: the utility of chevrons for determining the sex of crocodilians and non-avian dinosaurs. *Zoology* 108, 277–286.

Erickson, G.M., Currie, P.J., Inouye, B.D., Winn, A.A. (2006) Tyrannosaur life tables: an example of nonavian dinosaur population biology. *Science* 313, 213–217.

Erickson, G.M., Curry-Rogers, K.A., Varricchio, D. J., Norell, M.A., Xu, X. (2007) Growth patterns in brooding dinosaurs reveal the timing of sexual maturity in non-avian dinosaurs and the genesis of the avian condition. *Biology Letters* 3, 558–561.

Erickson, G.M., Rauhut, O.W.M., Zhou, Z. et al. (2009a) Was dinosaurian physiology inherited by birds? Reconciling slow growth in *Archaeopteryx*. *PLoS ONE* 4, e7390.

Erickson, G.M., Makovicky, P.J., Inouye, B.D., Zhou, C.-F., Gao, K.-Q. (2009b) A life table for *Psittacosaurus lujiatuensis*: initial insights into ornithischian dinosaur population biology. *Anatomical Record* 292, 1514–1521.

Erickson, G.M., Currie, P.J., Inouye, B.D., Winn, A. A. (2010) A revised life table and survivorship curve for *Albertosaurus sarcophagus* based on the Dry Island mass death assemblage. *Canadian Journal of Earth Sciences* 47, 1269–1275.

Erwin, D.H. (2006) *Extinction: How Life on Earth Nearly Ended 250 Million Years Ago*. Princeton University Press, Princeton, NJ.

Erwin, D.H. (2007) Disparity: morphological pattern and developmental context. *Palaeontology* 50, 57–73.

Evans, D.C. (2005) New evidence on brain–endocranial cavity relationships in ornithischian dinosaurs. *Acta Palaeontologica Polonica* 50, 617–622.

Evans, D.C. (2006) Nasal cavity homologies and cranial crest function in lambeosaurine dinosaurs. *Paleobiology* 32, 109–125.

Evans, D.C. (2010) Cranial anatomy and systematics of *Hypacrosaurus altispinus*, and a comparative analysis of skull growth in lambeosaurine hadrosaurids (Dinosauria: Ornithischia). *Zoological Journal of the Linnean Society* 159, 398–434.

Evans, D.C., Reisz, R.R. (2007) Anatomy and relationships of *Lambeosaurus magnicristatus*, a crested hadrosaurid dinosaur (Ornithischia) from the Dinosaur Park Formation, Alberta. *Journal of Vertebrate Paleontology* 27, 373–393.

Evans, D.C., Ridgely, R., Witmer, L.M. (2009) Endocranial anatomy of lambeosaurine hadrosaurids (Dinosauria: Ornithischia): a sensorineural perspective on cranial crest

function. *Anatomical Record* 292, 1315–1337.

Ewer, R.F. (1965) The anatomy of the thecodont reptile *Euparkeria capensis* Broom. *Philosophical Transactions of the Royal Society of London, Series B* 248, 379–435.

Ezcurra, M.D. (2006) A review of the systematic position of the dinosauriform archosaur *Eucoelophysis baldwini* Sullivan and Lucas, 1999 from the Upper Triassic of New Mexico, USA. *Geodiversitas* 28, 649–684.

Ezcurra, M.D. (2010a) A new early dinosaur (Saurischia: Sauropodomorpha) from the Late Triassic of Argentina: a reassessment of dinosaur origin and phylogeny. *Journal of Systematic Palaeontology* 8, 371–425.

Ezcurra, M.D. (2010b) Biogeography of Triassic tetrapods: evidence for provincialism and driven sympatric cladogenesis in the early evolution of modern tetrapod lineages. *Proceedings of the Royal Society of London, Series B* 277, 2547–2552.

Ezcurra M.D., Brusatte, S.L. (2011) Taxonomic and phylogenetic reassessment of the early neotheropod dinosaur *Camposaurus arizonensis* from the Late Triassic of North America. *Palaeontology* 54, 763–772.

Falkingham, P.L., Margetts, L., Manning, P.L. (2010) Fossil vertebrate tracks as paleopenetrometers: confounding effects of foot morphology. *Ichnos* 25, 356–360.

Falkingham, P.L., Bates, K.T., Margetts, L., Manning, P.L. (2011) Simulating sauropod manus-only trackway formation using finite-element analysis. *Biology Letters* 7, 142–145.

Farke, A.A. (2004) Horn use in *Triceratops* (Dinosauria: Ceratopsidae): testing behavioral hypotheses using scale models. *Palaeontologia Electronica* 7(1), Article 1.

Farke, A.A. (2011) Anatomy and taxonomic status of the chasmosaurine ceratopsid *Nedoceratops hatcheri* from the Upper Cretaceous Lance Formation of Wyoming, U.S.A. *PLoS ONE* 6, e16196.

Farke, A.A., Wolff, E.D.S., Tanke, D.H. (2009) Evidence of combat in *Triceratops*. *PLoS ONE* 4, e4252.

Farke, A.A., Chapman, R.E., Andersen, A. (2010) Modeling structural properties of the frill of *Triceratops*. In: Ryan, M.J., Chinnery-Allgeier, B.

J., Eberth, D.A. (eds.), *New Perspectives on Horned Dinosaurs*. Indiana University Press, Bloomington, IN, pp. 264–270.

Farlow, J.O. (1981) Estimates of dinosaur speeds from a new trackway site in Texas. *Nature* 294, 747–748.

Farlow, J.O. (1987) Speculations about the diet and digestive physiology of herbivorous dinosaurs. *Paleobiology* 13, 60–72.

Farlow, J.O. (1990) Dinosaur energetics and thermal biology. In: Weishampel, D.B., Dodson, P., Osmólska, H. (eds.), *The Dinosauria*. University of California Press, Berkeley, CA, pp. 43–55.

Farlow, J.O., Brinkman, D.L. (1994) Wear surfaces on the teeth of tyrannosaurs. In: Rosenberg, G.D., Wolberg, D.L. (eds.), *Dino Fest: Proceedings of a Conference for the General Public*. Paleontological Society Special Publication No. 7, pp. 165–175. Paleontological Society, Knoxville, TN.

Farlow, J.O., Dodson, P. (1975) The behavioral significance of frill and horn morphology in ceratopsian dinosaurs. *Evolution* 29, 353–361.

Farlow, J.O., Thompson, C.V., Rosner, D.E. (1976) Plates of the dinosaur *Stegosaurus*: forced convection heat loss fins? *Science* 192, 1123–1125.

Farlow, J.O., Brinkman, D.L., Abler, W.L., Currie, P. J. (1991) Size, shape and serration density of theropod dinosaur lateral teeth. *Modern Geology* 16, 161–198.

Farlow, J.O., Smith, M.B., Robinson, J.M. (1995a) Body mass, bone "strength indicator," and cursorial potential of *Tyrannosaurus rex*. *Journal of Vertebrate Paleontology* 15, 713–725.

Farlow, J.O., Dodson, P., Chinsamy, A. (1995b) Dinosaur biology. *Annual Review of Ecology and Systematics* 26, 445–471.

Farlow, J.O., Gatesy, S.M., Holtz, T.R., Hutchinson, J.R., Robinson, J.M. (2000) Theropod locomotion. *American Zoologist* 40, 640–663.

Farlow, J.O., Hayashi, S., Tattersall, G.J. (2010) Internal vascularity of the dermal plates of *Stegosaurus* (Ornithischia, Thyreophora). *Swiss Journal of Geosciences* 103, 173–185.

Farmer, C.G., Sanders, K. (2010) Unidirectional airflow in the lungs of alligators. *Science* 327, 338–340.

Fastovsky, D.E., Sheehan, P.M. (2005) The extinction of the dinosaurs in North America. *GSA Today* 15, 4–10.

Fastovsky, D.E., Smith, J.B. (2004) Dinosaur paleoecology. In: Weishampel, D.B., Dodson, P., Osmólska, H. (eds.), *The Dinosauria*, 2nd edn. University of California Press, Berkeley, CA, pp. 614–626.

Fastovsky, D.E., Weishampel, D.B. (2005) *The Evolution and Extinction of the Dinosaurs*. Cambridge University Press, Cambridge.

Fastovsky, D.E., Huang, Y., Hsu, J., Martin-McNaughton, J., Sheehan, P.M., Weishampel, D.B. (2004) The shape of Mesozoic dinosaur richness. *Geology* 32, 877–880.

Felsenstein, J. (2003) *Inferring Phylogenies*. Sinauer Associates, Sunderland, MA.

Ferigolo, J., Langer, M.C. (2007) A Late Triassic dinosauriform from south Brazil and the origin of the ornithischian predentary bone. *Historical Biology* 19, 23–33.

Fiorillo, A.R. (1991) Prey bone utilization by predatory dinosaurs. *Palaeogeography, Palaeoclimatology, Palaeoecology* 88, 157–166.

Fiorillo, A.R. (1998) Dental microwear patterns of the sauropod dinosaurs *Camarasaurus* and *Diplodocus*: evidence for resource partitioning in the Late Jurassic of North America. *Historical Biology* 13, 1–16.

Fiorillo, A.R., Eberth, D.A. (2004) Dinosaur taphonomy. In: Weishampel, D.B., Dodson, P., Osmólska, H. (eds.), *The Dinosauria*, 2nd edn. University of California Press, Berkeley, CA, pp. 607–613.

Fisher, P.E., Russell, D.A., Stoskopf, M.K., Barrick, R.E., Hammer, M., Kuzmitz, A. (2000) Cardiovascular evidence for an intermediate or higher metabolic rate in an ornithischian dinosaur. *Science* 288, 503–505.

Fletcher, B.J., Brentnall, S.J., Anderson, C.W., Berner, R.A., Beerling, D.J. (2008) Atmospheric carbon dioxide linked with Mesozoic and early Cenozoic climate change. *Nature Geoscience* 1, 43–48.

Foote, M. (1997) The evolution of morphological diversity. *Annual Review of Ecology and Systematics* 28, 129–152.

Forster, C.A., Sampson, S.D., Chiappe, L.M., Krause, D.W. (1998) The theropod ancestry of birds: new evidence from the Late Cretaceous of Madagascar. *Science* 279, 1915–1919.

Foster, J.R. (2003) Paleoecological analysis of the vertebrate fauna of the Morrison Formation (Upper Jurassic), Rocky Mountain Region, USA. *New Mexico Museum of Natural History and Science Bulletin* 23, 1–95.

Foster, J.R. (2007) *Jurassic West: The Dinosaurs of the Morrison Formation and their World*. Indiana University Press, Bloomington, IN.

Fox, L.R. (1975) Cannibalism in natural populations. *Annual Review of Ecology and Systematics* 6, 87–106.

Franz, R., Hummel, J., Kienzle, E., Kölle, P., Gunga, H.-C., Clauss, M. (2009) Allometry of visceral organs in living amniotes and its implications for sauropod dinosaurs. *Proceedings of the Royal Society of London, Series B* 276, 1731–1736.

Franzosa, J.W., Rowe, T.B. (2005) Cranial endocast of the Cretaceous theropod dinosaur *Acrocanthosaurus atokensis*. *Journal of Vertebrate Paleontology* 25, 859–864.

Fraser, N.C. (2006) *Dawn of the Dinosaurs: Life in the Triassic*. Indiana University Press, Bloomington, IN.

Fraser, N.C., Padian, K., Walkden, G.M., Davis, A.L.M. (2002) Basal dinosauriform remains from Britain and the diagnosis of the Dinosauria. *Palaeontology* 45, 79–95.

Fricke, H.C., Pearson, D.A. (2008) Stable isotope evidence for changes in dietary niche partitioning among hadrosaurian and ceratopsian dinosaurs of the Hell Creek Formation, North Dakota. *Paleobiology* 34, 534–552.

Fricke, H.C., Rogers, R.R. (2000) Multiple taxon–multiple locality approach to providing oxygen isotope evidence for warm-blooded theropod dinosaurs. *Geology* 28, 799–802.

Fujiwara, S.-I. (2009) A reevauation of manus structure in *Triceratops* (Ceratopsia: Ceratopsidae). *Journal of Vertebrate Paleontology* 29, 1136–1147.

Galton, P.M. (1970) Pachycephalosaurids: dinosaurian battering rams. *Discovery* 6, 23–32.

Galton, P.M. (1974) The ornithischian dinosaur *Hypsilophodon* from the Wealden of the Isle of Wight. *Bulletin of the British Museum of Natural History (Geology)* 25, 1–152.

Galton, P.M. (1985) Diet of prosauropod dinosaurs from the Late Triassic and Early Jurassic. *Lethaia* 18, 105–123.

Galton, P.M. (1986) Herbivorous adaptations of Late Triassic and Early Jurassic dinosaurs. In: Padian, K. (ed.), *The Beginning of the Age of Dinosaurs*. Cambridge University Press, Cambridge, pp. 203–221.

Galton, P.M. (1990) Basal Sauropodomorpha–Prosauropoda. In: Weishampel, D.B., Dodson, P., Osmólska, H. (eds.), *The Dinosauria*. University of California Press, Berkeley, CA, pp. 320–344.

Galton, P.M., Upchurch, P. (2004a) Stegosauria. In: Weishampel, D.B., Dodson, P., Osmólska, H. (eds.), *The Dinosauria*, 2nd edn. University of California Press, Berkeley, CA, pp. 343–362.

Galton, P.M., Upchurch, P. (2004b) Prosauropoda. In: Weishampel, D.B., Dodson, P., Osmólska, H. (eds.), *The Dinosauria*, 2nd edn. University of California Press, Berkeley, CA, pp. 232–258.

Gand, G., Demathieu, G. (2005) Les pistes dinosauroides du Trias moyen francais: interprétation et réévaluation de la nomenclature. *Geobios* 38, 725–749.

Gatesy, S.M. (1990) Caudofemoral musculature and the evolution of theropod locomotion. *Paleobiology* 16, 170–186.

Gatesy, S.M. (2002) Locomotor evolution on the line to modern birds. In: Chiappe, L.M., Witmer, L.M. (eds.), *Mesozoic Birds: Above the Heads of Dinosaurs*. University of California Press, Berkeley, CA, pp. 432–447.

Gatesy, S.M., Dial, K.P. (1996) Locomotor modules and the evolution of avian flight. *Evolution* 50, 331–340.

Gatesy, S.M., Middleton, K.M. (1997) Bipedalism, flight, and the evolution of theropod diversity. *Journal of Vertebrate Paleontology* 17, 308–329.

Gatesy, S.M., Middleton, K.M., Jenkins, F.A., Shubin, N.H. (1999) Three-dimensional preservation of foot movements in Triassic theropod dinosaurs. *Nature* 399, 141–144.

Gatesy, S.M., Shubin, N.H., Jenkins, F.A. (2005) Anaglyph stereo imaging of dinosaur track morphology and microtopography. *Palaeontologia Electronica* 8(1), Article 8.1.10A.

Gatesy, S.M., Bäker, M., Hutchinson, J.R. (2009) Constraint-based exclusion of limb poses for reconstructing theropod dinosaur locomotion. *Journal of Vertebrate Paleontology* 29, 535–544.

Gauthier, J.A. (1986) Saurischian monophyly and the origin of birds. *Memoirs of the California Academy of Sciences* 8, 1–55.

Gauthier, J.A., Padian, K. (1985) Phylogenetic, functional, and aerodynamic analyses of the origin of birds and their flight. In: Hecht, M.K., Ostrom, J. H., Viohl, G., Wellnhofer, P. (eds.), *The Beginnings of the Birds*. Freunde des Jura-Museums, Eichstätt, Germany, pp. 185–197.

Ghosh, P., Bhattacharya, S.K., Sahni, A., Kar, R.K., Mohabey, D.M., Ambwani, K. (2003) Dinosaur coprolites from the Late Cretaceous (Maastrichtian) Lameta Formation of India and other markers suggesting a C3 plant diet. *Cretaceous Research* 24, 743–750.

Gignac, P.M., Makovicky, P.J., Erickson, G.M., Walsh, R.P. (2010) A description of *Deinonychus antirrhopus* bite marks and estimates of bite force using tooth indentation simulations. *Journal of Vertebrate Paleontology* 30, 1169–1177.

Gillooly, J.F., Allen, A.P., Charnov, E.L. (2006) Dinosaur fossils predict body temperatures. *PLoS Biology* 4, e248.

Gleich, O., Dooling, R.J., Manley, G. (2005) Audiogram, body mass, and basilar papilla length: correlations in birds and predictions for extinct archosaurs. *Naturwissenschaften* 92, 595–598.

Godefroit, P., Golovneva, L., Shchepetov, S., Garcia, G., Alekseev, P. (2009) The last polar dinosaurs: high diversity of latest Cretaceous arctic dinosaurs in Russia. *Naturwissenschaften* 96, 495–501.

Goloboff, P.A., Farris, J.S., Nixon, K.C. (2008) TNT, a free program for phylogenetic analysis. *Cladistics* 24, 774–786.

Goodwin, M.B., Horner, J.R. (2004) Cranial histology of pachycephalosaurs (Ornithischia: Marginocephalia) reveals transitory structures inconsistent with head-butting behavior. *Paleobiology* 30, 253–267.

Goodwin, M.B., Horner, J.R. (2008) Ontogeny of cranial epi-ossifications in *Triceratops*. *Journal of Vertebrate Paleontology* 28, 134–144.

Goodwin, M.B., Buchholz, E.A., Johnson, R.E. (1998) Cranial anatomy and diagnosis of *Stygimoloch spinifer* (Ornithischia:

Pachycephalosauria) with comments on cranial display structures in agonistic behavior. *Journal of Vertebrate Paleontology* 18, 363–375.

Gould, S.J. (1991) The disparity of the Burgess shale arthropod fauna and the limits of cladistic analysis: why we must strive to quantify morphospace. *Paleobiology* 17, 411–423.

Gould, S.J. (2002) *The Struture of Evolutionary Theory*. Harvard University Press, Cambridge, MA.

Gower, D.J. (2001) Posible postcranial pneumaticity in the last common ancestor of birds and crocodilians: evidence from *Erythrosuchus* and other Mesozoic archosaurs. *Naturwissenschaften* 88, 119–122.

Gregory, T.R. (2001) The bigger the C-value, the larger the cell: genome size and red blood cell size in vertebrates. *Blood Cells, Molecules, and Diseases* 27, 830–843.

Gregory, W.K. (1905) The weight of the *Brontosaurus*. *Science* 22, 572.

Grellet-Tinner, G., Fiorelli, L.E. (2010) A new Argentinean nesting site showing neosauropod dinosaur reproduction in a Cretaceous hydrothermal environment. *Nature Communications* 1, Article 32, doi: 10.1038/ncomms1031.

Grellet-Tinner, G., Makovicky, P.J. (2006) A possible egg of the dromaeosaur *Deinonychus antirrhopus*: phylogenetic and biological implications. *Canadian Journal of Earth Sciences* 43, 705–719.

Grellet-Tinner, G., Chiappe, L., Norell, M., Bottjer, D. (2006) Dinosaur eggs and nesting ecology: a paleobiological investigation. *Palaeogeography, Palaeoclimatology, Palaeoecology* 232, 294–321.

Gunga, H.C., Kirsch, K., Baartz, F. et al. (1995) New data on the dimensions of *Brachiosaurus brancai* and their physiological implications. *Naturwissenschaften* 82, 190–192.

Gunga, H.C., Kirsch, K., Rittweger, J. et al. (1999) Body size and body volume distribution in two sauropods from the Upper Jurassic of Tendaguru/Tansania (East Africa). *Mitteilungen aus dem Museum für Naturkunde in Berlin. Geowissenschaftliche Reihe* 2, 91–102.

Gunga, H.C., Suthau, T., Bellmann, A. et al. (2007) Body mass estimations for *Plateosaurus engelhardti* using laser scanning and 3D reconstruction methods. *Naturwissenschaften* 94, 623–630.

Gunga, H.C., Suthau, T., Bellmann, A. et al. (2008) A new body mass estimation of *Brachiosaurus brancai* Janensch 1914 mounted and exhibited at the Museum of Natural History (Berlin, Germany). *Fossil Record* 11, 28–33.

Happ, J.W. (2008) An analysis of predator–prey behavior in a head-to-head encounter between *Tyrannosaurus rex* and *Triceratops*. In: Larson, P., Carpenter, K. (eds.), *Tyrannosaurus rex: The Tyrant King*. Indiana University Press, Bloomington, IN, pp. 355–368.

Happ, J.W. (2010) New evidence regarding the structure and function of the horns in *Triceratops*. In: Ryan, M.J., Chinnery-Allgeier, B.J., Eberth, D. A. (eds.), *New Perspectives on Horned Dinosaurs*. Indiana University Press, Bloomington, IN, pp. 271–281.

Haq, B.U., Hardenbol, J., Vail, P.R. (1987) Chronology of fluctuating sea levels since the Triassic. *Science* 235, 1156–1167.

Harris, J.D. (2006) The significance of *Suuwassea emilieae* (Dinosauria: Sauropoda) for flagellicaudatan intrarelationships and evolution. *Journal of Systematic Palaeontology* 4, 185–198.

Haubold, H. (1999) Tracks of the Dinosauromorpha from the Early Triassic. *Zentralblatt fur Geologie und Paläontologie* Teil I 1998, 783–795.

Hayashi, S., Carpenter, K., Suzuki, D. (2009) Different growth patterns between the skeleton and osteoderms of *Stegosaurus* (Ornithischia: Thyreophora). *Journal of Vertebrate Paleontology* 29, 123–131.

Hayashi, S., Carpenter, K., Scheyer, T.M., Watabe, M., Suzuki, D. (2010) Function and evolution of ankylosaur dermal armor. *Acta Palaeontologica Polonica* 55, 213–228.

Heilmann, G. (1926) *The Origin of Birds*. Witherby, London.

Henderson, D.M. (1999) Estimating the masses and centers of masses of extinct animals by 3-D mathematical slicing. *Paleobiology* 25, 88–106.

Henderson, D.M. (2002) The eyes have it: sizes, shapes, and orientations of theropod orbits as indicators of skull strength and bite force. *Journal of Vertebrate Paleontology* 22, 766–778.

Henderson, D.M. (2003) Footprints, trackways, and hip heights of bipedal dinosaurs: testing hip height predictions with computer models. *Ichnos* 10, 99–114.

Henderson, D.M. (2004) Tipsy punters: sauropod dinosaur pneumaticity, buoyancy, and aquatic habits. *Proceedings of the Royal Society of London, Series B* 271, S180–S183.

Henderson, D.M. (2006a) Simulated weathering of dinosaur tracks and the implications for their characterization. *Canadian Journal of Earth Sciences* 43, 691–704.

Henderson, D.M. (2006b) Burly gaits: centers of mass, stability, and the trackways of sauropod dinosaurs. *Journal of Vertebrate Paleontology* 6, 907–921.

Henderson, D.M. (2010) Skull shapes as indicators of niche partitioning by sympatric chasmosaurine and centrosaurine dinosaurs. In: Ryan, M.J., Chinnery-Allgeier, B.J., Eberth, D.A. (eds.), *New Perspectives on Horned Dinosaurs*. Indiana University Press, Bloomington, IN, pp. 293–307.

Henderson, D.M., Snively, E. (2004) *Tyrannosaurus* en pointe: allometry minimized rotational inertia of large carnivorous dinosaurs. *Proceedings of the Royal Society of London, Series B* 271, S57–S60.

Hennig, W. (1965) Phylogenetic systematics. *Annual Review of Entomology* 10, 97–116.

Hennig, W. (1966) *Phylogenetic Systematics*. University of Illinois Press, Urbana, IL.

Hieronymus, T.L., Witmer, L.M., Tanke, D.H., Currie, P.J. (2009) The facial integument of centrosaurine ceratopsids: morphological and histological correlates of novel skin structures. *Anatomical Record* 292, 1370–1396.

Hildebrand, A.R., Penfield, G.T., Kring, D.A. et al. (1991) Chicxulub crater: a possible Cretaceous Tertiary boundary impact crater on the Yucatan Peninsula, Mexico. *Geology* 19, 867–871.

Hildebrand, M., Goslow, G. (1998) *Analysis of Vertebrate Structure*. John Wiley & Sons Inc., Hoboken, NJ.

Hill, R.V., Witmer, L.M., Norell, M.A. (2003) A new specimen of *Pinacosaurus grangeri* (Dinosauria: Ornithischia) from the Late Cretaceous of Mongolia: ontogeny and phylogeny of ankylosaurs. *American Museum Novitates* 3395, 1–29.

Hillenius, W.J., Ruben, J. (2004) The evolution of endothermy in terrestrial vertebrates: who? when? why? *Physiological and Biochemical Zoology* 77, 1019–1042.

Holliday, C.M. (2009) New insights into dinosaur jaw muscle anatomy. *Anatomcial Record* 292, 1246–1265.

Holliday, C.M., Witmer, L.M. (2007) Archosaur adductor chamber evolution: integration of musculoskeletal and topological criteria in jaw muscle homology. *Journal of Morphology* 268, 457–484.

Holliday, C.M., Witmer, L.M. (2008) Cranial kinesis in dinosaurs: intracranial joints, protractor muscles, and their significance for cranial evolution and function in diapsids. *Journal of Vertebrate Paleontology* 28, 1073–1088.

Holliday, C.M., Ridgely, R.C., Sedlmayr, J.C., Witmer, L.M. (2010) Cartilaginous epiphyses in extant archosaurs and their implications for reconstructing limb function in dinosaurs. *PLoS ONE* 5, e13120.

Holtz, T.R. (1994) The phylogenetic position of Tyrannosauridae: implications for theropod systematics. *Journal of Paleontology* 68, 1110–1117.

Holtz, T.R. (2000) A new phylogeny of the carnivorous dinosaurs. *Gaia* 15, 5–61.

Holtz, T.R. (2004) Tyrannosauroidea. In: Weishampel, D.B., Dodson, P., Osmólska, H. (eds.), *The Dinosauria*, 2nd edn. University of California Press, Berkeley, CA, pp. 111–136.

Holtz, T.R. (2008) A critical reappraisal of the obligate scavenging hypothesis for *Tyrannosaurus rex* and other tyrant dinosaurs. In: Larson, P., Carpenter, K. (eds.), *Tyrannosaurus rex: The Tyrant King*. Indiana University Press Press, Bloomington, IN, pp. 371–396.

Holtz, T.R., Brett-Surman, M.K. (1997) The osteology of the dinosaurs. In: Farlow, J.O., Brett-Surman, M.K. (eds.), *The Complete Dinosaur*. Indiana University Press, Bloomington, IN, pp. 78–91.

Holtz, T.R., Brinkman, D.L., Chandler, C.L. (2000) Denticle morphometrics and a possibly omnivorous feeding habit for the theropod dinosaur *Troodon*. *Gaia* 15, 159–166.

Hone, D.W.E., Rauhut, O.W.M. (2010) Feeding behaviour and bone utilization by theropod dinosaurs. *Lethaia* 43, 232–244.

Hone, D.W.E., Watabe, M. (2010) New information on scavenging and selective feeding behaviour of tyrannosaurs. *Acta Palaeontologica Polonica* 55, 627–634.

Hone, D.W.E., Keesey, T.M., Pisani, D., Purvis, A. (2005) Macroevolutionary trends in the Dinosauria: Cope's Rule. *Journal of Evolutionary Biology* 18, 587–595.

Hone, D.W.E., Choiniere, J., Sullivan, C., Xu, X., Pittman, M., Tan, Q. (2010) New evidence for a trophic relationship between the dinosaurs *Velociraptor* and *Protoceratops*. *Palaeoeography, Palaeoclimatology, Palaeoecology* 291, 488–492.

Hopson, J.A. (1975) The evolution of cranial display structures in hadrosaurian dinosaurs. *Paleobiology* 1, 21–43.

Hopson, J.A. (1977) Relative brain size and behavior in archosaurian reptiles. *Annual Review of Ecology and Systematics* 8, 429–448.

Horner, J.R. (1982) Evidence of colonial nesting and "site fidelity" among ornithischian dinosaurs. *Nature* 297, 675–676.

Horner, J.R. (1994) Steak knives, beady eyes, and tiny little arms (a portrait of *T. rex* as a scavenger). In: Rosenberg, G.D., Wolberg, D.L. (eds.), *Dino Fest: Proceedings of a Conference for the General Public*. Paleontological Society Special Publication No. 7, pp. 157–164. Paleontological Society, Knoxville, TN.

Horner, J.R. (1999) Egg clutches and embryos of two hadrosaurian dinosaurs. *Journal of Vertebrate Paleontology* 19, 607–611.

Horner, J.R. (2000) Dinosaur reproduction and parenting. *Annual Review of Earth and Planetary Sciences* 28, 19–45.

Horner, J.R., Currie, P.J. (1994) Embryonic and neonatal morphology and ontogeny of a new species of *Hypacrosaurus* (Ornithischia, Lambeosauridae) from Montana and Alberta. In: Carpenter, K., Hirsch, K.F., Horner, J.R. (eds.), *Dinosaur Eggs and Babies*. Cambridge University Press, Cambridge, pp. 312–322.

Horner, J.R., Goodwin, M.B. (2006) Major cranial changes during *Triceratops* ontogeny. *Proceedings of the Royal Society of London, Series B* 273, 2757–2761.

Horner, J.R., Goodwin, M.B. (2009) Extreme cranial ontogeny in the Upper Cretaceous dinosaur *Pachycephalosaurus*. *PLoS ONE* 4, e7626.

Horner, J.R., Makela, R. (1979) Nest of juveniles provides evidence of family structure among dinosaurs. *Nature* 282, 296–298.

Horner, J.R., Marshall, C.L. (2002) Keratin-covered dinosaur skulls. *Journal of Vertebrate Paleontology* 22 (3 suppl.), 67A.

Horner, J.R., Padian, K. (2004) Age and growth dynamics of *Tyrannosaurus rex*. *Proceedings of the Royal Society of London, Series B* 271, 1875–1880.

Horner, J.R., Weishampel, D.B. (1988) A comparative embryological study of two ornithischian dinosaurs. *Nature* 332, 256–257.

Horner, J.R., de Ricqlès, A., Padian, K. (1999) Variation in skeletochronological indicators of the hadrosaurid dinosaur *Hypacrosaurus*: implications for age assessment of dinosaurs. *Paleobiology* 25, 295–304.

Horner, J.R., de Ricqlès, A., Padian, K. (2000) The bone histology of the hadrosaurid dinosaur *Maiasaura peeblesorum*: growth dynamics and physiology based on an ontogenetic series of skeletal elements. *Journal of Vertebrate Paleontology* 20, 109–123.

Horner, J.R., Weishampel, D.B., Forster, C.A (2004) Hadrosauridae. In: Weishampel, D.B., Dodson, P., Osmólska, H. (eds.), *The Dinosauria*, 2nd edn. University of California Press, Berkeley, CA, pp. 438–463.

Horner, J.R., de Ricqlès, A., Padian, K., Sheetz, R.D. (2009) Comparative long bone histology and growth of the "hypsilophodontid" dinosaurs *Orodromeus makelai*, *Dryosaurus altus*, and *Tenontosaurus tilletti* (Ornithischia: Euornithopoda). *Journal of Vertebrate Paleontology* 29, 734–747.

Hu, D., Hou, L., Zhang, L., Xu, X. (2009) A pre-*Archaeopteryx* troodontid theropod from China with long feathers on the metatarsus. *Nature* 461, 640–643.

Hu, Y., Meng, J., Wang, Y., Li, C. (2005) Large Mesozoic mammals fed on young dinosaurs. *Nature* 433, 149–152.

Huelsenbeck, J.P., Ronquist, F., Nielsen, R., Bollback, J. (2001) Bayesian influence of phylogeny and its impact on evolutionary biology. *Science* 294, 2310–2314.

Huene, F. von. (1934) Ein neurer Coelurosaurier in der thüringischen Trias. *Paläontologische Zeitschrift* 16, 145–170.

Hullar, T.E. (2006) Semicircular canal geometry, afferent sensitivity, and animal behavior. *Anatomical Record* 288A, 466–472.

Hummel, J., Gee, C.T., Südekum, K.-H., Sander, P.M., Nogge, G., Clauss, M. (2008) *In vitro* digestibility of fern and gymnosperm foliage: implications for sauropod feeding ecology and diet selection. *Proceedings of the Royal Society of London, Series B* 275, 1015–1021.

Humphries, C.J., Parenti, L.R. (1999) *Cladistic Biogeography*, 2nd edn. Oxford University Press, Oxford.

Hunt, R.M. (1978) Depositional setting of a Miocene mammal assemblage, Sioux County, Nebraska (U.S.A.). *Palaeogeography, Palaeoclimatology, Palaeoecology* 24, 1–52.

Hutchinson, J.R. (2001a) The evolution of pelvic osteology and soft tissues on the line to extant birds (Neornithes). *Zoological Journal of the Linnean Society* 131, 123–168.

Hutchinson, J.R. (2001b) The evolution of femoral osteology and soft tissues on the line to extant birds (Neornithes). *Zoological Journal of the Linnean Society* 131, 169–197.

Hutchinson, J.R. (2002) The evolution of hindlimb tendons and muscles on the line to crown-group birds. *Comparative Biochemistry and Physiology Part A* 133, 1051–1086.

Hutchinson, J.R. (2004) Biomechanical modeling and sensitivity analysis of bipedal running ability. II. Extinct taxa. *Journal of Morphology* 262, 441–461.

Hutchinson, J.R. (2005) Dinosaur locomotion. *Encyclopedia of Life Sciences* doi: 10.1038/npg. els.0003320.

Hutchinson, J.R. (2006) The evolution of locomotion in archosaurs. *Comptes Rendus Palevol* 5, 519–530.

Hutchinson, J.R. (2011) On the influence of function from structure using biomechanical modelling and simulation of extinct organisms. *Biology Letters* doi: 10.1098/rsbl.2011.0399.

Hutchinson, J.R., Allen, V. (2009) The evolutionary continuum of limb function from early theropods to birds. *Naturwissenschaften* 96, 423–448.

Hutchinson, J.R., Garcia, M. (2002) *Tyrannosaurus* was not a fast runner. *Nature* 415, 1018–1021.

Hutchinson, J.R., Gatesy, S.M. (2000) Adductors, abductors, and the evolution of archosaur locomotion. *Paleobiology* 26, 734–751.

Hutchinson, J.R., Gatesy, S.M. (2006) Dinosaur locomotion: beyond the bones. *Nature* 440, 292–294.

Hutchinson, J.R., Anderson, F.C., Blemker, S., Delp, S.L. (2005) Analysis of hindlimb muscle moment arms in *Tyrannosaurus rex* using a three-dimensional musculoskeletal computer model. *Paleobiology* 31, 676–701.

Hutchinson, J.R., Ng-Thow-Hing, V., Anderson, F.C. (2007) A 3D interactive method for estimating body segmental parameters in animals: application to the turning and running performance of *Tyrannosaurus rex*. *Journal of Theoretical Biology* 246, 660–680.

Huxley, T.H. (1868) On the animals which are most nearly intermediate between the birds and reptiles. *Geological Magazine* 5, 357–365.

Huxley, T.H. (1870a) Further evidence of the affinity between the dinosaurian reptiles and birds. *Quarterly Journal of the Geologial Society of London* 26, 12–31.

Huxley, T.H. (1870b) On the classification of the Dinosauria with observations on the Dinosauria of the Trias. I. The classification and affinities of the Dinosauria. *Quarterly Journal of the Geological Society of London* 26, 32–38.

Hwang, S.H. (2005) Phylogenetic patterns of enamel microstructure in dinosaur teeth. *Journal of Morphology* 266, 208–240.

Hwang, S.H. (2010) The utility of tooth enamel microstructure in identifying isolated dinosaur teeth. *Lethaia* 43, 307–322.

Hwang, S.H. (2011) The evolution of dinosaur tooth enamel microstructure. *Biological Reviews* 86, 183–216.

Hwang, S.H., Norell, M.A., Ji, Q., Gao, K.Q. (2004) A large compsognathid from the Early Cretaceous Yixian Formation of China. *Journal of Systematic Palaeontology* 2, 13–30.

Iordansky, N.N. (1973) The skull of the Crocodilia. In: Gans, C., Parsons, T.S. (eds.), *Biology of the Reptilia*, vol. 4. Academic Press, New York, pp. 201–262.

Irby, G.V. (1996) Paleoichnological evidence for running dinosaurs worldwide. *Museum of Northern Arizona Bulletin* 60, 109–112.

Irmis, R.B. (2011) Evaluating hypotheses for the early diversification of dinosaurs. *Earth and Environmental Science Transactions of the Royal Society of Edinburgh* 101, 397–426.

Irmis, R.B., Nesbitt, S.J., Padian, K. et al. (2007a) A Late Triassic dinosauromorph assemblage from New Mexico and the rise of dinosaurs. *Science* 317, 358–361.

Irmis, R.B., Parker, W.G., Nesbitt, S.J., Liu, J. (2007b) Early ornithischian dinosaurs: the Triassic record. *Historical Biology* 19, 3–22.

Jacobsen, A.R. (1998) Feeding behavior of carnivorous dinosaurs as determined by tooth marks on dinosaur bones. *Historical Biology* 13, 17–26.

Jasinoski, S.C., Russell, A.P., Currie P.J. (2006) An integrative phylogenetic and extrapolatory approach to the reconstruction of dromaeosaur (Theropoda: Eumaniraptora) shoulder musculature. *Zoological Journal of the Linnean Society* 146, 301–344.

Jenkins, F.A., Dial, K.P., Goslow, G.E. (1988) A cineradiographic analysis of bird flight: the wishbone in starlings is a spring. *Science* 241, 1495–1498.

Jenkyns, H.C. (1980) Cretaceous anoxic events: from continents to oceans. *Journal of the Geological Society London* 137, 171–188.

Jerison, H.J. (1973) *Evolution of the Brain and Intelligence*. Academic Press, New York.

Ji, Q., Ji, S.-A. (1996) On discovery of the earliest bird fossil in China and the origin of birds. *Chinese Geology* 10, 30–33.

Ji, Q., Currie, P.J., Norell, M.A., Ji, S.-A. (1998) Two feathered dinosaurs from northeastern China. *Nature* 393, 753–761.

Juul, L. (1994) The phylogeny of basal archosaurs. *Palaeontologia Africana* 31, 1–38.

Kardong, K. (2008) *Vertebrates: Comparative Anatomy, Function, Evolution*. McGraw-Hill, Columbus, OH.

Kaye, T.G., Gaugler, G., Sawlowicz, Z. (2008) Dinosaurian soft tissues interpreted as bacterial biofilms. *PLoS ONE* 3, e2808.

Keller, G., Adatte, T., Gardin, S., Bartolini, A., Bajpai, S. (2008) Main Deccan volcanism phase ends near the K-T boundary: evidence from the Krishna-Godavari Basin, SE India. *Earth and Planetary Science Letters* 268, 293–311.

Kellner, A.W.A. (1996) Fossilized theropod soft tissue. *Nature* 379, 32.

Kellner, A.W.A., Wang, X., Tischlinger, H., Campos, D.A., Hone, D.W.E., Meng, X. (2009) The soft tissue of *Jeholopterus* (Pterosauria, Anurognathidae, Batrachognathidae) and the structure of the pterosaur wing membrane. *Proceedings of the Royal Society of London, Series B* 277, 321–329.

Kielan-Jaworowska, Z., Cifelli, R.L., Luo, X.-L. (2004) *Mammals from the Age of Dinosaurs: Origins, Evolution, and Structure*. Columbia University Press, New York.

Kilbourne, B.M., Makovicky, P.J. (2010) Limb bone allometry during postnatal ontogeny in non-avian dinosaurs. *Journal of Anatomy* 217, 135–152.

Kim, J.Y., Kim, K.S., Lockley, M.G., Seo, S.J. (2010) Dinosaur skin impressions from the Cretaceous of Korea: new insights into modes of preservation. *Palaeogeography, Palaeoclimatology, Palaeoecology* 293, 167–174.

Kitching, I.J., Forey, P.L., Humphries, C.H., Williams, D.M. (1998) *Cladistics: The Theory and Practice of Parsimony Analysis*. The Systematics Association, London.

Klein, H., Haubold, H. (2007) Archosaur footprints: potential for biochronology of Triassic continental sequences. In: Lucas, S.G., Spielmann, J.A. (eds.), *The Global Triassic*. New Mexico Museum of Natural History and Science Bulletin No. 41, pp. 120–130.

Klein, N., Sander, P.M. (2008) Ontogenetic stages in the long bone histology of sauropod dinosaurs. *Paleobiology* 34, 247–263.

Knoll, F., Padian, K., de Ricqlès, A. (2010) Ontogenetic change and adult body size of the early ornithischian dinosaur *Lesothosaurus diagnosticus*: implications for basal ornithischian taxonomy. *Gondwana Research* 17, 171–179.

Kobayashi, Y., Barsbold, R. (2005) Reexamination of a primitive ornithomimosaur, *Garudimimus*

brevipes Barsbold, 1981 (Dinosauria: Theropoda), from the Late Cretaceous of Mongolia. *Canadian Journal of Earth Sciences* 42, 1501–1521.

Kobayashi, Y., Lü, J.-C. (2003) A new ornithomimid dinosaur with gregarious habits from the Late Cretaceous of China. *Acta Palaeontologica Polonica* 48, 235–259.

Kobayashi, Y., Lu, J.-C., Dong, Z.-M., Barsbold, R., Azuma, Y. (1999) Herbivorous diet in an ornithomimid dinosaur. *Nature* 402, 480–481.

Kundrát, M. (2007) Avian-like attributes of a virtual brain model of the oviraptorid theropod *Conchoraptor gracilis*. *Naturwissenschaften* 94, 499–504.

Kundrát, M., Cruickshank, A.R.I., Manning, T.W., Nudds, J. (2008) Embryos of therizinosauroid theropods from the Upper Cretaceous of China: diagnosis and analysis of ossification patterns. *Acta Zoologica* 89, 231–251.

Kurzanov, S.M. (1972) Sexual dimorphism in protoceratopsians. *Paleontological Journal* 1972, 91–97.

Langer, M.C. (2004) Basal Saurischia. In: Weishampel, D.B., Dodson, P., Osmólska, H. (eds.), *The Dinosauria*, 2nd edn. University of California Press, Berkeley, CA, pp. 25–46.

Langer, M.C., Benton, M.J. (2006) Early dinosaurs: a phylogenetic study. *Journal of Systematic Palaeontology* 4, 309–358.

Langer, M.C., Abdala, F., Richter, M., Benton, M.J. (1999) A sauropodomorph dinosaur from the Upper Triassic (Carnian) of southern Brazil. *Comptes Rendus de l'Académie des Sciences, Sciences de la Terre et des Planètes* 329, 511–517.

Langer, M.C., França, M.A.G., Gabriel, S. (2007) The pectoral girdle and forelimb anatomy of the stem-sauropodomorph *Saturnalia tupiniquim* (Upper Triassic, Brazil). *Special Papers in Palaeontology* 77, 113–137.

Langer, M.C., Ezcurra, M.D., Bittencourt, J.S., Novas, F.E. (2010) The origin and early evolution of dinosaurs. *Biological Reviews* 85, 55–110.

Langston, W. (1973) The crocodilian skull in historical perspective. In: Gans, C., Parsons, T.S. (eds.), *Biology of the Reptilia*, vol. 4. Academic Press, New York, pp. 263–284.

Langston, W.L. (1975) The ceratopsian dinosaurs and associated lower vertebrates from the

St. Mary River Formation (Maastrichtian) at Scabby Butte, southern Alberta. *Canadian Journal of Earth Sciences* 12, 1576–1608.

Larson, P.L. (1994) *Tyrannosaurus* sex. In: *Dino Fest: Proceedings of a Conference for the General Public*. Paleontological Society Special Publication No. 7, pp. 139–155. Paleontological Society, Knoxville, TN.

Larsson, H.C.E., Sereno, P.C., Wilson, J.A. (2000) Forebrain enlargement among nonavian dinosaurs. *Journal of Vertebrate Paleontology* 20, 615–618.

Lee, A.H., Werning, S. (2008) Sexual maturity in growing dinosaurs does not fit reptilian growth models. *Proceedings of the National Academy of Sciences USA* 105, 582–587.

Lee, Y.N., Huh, M. (2002) Manus-only sauropod tracks in the Uhangri Formation (upper Cretaceous), Korea, and their paleobiological implications. *Journal of Paleontology* 76, 558–564.

Lehman, T.M. (1987) Late Maastrichtian paleoenvironments and dinosaur biogeography in the Western Interior of North America. *Palaeogeography, Palaeoclimatology, Palaeoecology* 60, 189–217.

Lehman, T.M. (1990) The ceratopsian subfamily Chasmosaurinae: sexual dimorphism and systematics. In: Carpenter, K., Currie, P.J. (eds.), *Dinosaur Systematics: Approaches and Perspectives*. Cambridge University Press, Cambridge, pp. 211–230.

Lehman, T.M., Woodward, H.N. (2008) Modeling growth rates for sauropod dinosaurs. *Paleobiology* 34, 264–281.

Li, Q., Gao, K.-Q., Vinther, J. et al. (2010) Plumage color patterns of an extinct dinosaur. *Science* 327, 1369–1372.

Liem, K., Bemis, W., Walker, W.F., Grande, L. (2008) *Functional Anatomy of the Vertebrates: An Evolutionary Perspective*. Brooks Cole, Florence, KY.

Lillegraven, J.A., Eberle, J.J. (1999) Vertebrate faunal changes through Lancian and Puercan time in southern Wyoming. *Journal of Paleontology* 73, 691–710.

Lingham-Soliar, T. (2003) Evolution of birds: ichthyosaur integumental fibers conform to dromaeosaur protofeathers. *Naturwissenschaften* 90, 428–432.

Lingham-Soliar, T., Feduccia, A., Wang, X. (2007) A new Chinese specimen indicates that "protofeathers" in the Early Cretaceous theropod dinosaur *Sinosauropteryx* are degraded collagen fibers. *Proceedings of the Royal Society of London, Series B* 274, 1823–1829.

Lipkin, C., Carpenter, K. (2008) Looking again at the forelimb of *Tyrannosaurus rex*. In: Larson, P., Carpenter, K. (eds.), *Tyrannosaurus rex: The Tyrant King*. Indiana University Press, Bloomington, IN, pp. 167–190.

Lloyd, G.T. (2011) A refined modelling approach to assess the influence of sampling on palaeobiodiversity curves: new support for declining Cretaceous dinosaur richness. *Biology Letters* doi: 10.1098/rsbl.2011.0210.

Lloyd, G.T., Davis, K.E., Pisani, D. et al. (2008) Dinosaurs and the Cretaceous Terrestrial Revolution. *Proceedings of the Royal Society of London, Series B* 275, 2483–2490.

Lockley, M.G. (1992) A quadrupedal ornithopod trackway from the Lower Cretaceous of La Rioja (Spain): inferences on gait and hand structure. *Journal of Vertebrate Paleontology* 12, 150–157.

Lockley, M.G., Hunt, A.P. (1995) *Dinosaur Tracks and Other Fossil Footprints of the Western United States*. Columbia University Press, New York.

Lockley, M.G., Matsukawa, M. (1999) Some observations on trackway evidence for gregarious behavior among small bipedal dinosaurs. *Palaeogeography, Palaeoclimatology, Palaeoecology* 150, 25–31.

Lockley, M.G., Houck, K.J., Prince, N.K. (1986) North America's largest dinosaur trackway site: implications for Morrison Formation paleoecology. *Bulletin of the Geological Society of America* 97, 1163–1176.

Lockley, M.G., Meyer, C.A., Hunt, A.P., Lucas, S.G. (1994) The distribution of sauropod tracks and trackmakers. *Gaia* 10, 233–248.

Lonardelli, I., Wenk, H.-R., Lutterotti, L., Goodwin, M. (2005) Texture analysis from synchrotron diffraction images with the Rietveld method: dinosaur tendon and salmon scale. *Journal of Synchrotron Radiation* 12, 354–360.

Longrich, N. (2006) Structure and function of hindlimb feathers in *Archaeopteryx lithographica*. *Paleobiology* 32, 417–431.

Longrich, N. (2010) The function of large eyes in *Protoceratops*: a nocturnal ceratopsian? In: Ryan, M.J., Chinnery-Allgeier, B.J., Eberth, D.A. (eds.), *New Perspectives on Horned Dinosaurs*. Indiana University Press, Bloomington, IN, pp. 307–327.

Longrich, N.R., Currie, P.J. (2009) *Albertonykus borealis*, a new alvarezsaur (Dinosauria: Theropoda) from the Early Maastrichtian of Alberta, Canada: implications for the systematics and ecology of the Alvarezsauridae. *Cretaceous Research* 30, 239–252.

Longrich, N.R., Sankey, J., Tanke, D. (2010a) *Texacephale langstoni*, a new genus of pachycephalosaurid (Dinosauria: Ornithischia) from the upper Campanian Aguja Formation, southern Texas, USA. *Cretaceous Research* 31, 274–284.

Longrich, N.R., Horner, J.R., Erickson, G.M., Currie, P.J. (2010b) Cannibalism in *Tyrannosaurus rex*. *PLoS ONE* 5, e13419.

Loope, D.B., Dingus, L., Swisher, C.C., Mingin, C. (1998) Life and death in a Late Cretaceous dunefield, Nemegt Basin, Mongolia. *Geology* 26, 27–30.

López-Martínez, N., Moratalla, J.J., Sanz, J.L. (2000) Dinosaurs nesting on tidal flats. *Palaeogeography, Palaeoclimatology, Palaeoecology* 160, 153–163.

Luo, Z.-X. (2007) Transformation and diversification in early mammal evolution. *Nature* 450, 1011–1019.

Lyson, T.R., Longrich, N.R. (2011) Spatial niche partitioning in dinosaurs from the latest Cretaceous (Maastrichtian) of North America. *Proceedings of the Royal Society of London, Series B* 278, 1158–1164.

McDonald, A.T., Horner, J.R. (2010) New material of "*Styracosaurus*" *ovatus* from the Two Medicine Formation of Montana. In: Ryan, M.J., Chinnery-Allgeier, B.J., Eberth, D.A. (eds.), *New Perspectives on Horned Dinosaurs*. Indiana University Press, Bloomington, IN, pp. 156–168.

McDonald, A.T., Barrett, P.M., Chapman, S.D. (2010) A new basal iguanodont (Dinosauria: Ornithischia) from the Wealden (Lower Cretaceous) of England. *Zootaxa* 2569: 1–43.

MacDougall, J.D. (1988) Seawater strontium isotopes, acid rain, and the Cretaceous–Tertiary boundary. *Science* 239, 485–487.

McElwain, J.C., Beerling, D.J., Woodward, F.I. (1999) Fossil plants and global warming at the Triassic–Jurassic boundary. *Science* 285, 1386–1390.

McGowan, A.J., Smith, A.B. (2008) Are global Phanerozoic marine diversity curves truly global? A study of the relationship between regional rock records and global Phanerozoic marine diversity. *Paleobiology* 34, 80–103.

McNab, B.K. (2009) Resources and energetics determined dinosaur maximal size. *Proceedings of the National Academy of Sciences USA* 106, 12184–12188.

McWhinney, L.A., Rothschild, B.M., Carpenter, K. (2001) Posttraumatic chronic osteomyelitis in *Stegosaurus* dermal spikes. In: Carpenter, K (ed.), *The Armored Dinosaurs*. Indiana University Press, Bloomington, IN, pp. 141–156.

Maidment, S.C.R., Norman, D.B., Barrett, P.M., Upchurch, P. (2008) Systematics and phylogeny of Stegosauria (Dinosauria: Ornithischia). *Journal of Systematic Palaeontology* 6, 367–407.

Main, R.P., de Ricqlès, A., Horner, J.R., Padian, K. (2005) The evolution and function of thyreophoran dinosaur scutes: implications for plate function in stegosaurs. *Paleobiology* 31, 293–316.

Maina, J.N. (2000) Comparative respiratory morphology: themes and principles in the design and construction of the gas exchangers. *Anatomical Record* 261, 25–44.

Makovicky, P.J. (1997) Postcranial axial skeleton. In: Currie, P.J., Padian, K. (eds.), *Encyclopedia of Dinosaurs*. Academic Press, New York, pp. 579–590.

Makovicky, P.J. (2001) A *Montanoceratops cerorhynchus* (Dinosauria: Ceratopsia) braincase from the Horseshoe Canyon Formation of Alberta. In: Tanke, D.H., Carpenter, K. (eds.), *Mesozoic Vertebrate Life*. Indiana University Press, Bloomington, IN, pp. 243–262.

Makovicky, P.J., Norell, M.A. (2004) Troodontidae. In: Weishampel, D.B., Dodson, P., Osmólska, H. (eds.), *The Dinosauria*, 2nd edn. University of California Press, Berkeley, CA, pp. 184–195.

Makovicky, P.J., Norell, M.A. (2006) *Yamaceratops dorngobiensis*, a new primitive ceratopsian (Dinosauria: Ornithischia) from the Cretaceous of Mongolia. *American Museum Novitates* 3530, 1–42.

Makovicky, P.J., Kobayashi, Y., Currie, P.J. (2004) Ornithomimosauria. In: Weishampel, D.B., Dodson, P., Osmólska, H. (eds.), *The Dinosauria*, 2nd edn. University of California Press, Berkeley, CA, pp. 137–150.

Makovicky, P.J., Apesteguía, S., Angolín, F.L. (2005) The earliest dromaeosaurid theropod from South America. *Nature* 437, 1007–1011.

Makovicky, P.J., Li, D., Gao, K.-Q., Lewin, M., Erickson, G.M., Norell, M.A. (2010) A giant ornithomimosaur from the Early Cretaceous of China. *Proceedings of the Royal Society of London, Series B* 277, 191–198.

Mallison, H. (2010a) The digital *Plataeosaurus* I: body mass, mass distribution and posture assessed using CAD and CAE on a digitally mounted complete skeleton. *Palaeontologia Electronica* 13 (2), Article 13.2.8A.

Mallison, H. (2010b) The digital *Plateosaurus* II: an asscssmcnt of the range of motion of the limbs and vertebral column and of previous reconstructions using a digital skeletal mount. *Acta Palaeontologica Polonica* 55, 433–458.

Mallison, H. (2010c) CAD assessment of the posture and range of motion of *Kentrosaurus aethiopicus* Hennig 1915. *Swiss Journal of Geosciences* 103, 211–233.

Mallison, H. (2011) Defense capabilities of *Kentrosaurus aethiopicus* Hennig, 1915. *Palaeontologia Electronica* 14(2), Article 14.2.10A.

Manning, P.L., Payne, D., Pennicott, J., Barrett, P.M., Ennos, R.A. (2006) Dinosaur killer claws or climbing crampons? *Biology Letters* 2, 110–112.

Manning, P.L., Morris, P.M., McMahon, A. et al. (2009) Mineralized soft-tissue structure and chemistry in a mummified hadrosaur from the Hell Creek Formation, North Dakota (USA). *Proceedings of the Royal Society of London, Series B* 276, 3429–3437.

Mannion, P.D., Upchurch, P. (2010a) A quantitative analysis of environmental associations in sauropod dinosaurs. *Paleobiology* 36, 253–282.

Mannion, P.D., Upchurch, P. (2010b) Completeness metrics and the quality of the sauropodomorph fossil record through geological and historical time. *Paleobiology* 36, 283–302.

Mannion, P.D., Upchurch, P., Carrano, M.T., Barrett, P.M. (2011a) Testing the effect of the rock record on diversity: a multidisciplinary approach to elucidating the generic richness of sauropodomorph dinosaurs through time. *Biological Reviews* 86, 157–181.

Mannion, P.D., Benson, R.B.J., Upchurch, P., Butler, R.J., Carrano, M.T., Barrett, P.M. (2011b) A temperate palaeodiversity peak in Mesozoic dinosaurs and evidence for Late Cretaceous geographical partitioning. *Global Ecology and Biogeography* (in press).

Martill, D.M. (1991) Organically preserved dinosaur skin: taphonomic and biological implications. *Modern Geology* 16, 61–68.

Martill, D.M., Batten, D.J., Loydell, D.K. (2000) A new specimen of the thyreophoran dinosaur cf. *Scelidosaurus* with soft tissue preservation. *Palaeontology* 43, 549–559.

Martinez, R.N., Alcober, O.A. (2009) A basal sauropodomorph (Dinosauria: Saurischia) from the Ischigualasto Formation (Triassic, Carnian) and the early evolution of Sauropodomorpha. *PLoS One* 4, e4397.

Martinez, R.N., Sereno, P.C., Alcober, O.A. et al. (2011) A basal dinosaur from the dawn of the dinosaur era in southwestern Pangaea. *Science* 331, 206–210.

Maryanska, T., Chapman, R.E., Weishampel, D.B. (2004) Pachycephalosauria. In: Weishampel, D.B., Dodson, P., Osmólska, H. (eds.), *The Dinosauria*, 2nd edn. University of California Press, Berkeley, CA, pp. 494–513.

Mateus, I., Mateus, H., Telles-Antunes, M. et al. (1997) Couvée, oeufs et embryons d'un dinosaure Theropode du Jurassique supérieur de Lourinhã (Portugal). *Comptes Rendus de l'Académie des Sciences II* 325, 71–78.

Mateus, O., Maidment, S.C.R., Christiansen, N.A. (2009) A new long-necked "sauropod-mimic" stegosaur and the evolution of plated dinosaurs. *Proceedings of the Royal Society of London, Series B* 276, 1815–1821.

Mathews, J.C., Brusatte, S.L., Williams, S.A., Henderson, M.D. (2009) The first *Triceratops* bonebed and its implications for gregarious behavior. *Journal of Vertebrate Paleontology* 29, 286–290.

Maurrasse, F.J.-M.R., Sen, G. (1991) Impacts, tsunamis, and the Haitian Cretaceous–Tertiary boundary layer. *Science* 252, 1690–1693.

Maxwell, D.D., Ostrom, J.A. (1995) Taphonomy and paleobiological implications of *Tenontosaurus–Deinonychus* associations. *Journal of Vertebrate Paleontology* 15, 707–712.

Mayr, G., Peters, D.S., Plodowski, G., Vogel, O. (2002) Bristle-like integumentary structures at the tail of the horned dinosaur *Psittacosaurus*. *Naturwissenschaften* 89, 361–365.

Mazzetta, G.V., Blanco, R.E., Cisilino, A.P. (2004) Modelizacio'n con elementos finitos de un diente referido al ge'nero *Giganotosaurus* Coria y Salgado, 1995 (Theropoda: Carcharodontosauridae). *Ameghiniana* 41, 619–626.

Mazzetta, G.V., Cisilino, A.P., Blanco, R.E., Calvo, N. (2009) Cranial mechanics and functional interpretation of the horned carnivorous dinosaur *Carnotaurus sastrei*. *Journal of Vertebrate Paleontology* 29, 822–830.

Meers, M.B. (2002) Maximum bite force and prey size of *Tyrannosaurus rex* and their relationship to the inference of feeding behaviour. *Historical Biology* 16, 1–22.

Melchor, R.N., de Valais, S. (2006) A review of Triassic tetrapod track assemblages from Argentina. *Palaeontology* 49, 355–379.

Meyer, C.A. (1993) A sauropod dinosaur megatracksite from the Late Jurassic of northern Switzerland. *Ichnos* 3, 29–38.

Middleton, K.M., Gatesy, S.M. (2000) Theropod forelimb design and evolution. *Zoological Journal of the Linnean Society* 128, 149–187.

Mikhailov, K. (1991) Classification of fossil eggshells of amniotic vertebrates. *Acta Palaeontologica Polonica* 36, 193–238.

Mikhailov, K. (1997) Fossil and recent eggshells in amniotic vertebrates: fine structure, comparative morphology, and classification. *Special Papers in Palaeontology* 56, 1–80.

Milner, A.C. (2003) Fish-eating theropods: a short review of the systematics, biology and

palaeobiology of spinosaurs. *Journadas Internacionales sobre paleontologiá de Dinosaurios y su Entoro* 2, 129–138.

Milner, A.R.C., Harris, J.D., Lockley, M.G., Kirkland, J.I., Matthews, N.A. (2009) Bird-like anatomy, posture, and behavior revealed by an Early Jurassic theropod dinosaur resting trace. *PLoS ONE* 4, e4591.

Molnar, R.E. (2000) Mechanical factors in the design of the skull of *Tyrannosaurus rex* (Osborn, 1905). *Gaia* 15, 193–218.

Molnar, R.E., Clifford, H.T. (2000) Gut contents of a small ankylosaur. *Journal of Vertebrate Paleontology* 20, 194–196.

Montanari, S., Brusatte, S.L., De Wolf, W., Norell, M.A. (2011) Variation of osteocyte lacunae size within the tetrapod skeleton: implications for palaeogenomics. *Biology Letters* 7, 751–754.

Moreno, K., Carrano, M.T., Snyder, R. (2007) Morphological changes in pedal phalanges through ornithopod dinosaur evolution: a biomechanical approach. *Journal of Morphology* 268, 50–63.

Müller, J., Scheyer, T.M., Head, J.J. et al. (2010) Homeotic effects, somitogenesis and the evolution of vertebral numbers in recent and fossil amniotes. *Proceedings of the National Academy of Sciences USA* 107, 2118–2123.

Myers, T.S., Fiorillo, A.R. (2009) Evidence for gregarious behavior and age segregation in sauropod dinosaurs. *Palaeogeography, Palaeoclimatology, Palaeoecology* 274, 96–104.

Myhrvold, N.P., Currie, P.J. (1997) Supersonic sauropods? Tail dynamics in diplodocids. *Paleobiology* 23, 393–409.

Nee, S. (2006) Birth–death models in macroevolution. *Annual Review of Ecology, Evolution, and Systematics* 37, 1–17.

Nesbitt, S.J. (2003) *Arizonasaurus* and its implications for archosaur divergence. *Proceedings of the Royal Society of London, Series B* 270, S234–S237.

Nesbitt, S.J. (2007) The anatomy of *Effigia okeeffeae* (Archosauria, Suchia), theropod-like convergence, and the distribution of related taxa. *Bulletin of the American Museum of Natural History* 302, 1–84.

Nesbitt, S.J. (2011) The early evolution of archosaurs: relationships and the origin of major clades. *Bulletin of the American Museum of Natural History* 352, 1–292.

Nesbitt, S.J., Norell, M.A. (2006) Extreme convergence in the body plans of an early suchian (Archosauria) and ornithomimid dinosaurs (Theropoda). *Proceedings of the Royal Society of London, Series B* 273, 1045–1048.

Nesbitt, S.J., Turner, A.H., Erickson, G.M., Norell, M.A. (2006) Prey choice and cannibalistic behavior in the theropod *Coelophysis*. *Biology Letters* 2, 611–614.

Nesbitt, S.J., Irmis, R.B., Parker, W.G. (2007) A critical re-evaluation of the Late Triassic dinosaur taxa of North America. *Journal of Systematic Palaeontology* 5, 209–243.

Nesbitt, S.J., Stocker, M.R., Small, B.J., Downs, A. (2009a) The osteology and relationships of *Vancleavea campi* (Reptilia: Archosauriformes). *Zoological Journal of the Linnean Society* 157, 814–864.

Nesbitt, S.J., Smith, N.D., Irmis, R.B., Turner, A.H., Downs, A., Norell, M.A. (2009b) A complete skeleton of a Late Triassic saurischian and the early evolution of dinosaurs. *Science* 326, 1530–1533.

Nesbitt, S.J., Turner, A.H., Spaulding, M., Conrad, J. L., Norell, M.A. (2009c) The theropod furcula. *Journal of Morphology* 270, 856–879.

Nesbitt, S.J., Sidor, C.A., Irmis, R.B., Angielczyk, K. D., Smith, R.M.H., Tsuji, L.A. (2010) Ecologically distinct dinosaurian sister group shows early diversification of Ornithodira. *Nature* 464, 95–98.

Nesbitt, S.J., Liu, J., Li, C. (2011) A sail-backed suchian from the Heshanggou Formation (Early Triassic: Olenekian) of China. *Earth and Environmental Science Transactions of the Royal Society of Edinburgh* 101, 271–284.

Nicholls, E.L., Russell, A.P. (1985) Structure and function of the pectoral girdle and forelimb of *Struthiomimus altus* (Theropoda: Ornithomimidae). *Palaeontology* 28, 643–677.

Niedźwiedzki, G., Szrek, P., Narkiewicz, K., Narkiewicz, M., Ahlberg, P.E. (2010) Tetrapod trackways from the early Middle Devonian period of Poland. *Nature* 463, 43–48.

Nopcsa, F.B. (1929) Sexual differences in ornithopodous dinosaurs. *Palaeobiologica* 2, 187–200.

Norell, M.A. (1992) Taxic origin and temporal diversity: the effect of phylogeny. In: Novacek, M.D., Wheeler, Q.D. (eds.), *Extinction and Phylogeny*. Columbia University Press, New York, pp. 89–118.

Norell, M.A. (1993) Tree-based approaches to understanding history: comments on ranks, rules and the quality of the fossil record. *American Journal of Science* 293, 407–417.

Norell, M.A., Makovicky, P.J. (1997) Important features of the dromaeosaur skeleton: information from a new specimen. *American Museum Novitates* 3215, 1–28.

Norell, M.A., Makovicky, P.J. (2004) Dromaeosauridae. In: Weishampel, D.B., Dodson, P., Osmólska, H. (eds.), *The Dinosauria*, 2nd edn. University of California Press, Berkeley, CA, pp. 196–209.

Norell, M.A., Xu, X. (2005) Feathered dinosaurs. *Annual Review of Earth and Planetary Sciences* 33, 277–299.

Norell, M.A., Clark, J.M., Chiappe, L.M., Dashzeveg, D. (1995) A nesting dinosaur. *Nature* 378, 774–776.

Norell, M.A., Clark, J.M., Makovicky, P.J. (2001a) Relationships among Maniraptora: problems and prospects. In: Gauthier, J.A., Gall, L.F. (eds.), *New Perspectives on the Origin and Early Evolution of Birds*. Peabody Museum of Natural History, New Haven, CT, pp. 49–67.

Norell, M.A., Makovicky, P.J., Currie, P.J. (2001b) The beaks of ostrich dinosaurs. *Nature* 412, 874.

Norell, M.A., Clark, J.M., Chiappe, L.M. (2001c) An embryonic oviraptorid (Dinosauria: Theropoda) from the Upper Cretaceous of Mongolia. *American Museum Novitates* 3315, 1–20.

Norell, M.A., Ji, Q., Gao, K., Yuan, C., Zhao, Y., Wang, L. (2002) "Modern" feathers on a non-avian dinosaur. *Nature* 416, 36–37.

Norell, M.A., Makovicky, P.J., Bever, G.S. et al. (2009) A review of the Mongolian Cretaceous dinosaur *Saurornithoides* (Troodontidae: Theropoda). *American Museum Novitates* 3654, 1–63.

Norman, D.B. (1984) On the cranial morphology and evolution of ornithopod dinosaurs.

Symposium of the Zoological Society of London 52, 521–547.

Norman, D.B. (1986) On the anatomy of *Iguanodon atherfieldensis* (Ornithischia: Ornithopoda). *Bulletin de l'Institut Royale des Sciences Naturalles de Belgique* 56, 281–372.

Norman, D.B. (1998) On Asian ornithopods (Dinosauria: Ornithischia). 3. A new species of iguanodontid dinosaur. *Zoological Journal of the Linnean Society* 122, 291–348.

Norman, D.B. (2001) Dinosaur feeding. *Encyclopedia of Life Sciences* doi: 10.1038/npg. els.0003321.

Norman, D.B. (2004) Basal Iguanodontia. In: Weishampel, D.B., Dodson, P., Osmólska, H. (eds.), *The Dinosauria*, 2nd edn. University of California Press, Berkeley, CA, pp. 413–437.

Norman, D.B., Weishampel, D.B. (1985) Ornithopod feeding mechanisms: their bearing on the evolution of herbivory. *American Naturalist* 126, 151–164.

Norman, D.B., Weishampel, D.B. (1991) Feeding mechanisms in some small herbivorous dinosaurs: processes and patterns. In: Rayner, J.M. V., Wootton, R.J. (eds.), *Biomechanics and Evolution*. Cambridge University Press, Cambridge, pp. 161–181.

Norman, D.B., Sues, H.-D., Witmer, L.M., Coria, R.A. (2004a) Basal Ornithopoda. In: Weishampel, D.B., Dodson, P., Osmólska, H. (eds.), *The Dinosauria*, 2nd edn. University of California Press, Berkeley, CA, pp. 393–412.

Norman, D.B., Witmer, L.M., Weishampel, D.B. (2004b) Basal Thyreophora. In: Weishampel, D.B., Dodson, P., Osmólska, H. (eds.), *The Dinosauria*, 2nd edn. University of California Press, Berkeley, CA, pp. 335–342.

Norman, D.B., Witmer, L.M., Weishampel, D.B. (2004c) Basal Ornithischia. In: Weishampel, D.B., Dodson, P., Osmólska, H. (eds.), *The Dinosauria*, 2nd edn. University of California Press, Berkeley, CA, pp. 325–334.

Noto, C.R., Grossman, A. (2010) Broad-scale patterns of Late Jurassic dinosaur paleoecology. *PLoS ONE* 5, e12553.

Novas, F.E. (1989) The tibia and tarsus in the Herrerasauridae (Dinosauria, incertae sedis) and the origin and evolution of the dinosaurian tarsus. *Journal of Paleontology* 63, 677–690.

Novas, F.E. (1992) Phylogenetic relationships of the basal dinosaurs, the Herrerasauridae. *Palaeontology* 16, 51–62.

Novas, F.E. (1996) Dinosaur monophyly. *Journal of Vertebrate Paleontology* 16, 723–741.

Novas, F.E. (2009) *The Age of Dinosaurs in South America*. Indiana University Press, Bloomington, IN.

Nudds, R.L., Dyke, G.J. (2010) Narrow primary feather rachises in *Confuciusornis* and *Archaeopteryx* suggest poor flight ability. *Science* 328, 887–889.

O'Connor, M.P., Dodson, P. (1999) Biophysical constraints on the thermal ecology of dinosaurs. *Paleobiology* 25, 341–368.

O'Connor, P.M. (2004) Pulmonary pneumaticity in the postcranial skeleton of extant Aves: a case study examining Anseriformes. *Journal of Morphology* 261, 141–161.

O'Connor, P.M. (2006) Postcranial pneumaticity: an evaluation of soft-tissue influences on the postcranial skeleton and the reconstruction of pulmonary anatomy in archosaurs. *Journal of Morphology* 267, 1199–1226.

O'Connor, P.M. (2009) Evolution of archosaurian body plans: skeletal adaptations of an air-sac based breathing apparatus in birds and other archosaurs. *Journal of Experimental Zoology* 311A, 629–646.

O'Connor, P.M., Claessens, L.P.A.M. (2005) Basic avian pulmonary design and flow-through ventilation in non-avian theropod dinosaurs. *Nature* 436, 253–256.

O'Higgins, P. (2000) The study of morphological variation in the hominid fossil record: biology, landmarks and geometry. *Journal of Anatomy* 197, 103–120.

Olsen, P.E., Schlische, R.W., Gore, P.J.W. (1989) Tectonic, depositional and paleoecological history of Early Mesozoic rift basins, eastern North America. *American Geophysical Union Field Trip Guidebook* T351, 1–174.

Olsen, P.E., Kent, D.V., Sues, H.D. et al. (2002) Ascent of dinosaurs linked to an iridium anomaly at the Triassic–Jurassic boundary. *Science* 296, 1305–1307.

Olsen, P.E., Kent, D.V., Whiteside, J.H. (2011) Implications of the Newark Supergroup-based astrochronology and geomagnetic polarity time scale (Newark-APTS) for the tempo and mode of the early diversification of the Dinosauria. *Earth and Environmental Science Transactions of the Royal Society of Edinburgh* 101, 201–229.

Organ, C.L. (2006) Thoracic epaxial muscles in living archosaurs and ornithopod dinosaurs. *Anatomical Record* 288A, 782–793.

Organ, C.L., Adams, J. (2005) The histology of ossified tendon in dinosaurs. *Journal of Vertebrate Paleontology* 25, 602–613.

Organ, C.L., Shedlock, A.M., Meade, A., Pagel, M., Edwards, S.V. (2007) Origin of avian genome size and structure in non-avian dinosaurs. *Nature* 446, 180–184.

Organ, C.L., Schweitzer, M.H., Zheng, W., Freimark, L.M., Cantley, L.C., Asara, J.M. (2008) Molecular phylogenetics of mastodon and *Tyrannosaurus rex. Science* 320, 499.

Organ, C.L., Brusatte, S.L., Stein, K. (2009) Sauropod dinosaurs evolved moderately sized genomes unrelated to body size. *Proceedings of the Royal Society of London, Series B* 276, 4303–4308.

Ortega, F., Escaso, F., Sanz, J.L. (2010) A bizarre, humped Carcharodontosauria (Theropoda) from the Lower Cretaceous of Spain. *Nature* 467, 203–206.

Osborn, H.F. (1912) Crania of *Tyrannosaurus* and *Allosaurus. Memoirs of the American Museum of Natural History* 1, 33–54.

Ősi, A., Butler, R.J., Weishampel, D.B. (2010) A Late Cretaceous ceratopsian dinosaur from Europe with Asian affinities. *Nature* 465, 466–468.

Osmólska, H. (2004) Evidence on relation of brain to endocranial cavity in oviraptorid dinosaurs. *Acta Palaeontologica Polonica* 49, 321–324.

Osmólska, H., Currie, P.J., Barsbold, R. (2004) Oviraptorosauria. In: Weishampel, D.B., Dodson, P., Osmólska, H. (eds.), *The Dinosauria*, 2nd edn. University of California Press, Berkeley, CA, pp. 165–183.

Ostrom, J.H. (1962) The cranial crests of hadrosaurian dinosaurs. *Postilla* 62, 1–29.

Ostrom, J.H. (1964) A functional analysis of jaw mechanics in the dinosaur *Triceratops. Postilla* 88, 1–35.

Ostrom, J.H. (1966) Functional morphology and evolution of the ceratopsian dinosaurs. *Evolution* 20, 290–308.

Ostrom, J.H. (1969) A new theropod dinosaur from the Lower Cretaceous of Montana. *Postilla* 128, 1–17.

Ostrom, J.H. (1973) The ancestry of birds. *Nature* 242, 136.

Ostrom, J.H. (1974) The pectoral girdle and forelimb function of *Deinonychus* (Reptilia: Saurischia): a correction. *Postilla* 165, 1–11.

Ostrom, J.H. (1978) The osteology of *Compsognathus longipes* Wagner. *Zitteliana* 4, 73–118.

Ostrom, J.H. (1990) Dromaeosauridae. In: Weishampel, D.B., Dodson, P., Osmólska, H. (eds.), *The Dinosauria*. University of California Press, Berkeley, CA, pp. 269–279.

Ostrom, J.H. (1994) *Deinonychus*, the ultimate killing machine. In: Rosenberg, G.D., Wolberg, D.L. (eds.), *Dino Fest: Proceedings of a Conference for the General Public*. Paleontological Society Special Publication No. 7, pp. 127–138. Paleontological Society, Knoxville, TN.

Owen, R. (1842) Report on British Fossil Reptiles. Part II. *Reports of the British Association for the Advancement of Science* 11, 60–204.

Padian, K. (2003) Four-winged dinosaurs, bird precursors, or neither? *Bioscience* 53, 450–452.

Padian, K. (2004) Basal Avialae. In: Weishampel, D.B., Dodson, P., Osmólska, H. (eds.), *The Dinosauria*, 2nd edn. University of California Press, Berkeley, CA, pp. 210–231.

Padian, K., Chiappe, L.M. (1998) The origin and early evolution of birds. *Biological Reviews* 73, 1–42.

Padian, K., Horner, J.R. (2004) Dinosaur physiology. In: Weishampel, D.B., Dodson, P., Osmólska, H. (eds.), *The Dinosauria*, 2nd edn. University of California Press, Berkeley, CA, pp. 660–671.

Padian, K., Horner, J.R. (2011) The evolution of "bizarre structures" in dinosaurs: biomechanics, sexual selection, social selection or species recognition? *Journal of Zoology* 283, 3–17.

Padian, K., May, C.L. (1993) The earliest dinosaurs. In: Lucas, S.G., Morales, M. (eds.), *The Nonmarine Triassic*. New Mexico Museum of Natural History and Science Bulletin No. 3, pp. 379–380.

Padian, K., de Ricqlès, A.J., Horner, J.R. (2001) Dinosaurian growth rates and bird origins. *Nature* 412, 405–408.

Padian, K., Horner, J.R., de Ricqlès, A. (2004) Growth in small dinosaurs and pterosaurs: the evolution of archosaurian growth strategies. *Journal of Vertebrate Paleontology* 24, 555–571.

Page, R.D.M. (1988) Quantitative cladistic biogeography: constructing and comparing area cladograms. *Systematic Zoology* 37, 254–270.

Paik, I.S., Kim, H.J., Lim, J.D., Huh, M., Lee, H.I. (2011) Diverse tooth marks on an adult sauropod bone from the Early Cretaceous, Korea: implications in feeding behaviour of theropod dinosaurs. *Palaeogeography, Palaeoclimatology, Palaeoecology* 309, 342–346.

Paladino, F.V., O'Connor, M.P., Spotila, J.R. (1990) Metabolism of leatherback turtles, gigantothermy, and thermoregulation of dinosaurs. *Nature* 344, 858–860.

Parrish, J.M. (1987) The origin of crocodilian locomotion. *Paleobiology* 13, 396–414.

Paul, G.S. (1984) The segnosaurian dinosaurs: relics of the prosauropod–ornithischian transition? *Journal of Vertebrat Paleontology* 4, 507–515.

Paul, G.S. (1988) *Predatory Dinosaurs of the World.* Simon and Schuster, New York.

Paul, G.S. (2000) Limb design, function and running performance in ostrich-mimics and tyrannosaurs. *Gaia* 15, 257–270.

Paul, G.S. (2002) *Dinosaurs of the Air: The Evolution and Loss of Flight in Dinosaurs and Birds.* Johns Hopkins University Press, Baltimore, MD.

Paul, G.S. (2008) The extreme lifestyle and habits of the gigantic tyrannosaurid superpredators of the late Cretaceous of North America and Asia. In: Larson, P.L., Carpenter, K., (eds.), *Tyrannosaurus rex: The Tyrant King.* Indiana University Press, Bloomington, IN, pp. 307–351.

Paul, G.S., Christiansen, P. (2000) Forelimb posture in neoceratopsian dinosaurs: implications for gait and locomotion. *Paleobiology* 26, 450–465.

Peabody, F.E. (1948) Reptile and amphibian trackways from the Moenkopi Formation of Arizona and Utah. *University of California Publications, Bulletin of the Department of Geological Sciences* 27, 295–468.

Pereda-Suberbiola, X. (2009) Biogeographical affinities of Late Cretaceous continental tetrapods of Europe: a review. *Bulletin de la Societe Géologique de France* 180, 57–71.

Perle, A., Norell, M.A., Chiappe, L.M., Clark, J.M. (1993) Flightless bird from the Cretaceous of Mongolia. *Nature* 362, 623–626.

Perry, S.F. (1983) Reptilian lungs: functional anatomy and evolution. *Advances in Anatomy, Embryology, and Cell Biology* 79, 1–81.

Peters, S.E. (2005) Geologic constraints on the macroevolutionary history of marine animals. *Proceedings of the National Academy of Sciences USA* 102, 12326–12331.

Peters, S.E. (2008) Environmental determinants of extinction selectivity in the fossil record. *Nature* 454, 626–629.

Peters, S.E., Foote, M. (2001) Biodiversity in the Phanerozoic: a reinterpretation. *Paleobiology* 27, 583–601.

Piechowski, R., Dzik, J. (2010) The axial skeleton of *Silesaurus opolensis. Journal of Vertebrate Paleontology* 30, 1127–1141.

Platnick, N.L., Nelson, G. (1981) A method of analysis for historical biogeography. *Systematic Zoology* 27, 1–16.

Pol, D., Garrido, A., Cerda, I.A. (2011) A new sauropodomorph dinosaur from the Early Jurassic of Patagonia and the origin and evolution of the sauropod-type sacrum. *PLoS ONE* 6, e14572.

Pontzer, H., Allen, V., Hutchinson, J.R. (2009) Biomechanics of running indicates endothermy in bipedal dinosaurs. *PLoS ONE* 4, e7783.

Pope, K.O., Baines, K.H., Ocampo, A.C., Ivanov, B. A. (1997) Energy, volatile production, and climatic effects of the Chicxulub Cretaceous/Tertiary impact. *Journal of Geophysical Research* 102 (E9), 21645–21664.

Pope, K.O., D'Hondt, S.L., Marshall, C.R. (1998) Meteorite impact and the mass extinction of species at the Cretaceous/Tertiary boundary. *Proceedings of the National Academy of Sciences USA* 95, 11028–11029.

Pough, F.H., Andrews, R.M., Cadle, J.E., Crump, M. L., Savitsky, A.H., Wells, K.D. (2003) *Herpetology,* 3rd edn. Benjamin Cummings, New York.

Pough, F.H., Janis, C.M., Heiser, J.B. (2008) *Vertebrate Life.* Benjamin Cummings, New York.

Powell, J.L. (1998) *Night Comes to the Cretaceous.* W.H. Freeman and Company, New York.

Prasad, V., Strömberg, C.A.E., Alimohamamadian, H., Sahni, A. (2005) Dinosaur coprolites and the early evolution of grasses and grazers. *Science* 310, 1177–1180.

Prieto-Márquez, A. (2010a) Global phylogeny of Hadrosauridae (Dinosauria: Ornithopoda) using parsimony and Bayesian methods. *Zoological Journal of the Linnean Society* 159, 435–502.

Prieto-Márquez, A. (2010b) Global historical biogeography of hadrosaurid dinosaurs. *Zoological Journal of the Linnean Society* 159, 503–525.

Prieto-Márquez, A., Wagner, J.R. (2009) *Pararhabdodon isonensis* and *Tsintaosaurus spinorhinus*: a new clade of lambeosaurine hadrosaurids from Eurasia. *Cretaceous Research* 30, 1238–1246.

Prieto-Márquez, A., Gignac, P.M., Joshi, S. (2007) Neontological evaluation of pelvic skeletal attributes purported to reflect sex in extinct non-avian archosaurs. *Journal of Vertebrate Paleontology* 27, 603–609.

Prinn, R.G., Fegley, F. (1987) Bolide impacts, acid rain, and biospheric traumas at the Cretaceous–Tertiary boundary. *Earth and Planetary Science Letters* 83, 1–15.

Prum, R.O., Brush, A.H. (2002) The evolutionary origin and diversification of feathers. *Quarterly Review of Biology* 77, 261–295.

Ptaszynski, T. (2000) Lower Triassic vertebrate footprints from Wióry, Holy Cross Mountains, Poland. *Acta Paleontologica Polonica* 45, 151–194.

Purvis, A. (2008) Phylogenetic approaches to the study of extinction. *Annual Review of Ecology, Evolution, and Systematics* 39, 301–319.

Raath, M.A. (1990) Morphological variation in small theropods and its meaning in systematics: evidence from *Syntarsus rhodesiensis*. In: Carpenter, K., Currie, P.J. (eds.), *Dinosaur Systematics: Approaches and Perspectives*. Cambridge University Press, Cambridge, pp. 91–105.

Rannala, B., Yang, Z. (1996) Probability distribution of molecular evolutionary trees: a new method of phylogenetic inference. *Journal of Molecular Evolution* 43, 304–311.

Rauhut, O.W.M. (2003) The interrelationships and evolution of basal theropod dinosaurs. *Special Papers in Palaeontology* 69, 1–213.

Rauhut, O.W.M., Remes, K., Fechner, R., Cladera, G., Puerta, P. (2005) Discovery of a short-necked sauropod dinosaur from the Late Jurassic period of Patagonia. *Nature* 435, 670–672.

Rauhut, O.W.M., Milner, A.C., Moore-Fay, S. (2010) Cranial osteology and phylogenetic position of the theropod dinosaur *Proceratosaurus bradleyi* (Woodward, 1910) from the Middle Jurassic of England. *Zoological Journal of the Linnean Society* 158, 155–195.

Raup, D.M. (1979) Size of the Permo-Triassic bottleneck and its evolutionary implications. *Science* 206, 217–218.

Raup, D.M., Sepkoski, J.J. (1982) Mass extinctions in the marine fossil record. *Science* 215, 1501–1503.

Rayfield, E.J. (2004) Cranial mechanics and feeding in *Tyrannosaurus rex*. *Proceedings of the Royal Society of London, Series B* 271, 1451–1459.

Rayfield, E.J. (2005a) Aspects of comparative cranial mechanics in the theropod dinosaurs *Coelophysis*, *Allosaurus* and *Tyrannosaurus rex*. *Zoological Journal of the Linnean Society* 144, 309–316.

Rayfield, E.J. (2005b) Using finite-element analysis to investigate suture morphology: a case study using large carnivorous dinosaurs. *Anatomical Record* 283A, 349–365.

Rayfield, E.J. (2007) Finite element analysis and understanding the biomechancis and evolution of living and fossil organisms. *Annual Review of Earth and Planetary Sciences* 35, 541–576.

Rayfield, E.J., Norman, D.B., Horner, C.C. et al. (2001) Cranial design and function in a large theropod dinosaur. *Nature* 409, 1033–1037.

Rayfield, E.J., Milner, A.C., Xuan, V.B., Young, P.G. (2007) Functional morphology of spinosaur "crocodile-mimic" dinosaurs. *Journal of Vertebrate Paleontology* 27, 892–901.

Redelstorff, R., Sander, P.M. (2009) Long and girdle bone histology of *Stegosaurus*: implications for growth and life history. *Journal of Vertebrate Paleontology* 29, 1087–1099.

Rees, P.M., Zeigler, A.M., Valdes, P.J. (2000) Jurassic phytogeography and climates: new data and model comparisons. In: Huber, B.T., MacLeod, K. G., Wing, S.T. (eds.), *Warm Climates in Earth History*. Cambridge University Press, Cambridge, pp. 297–318.

Reichel, M. (2010) A model for the bite mechanics in the herbivorous dinosaur *Stegosaurus* (Ornithischia, Stegosauridae). *Swiss Journal of Geosciences* 103, 235–240.

Reid, R.E.H. (1981) Lamellar-zonal bone with zones of annuli in the pelvis of a sauropod dinosaur. *Nature* 292, 49–51.

Reid, R.E.H. (1984) Primary bone and dinosaur physiology. *Geological Magazine* 121, 589–598.

Reid, R.E.H. (1987) Bone and dinosaurian "endothermy." *Modern Geology* 11, 133–154.

Reid, R.E.H. (1997) Dinosauria physiology: the case for "intermediate" dinosaurs. In: Farlow, J.O., Brett-Surman, M.K. (eds.), *The Complete Dinosaur*. Indiana University Press, Bloomington, IN, pp. 449–473.

Reig, O.A. (1963) La presencia de dinosaurios saurisquios en los "Estratos de Ischigualasto" (Mesotriásico superior) de las Provincias de San Juan y La Rioja (Republica Argentina). *Ameghiniana* 3, 3–20.

Reisz, R.R., Scott, D., Sues, H.-D., Evans, D.C., Raath, M.A. (2005) Embryos of an Early Jurassic prosauropod dinosaur and their evolutionary significance. *Science* 309, 761–764.

Reisz, R.R., Evans, D.C., Sues, H.-D., Scott, D. (2010) Embryonic skeletal anatomy of the sauropodomorph dinosaur *Massospondylus* from the Lower Jurassic of South Africa. *Journal of Vertebrate Paleontology* 30, 1653–1665.

Remes, K., Ortega, F., Fierro, I. et al. (2009) A new basal sauropod dinosaur from the Middle Jurassic of Niger and the early evolution of Sauropoda. *PLoS ONE* 4, e6924.

Rich, T.H., Vickers-Rich, P., Gangloff, R.A. (2002) Polar dinosaurs. *Science* 295, 979–980.

Ricklefs, R.E. (2007a) Tyrannosaur ageing. *Biology Letters* 3, 214–217.

Ricklefs, R.E. (2007b) Estimating diversification rates from phylogenetic information. *Trends in Ecology and Evolution* 22, 601–610.

Rinehart, L.F., Lucas, S.G., Heckert, A.B., Spielmann, J.A., Celeskey, M.D. (2009) The paleobiology of *Coelophysis bauri* (Cope) from the Upper Triassic (Apachean) Whitaker quarry, New Mexico, with detailed analysis of a single quarry block. *New Mexico Museum of Natural History and Science Bulletin* 45, 1–260.

Roach, B.T., Brinkman, D.L. (2007) A reevaluation of cooperative pack hunting and gregariousness in *Deinonychus antirrhopus* and other nonavian theropod dinosaurs. *Bulletin of the Peabody Museum of Natural History* 48, 103–138.

Robertson, D.S., McKenna, M.C., Toon, O.B., Hope, S., Lillegraven, J.A. (2004) Survival in the first hours of the Cenozoic. *Geological Sociey of America Bulletin* 116, 760–768.

Rogers, R.R., Brady, M.E. (2010) Origins of microfossil bonebeds: insights from the Upper Cretaceous Judith River Formation of north-central Montana. *Paleobiology* 36, 80–112.

Rogers, R.R., Swisher, C.C., Sereno, P.C., Monetta, A.M., Forster, C.A., Martinez, R.N. (1993) The Ischigualasto tetrapod assemblage (Late Triassic, Argentina) and $^{40}Ar/^{39}Ar$ dating of dinosaur origins. *Science* 260, 794–797.

Rogers, R.R., Krause, D.W., Curry Rogers, K. (2003) Cannibalism in the Madagascan dinosaur *Majungatholus atopus*. *Nature* 422, 515–518.

Rogers, R.R., Eberth, D.A., Fiorillo, A.R. (2008) *Bonebeds: Genesis, Analysis, and Paleobiological Significance*. University of Chicago Press, Chicago, IL.

Rogers, S.W. (1998) Exploring dinosaur neuropaleobiology: computed tomography scanning and analysis of an *Allosaurus fragilis* endocast. *Neuron* 21, 673–679.

Rohlf, F.J., Marcus, L.F. (1993) A revolution in morphometrics. *Trends in Ecology and Evolution* 8, 129–132.

Romer, A.S. (1956) *Vertebrate Paleontology*. University of Chicago Press, Chicago, IL.

Romer, A.S. (1971) The Chañares (Argentina) Triassic reptile fauna. X. Two new but incompletely known long-limbed pseudosuchians. *Breviora* 378, 1–10.

Romer, A.S. (1972) The Chañares (Argentina) Triassic reptile fauna. XIV. *Lewisuchus admixtus*, gen. et sp. nov., a further thecodont from the Chañares beds. *Breviora* 390, 1–13.

Rose, K.D. (2006) *The Beginning of the Age of Mammals*. Johns Hopkins University Press, Baltimore, MD.

Rowe, T., Gauthier, J.A. (1990) Ceratosauria. In: Weishampel, D.B., Dodson, P., Osmólska, H. (eds.), *The Dinosauria*. University of California Press, Berkeley, CA, pp. 151–168.

Rowe, T., McBride, E.F., Sereno, P.C. (2001) Dinosaur with a heart of stone. *Science* 291, 783a.

Ruta, M., Wagner, P J., Coates, M.I. (2006) Evolutionary patterns in early tetrapods. I. Rapid initial diversification followed by decrease in rates of character change. *Proceedings of the Royal Society of London, Series B* 273, 2107–2111.

Ruxton, G.D., Houston, D.C. (2003) Could *Tyrannosaurus rex* have been a scavenger rather than a predator? An energetics approach. *Proceedings of the Royal Society of London, Series B* 270, 731–733.

Ruxton, G.D., Wilkinson, D.M. (2011) The energetics of low browsing in sauropods. *Biology Letters* 7, 779–781.

Ryan, M.J., Russell, A.P., Russell, D.A., Eberth, D.A., Currie, P.J. (2001) The taphonomy of a *Centrosaurus* (Ornithischia: Ceratopsidae) bone bed from the Dinosaur Park Formation (Upper Campanian), Alberta, Canada, with comments on cranial ontogeny. *Palaios* 16, 482–506.

Ryan, M.J., Chinnery-Allgeier, B.J., Eberth, D.A. (eds.) (2010a) *New Perspectives on Horned Dinosaurs*. Indiana University Press, Bloomington, IN.

Ryan, M.J., Eberth, D.A., Brinkman, D.B., Currie, P.J, Tanke, D.H. (2010b) A new *Pachyrhinosaurus*-like ceratopsid from the Upper Dinosaur Park Formation (Late Campanian) of southern Alberta, Canada. In: Ryan, M.J., Chinnery-Allgeier, B.J., Eberth, D.A. (eds.), *New Perspectives on Horned Dinosaurs*. Indiana University Press, Bloomington, IN, pp. 141–155.

Rybczynski, N., Vickaryous, M.K. (2001) Evidence of complex jaw movement in the Late Cretaceous ankylosaurid *Euoplocephalus tutus* (Dinosauria: Thyreophora). In: Carpenter, K. (ed.), *The Armored Dinosaurs*. Indiana University Press, Bloomington, IN, pp. 299–317.

Rybczynski, N., Tirabasso, A., Bloskie, P., Cuthbertson, R., Holliday, C. (2008) A three-dimensional animation model of *Edmontosaurus* (Hadrosauridae) for testing chewing hypotheses. *Palaeontologia Electronica* 11(2), Article 11.2.9A.

Sahney, S., Benton, M.J. (2008) Recovery from the most profound mass extinction of all time. *Proceedings of the Royal Society of London, Series B* 275, 759–765.

Sakamoto, M. (2010) Jaw biomechanics and the evolution of biting performance in theropod dinosaurs. *Proceedings of the Roayl Society of London, Series B* 277, 3327–3333.

Salgado, L., Coria, R.A., Calvo, J.O. (1997) Evolution of titanosaurid sauropods. I: phylogenetic analysis based on the postcranial evidence. *Ameghiniana* 34, 3–32.

Samman, T., Powell, G.L., Currie, P.J., Hills, L.V. (2005) Morphometry of the teeth of western North American tyrannosaurids and its applicability to quantitative classification. *Acta Palaeontologica Polonica* 50, 757–776.

Sampson, S.D. (1999) Sex and destiny: the role of mating signals in speciation and macroevolution. *Historical Biology* 13, 173–197.

Sampson, S.D., Witmer, L.M. (2007) Craniofacial anatomy of *Majungasaurus crenatissimus* (Theropoda: Abelisauridae) from the Late Cretaceous of Madagascar. *Society of Vertebrate Paleontology Memoir* 8, 32–102.

Sampson, S.D., Ryan, M.J., Tanke, D.H. (1997) Craniofacial ontogeny in centrosaurine dinosaurs (Ornithischia: Ceratopsidae): taxonomic and behavioral implications. *Zoological Journal of the Linnean Society* 121, 293–337.

Sampson, S.D., Witmer, L.M., Forster, C.A. et al. (1998) Predatory dinosaur remains from Madagascar: implications for Cretaceous biogeography of Gondwana. *Science* 280, 1048–1051.

Sampson, S.D., Carrano, M.T., Forster, C.A. (2001) A bizarre predatory dinosaur from the Late Cretaceous of Madagascar. *Nature* 409, 504–506.

Sampson, S.D., Loewen, M.A., Farke, A.A. et al. (2010) New horned dinosaurs from Utah provide evidence for intracontinental dinosaur endemism. *PLoS ONE* 5, e12292.

Sander, P.M. (1999) The microstructure of reptilian tooth enamel: terminology, function, and phylogeny. *Münchner Geowissenschaftliche Abhandlungen, Reihe A* 38, 1–102.

Sander, P.M. (2000) Long bone histology of the Tendaguru sauropods: implications for growth and biology. *Paleobiology* 36, 466–488.

Sander, P.M., Clauss, M. (2008) Perspective: Sauropod gigantism. *Science* 322, 200–201.

Sander, P.M., Klein, N. (2005) Developmental plasticity in the life history of a prosauropod dinosaur. *Science* 310, 1800–1802.

Sander, P.M., Klein, N., Buffetaut, E., Cuny, G., Suteethorn, V., Le Loeuff, J. (2004) Adaptive radiation in sauropod dinosaurs: bone histology indicates rapid evolution of giant body size through acceleration. *Organisms, Diversity, and Evolution* 4, 165–173.

Sander, P.M., Mateus, O., Laven, T., Knötschke, N. (2006) Bone histology indicates insular dwarfism in a new Late Jurassic sauropod dinosaur. *Nature* 441, 739–741.

Sander, P.M., Christian, A., Gee, C.T. (2009) Response to "Sauropods kept their heads down." *Science* 323, 1671–1672.

Sander, P.M., Gee, C.T., Hummel, J., Clauss, M. (2010) Mesozoic plants and dinosaur herbivory. In: Gee, C.T. (ed.), *Plants in Mesozoic Time*. Indiana University Press, Bloomington, IN, pp. 331–359.

Sander, P.M., Christian, A., Clauss, M. et al. (2011) Biology of the sauropod dinosaurs: the evolution of gigantism. *Biological Reviews* 86, 117–155.

Sanders, R.K., Smith, D.K. (2005) The endocranium of the theropod dinosaur *Ceratosaurus* studied with computed tomography. *Acta Palaeontologica Polonica* 50, 601–616.

Sanz, J.L., Moratalla, J.M., Diaz-Molina, M., Lopez-Martinez, N., Källn, O., Vianey-Liaud, M. (1995) Dinosaur nests at the sea shore. *Nature* 376, 731–732.

Sato, T., Cheng, Y., Wu, X., Zelenitsky, D.K., Hsiao, Y. (2005) A pair of shelled eggs inside a female dinosaur. *Science* 308, 375.

Scannella, J.B., Horner, J.R. (2010) *Torosaurus* Marsh, 1891, is *Triceratops* Marsh, 1889 (Ceratopsidae: Chasmosaurinae): synonymy through ontogeny. *Journal of Vertebrate Paleontology* 30, 1157–1168.

Schachner, E.R., Lyson, T.R., Dodson, P. (2009) Evolution of the respiratory system in nonavian theropods: evidence from rib and vertebral morphology. *Anatomical Record* 292, 1501–1513.

Schachner, E.R., Manning, P.L., Dodson, P. (2011a) Pelvis and hindlimb myology of the basal archosaur *Poposaurus gracilis* (Archosauria: Poposauroidea). *Journal of Morphology* 272, 1464–1491.

Schachner, E.R., Farmer, C.G., McDonald, A.T., Dodson, P. (2011b) Evolution of the dinosauriform respiratory apparatus: new evidence from the postcranial axial skeleton. *Anatomical Record* 294, 1532–1547.

Scheyer, T.M., Sander, P.M. (2004) Histology of anklosaur osteoderms: implications for systematics and function. *Journal of Vertebrate Paleontology* 24, 874–893.

Scheyer, T.M., Klein, N., Sander, P.M. (2010) Developmental palaeontology of Reptilia as revealed by histological studies. *Seminars in Cell and Developmental Biology* 21, 462–470.

Schmitz, L., Motani, R. (2011) Nocturnality in dinosaurs inferred from scleral ring and orbit morphology. *Science* 332, 705–708.

Schott, R.K., Evans, D.C., Williamson, T.E., Carr, T.D., Goodwin, M.B. (2009) The anatomy and systematics of *Colepiocephale lambei* (Dinosauria: Pachycephalosauridae). *Journal of Vertebrate Paleontology* 29, 771–786.

Schubert, B.W., Ungar, P.S. (2005) Wear facets and enamel spalling in tyrannosaurid dinosaurs. *Acta Palaeontologica Polonica* 50, 93–99.

Schuh, R.T., Brower, A.V.Z. (2009) *Biological Systematics: Principles and Applications*, 2nd edn. Cornell University Press, Ithaca, NY.

Schulte, P. Alegret. L., Arenillas, I. et al. (2010) The Chicxulub asteroid impact and mass extinction at the Cretaceous–Paleogene boundary. *Science* 327, 1214–1218.

Schwarz, D., Frey, E., Meyer, C.A. (2007) Pneumaticity and soft-tissue reconstructions in the neck of diplodocid and dicraeosaurid sauropods. *Acta Palaeontologica Polonica* 52, 167–188.

Schwarz-Wings, D. (2009) Reconstruction of the thoracic epaxial musculature of diplodocid and dicraeosaurid sauropods. *Journal of Vertebrate Paleontology* 29, 517–534.

Schweitzer, M.H. (2011) Soft tissue preservation in terrestrial Mesozoic vertebrates. *Annual Review of Earth and Planetary Sciences* 39, 187–216.

Schweitzer, M.H., Wittmeyer, J.L., Horner, J.R., Toporski, J.K. (2005a) Soft-tissue vessels and cellular preservation in *Tyrannosaurus rex*. *Science* 307, 1952–1955.

Schweitzer, M.H., Wittmeyer, J.L., Horner, J.R. (2005b) Gender-specific reproductive tissue in ratites and *Tyrannosaurus rex*. *Science* 308, 1456–1460.

Schweitzer, M.H., Suo, Z., Avci, R. et al. (2007a) Analyses of soft tissue from *Tyrannosaurus rex* suggest the presence of protein. *Science* 316, 277–280.

Schweitzer, M.H., Wittmeyer, J.L., Horner, J.R. (2007b) Soft tissue and cellular preservation in vertebrate skeletal elements from the Cretaceous to the present. *Proceedings of the Royal Society of London, Series B* 274, 183–197.

Schwimmer, D.R. (2002) *King of the Crocodylians: The Paleobiology of Deinosuchus*. Indiana University Press, Bloomington, IN.

Scotese, C.R. (2004) Cenozoic and Mesozoic paleogeography: changing terrestrial biogeographic pathways. In: Lomolino, M.V., Heaney, L.R. (eds.), *Frontiers of Biogeography*. Sinauer, Sunderland, MA, pp. 9–26.

Seebacher, F. (2001) A new method to calculate allometric length–mass relationships of dinosaurs. *Journal of Vertebrate Paleontology* 21, 51–60.

Seebacher, F. (2003) Dinosaur body temperatures: the occurrence of endothermy and ectothermy. *Paleobiology* 29, 105–122.

Seeley, H.G. (1887) On the classification of the fossil animals commonly named Dinosauria. *Proceedings of the Royal Society of London* 43, 165–171.

Sellers, W.I., Manning, P.M. (2007) Estimating dinosaur maximum running speeds using evolutionary robotics. *Proceedings of the Royal Society of London, Series B* 274, 2711–2716.

Sellers, W.I., Manning, P.L., Lyson, T., Stevens, K., Margetts, L. (2009) Virtual palaeontology: gait reconstruction of extinct vertebrates using high performance computing. *Palaeontologia Electronica* 12(3), Article 12.3.11A.

Sellwood, B.W., Valdes, P.J. (2006) Mesozoic climates: general circulation models and the rock record. *Sedimentary Geology* 190, 269–287.

Senter, P. (2005) Function in the stunted forelimbs of *Mononykus olecranus* (Theropoda), a dinosaurian anteater. *Paleobiology* 31, 373–381.

Senter, P. (2006a) Forelimb function in *Ornitholestes hermanni* Osborn (Dinosauria, Theropoda). *Palaeontology* 49, 1029–1034.

Senter, P. (2006b) Scapular orientation in theropods and basal birds, and the origin of flapping flight. *Acta Palaeontologica Polonica* 51, 305–313.

Senter, P. (2007) A new look at the phylogeny of Coelurosauria (Dinosauria: Theropoda). *Journal of Systematic Palaeontology* 5, 429–463.

Senter, P. (2010) Evidence for a sauropod-like metacarpal configuration in stegosaurian dinosaurs. *Acta Palaeontologica Polonica* 55, 427–432.

Sereno, P.C. (1986) Phylogeny of the bird-hipped dinosaurs. *National Geographic Research* 2, 234–256.

Sereno, P.C. (1990) Psittacosauridae. In: Weishampel, D.B., Dodson, P., Osmólska, H. (eds.), *The Dinosauria*. University of California Press, Berkeley, CA, pp. 579–592.

Sereno, P.C. (1991a) Basal archosaurs: phylogenetic relationships and functional implications. *Society of Vertebrate Paleontology Memoir* 2, 1–53.

Sereno, P.C. (1991b) *Lesothosaurus*, "fabrosaurids," and the early evolution of Ornithischia. *Journal of Vertebrate Paleontology* 11, 168–197.

Sereno, P.C. (1997) The origin and evolution of dinosaurs. *Annual Review of Earth and Planetary Science* 25, 435–489.

Sereno, P.C. (1998) A rationale for phylogenetic definitions, with application to the higher-level taxonomy of Dinosauria. *Neues Jahrbuch für Geologie und Paläontologie Abhandlungen* 210, 41–83.

Sereno, P.C. (1999) The evolution of dinosaurs. *Science* 284, 2137–2147.

Sereno, P.C. (2000) The fossil record, systematics and evolution of pachycephalosaurs and ceratopsians from Asia. In: Benton, M.J., Shishkin, M.A., Unwin, D.M., Kurochkin, E.N. (eds.), *The Age of Dinosaurs in Russia and Mongolia*. Cambridge University Press, Cambridge, pp. 480–516.

Sereno, P.C. (2001) Alvarezsaurids: birds or ornithomimosaurs? In: Gauthier, J., Gall, L.F. (eds.), *New Perspectives on the Origin and Early*

Evolution of Birds. Peabody Museum of Natural History, New Haven, CT, pp. 69–98.

Sereno, P.C. (2005) The logical basis of phylogenetic taxonomy. *Systematic Biology* 54, 595–619.

Sereno, P.C. (2010) Taxonomy, cranial morphology, and relationships of parrot-beaked dinosaurs (Ceratopsia: *Psittacosaurus*). In: Ryan, M.J., Chinnery-Allgeier, B.J., Eberth, D.A. (eds.), *New Perspectives on Horned Dinosaurs*. Indiana University Press, Bloomington, IN, pp. 21–58.

Sereno, P.C., Arcucci, A.B. (1990) The monophyly of crurotarsal archosaurs and the origin of bird and crocodile ankle joints. *Neues Jahrbuch für Geologie und Paläontologie Abhandlungen* 180, 21–52.

Sereno, P.C., Arcucci, A.B. (1993) Dinosaurian precursors from the Middle Triassic of Argentina: *Lagerpeton chanarensis*. *Journal of Vertebrate Paleontology* 13, 385–399.

Sereno, P.C., Arcucci, A.B. (1994) Dinosaurian precursors from the Middle Triassic of Argentina: *Marasuchus lilloensis*, gen. nov. *Journal of Vertebrate Paleontology* 14, 53–73.

Sereno, P.C., Brusatte, S.L. (2008) Basal abelisaurid and carcharodontosaurid theropods from the Lower Cretaceous Elrhaz Formation of Niger. *Acta Palaeontologica Polonica* 53, 15–46.

Sereno, P.C., Brusatte, S.L. (2009) Comparative assessment of tyrannosaurid interrelationships. *Journal of Systematic Palaeontology* 7, 455–470.

Sereno, P.C., Dong, Z.-M. (1992) The skull of the basal stegosaur *Huayangosaurus taibaii* and a cladistic diagnosis of Stegosauria. *Journal of Vertebrate Paleontology* 12, 318–343.

Sereno, P.C., Novas, F.E. (1992) The complete skull and skeleton of an early dinosaur. *Science* 258, 1137–1140.

Sereno, P.C., Novas F.E. (1994) The skull and neck of the basal theropod *Herrerasaurus ischigualastensis*. *Journal of Vertebrate Paleontology* 13, 451–476.

Sereno, P.C., Wilson, J.A. (2005) Structure and evolution of a sauropod tooth battery. In: Curry-Rogers, K.A., Wilson, J.A. (eds.), *The Sauropods: Evolution and Paleobiology*. University of California Press, Berkeley, CA, pp. 157–177.

Sereno, P.C., Forster, C.A., Rogers, R.R., Monetta, A.M. (1993) Primitive dinosaur skeleton from Argentina and the early evolution of Dinosauria. *Nature* 361, 64–66.

Sereno, P.C., Dutheil, D.B., Iarochene, M. et al. (1996) Predatory dinosaurs from the Sahara and Late Cretaceous faunal differentiation. *Science* 272, 986–991.

Sereno, P.C., Beck, A.L., Dutheil, D.B. et al. (1998) A long-snouted predatory dinosaur from Africa and the evolution of spinosaurids. *Science* 282, 1298–1302.

Sereno, P.C., Beck, A.L., Dutheil, D.B. et al. (1999) Cretaceous sauropods from the Sahara and the uneven rate of skeleton evolution among dinosaurs. *Science* 286, 1342–1347.

Sereno, P.C., Wilson, J.A., Conrad, J.L. (2004) New dinosaurs link southern landmasses in the mid-Cretaceous. *Proceedings of the Royal Society of London, Series B* 271, 1325–1330.

Sereno, P.C., McAllister, S., Brusatte, S.L. (2005) TaxonSearch: a relational database for suprageneric taxa and phylogenetic definitions. *PhyloInformatics* 8, 1–21.

Sereno, P.C., Wilson, J.A., Witmer, L.M. et al. (2007) Structural extremes in a Cretaceous dinosaur. *PLoS ONE* 2, e1230.

Sereno, P.C., Martinez, R.N., Wilson, J.A., Varricchio, D.J., Alcober, O.A., Larsson, H.C.E. (2008) Evidence for avian intrathoracic air sacs in a new predatory dinosaur from Argentina. *PLoS ONE* 3, e3303.

Sereno, P.C., Tan, L., Brusatte, S.L., Kriegstein, H.J., Zhao, X.-J., Cloward, K. (2009) Tyrannosaurid skeletal design first evolved at small body size. *Science* 326, 418–422.

Sertich, J.J.W., Loewen, M.A. (2010) A new basal sauropodomorph dinosaur from the Lower Jurassic Navajo Sandstone of southern Utah. *PLoS ONE* 5, e9789.

Seymour, R.S. (2009) Raising the sauropod neck: it costs more to get less. *Biology Letters* 5, 317–319.

Seymour, R.S., Lillywhite, H.B. (2000) Hearts, neck posture and metabolic intensity of sauropod dinosaurs. *Proceedings of the Royal Society of London, Series B* 267, 1883–1887.

Seymour, R.S., Smith, S.L., White, C.R., Henderson, D.M., Schwarz-Wings, D. (2011) Blood flow to long bones indicates activity metabolism in mammals, reptiles and dinosaurs. *Proceedings of the Royal Society of London, Series B* doi: 10.1098/rspb.2011.0968.

Sheehan, P.M., Fastovsky, D.E., Hoffmann, R.G., Berghaus, C.B., Gabriel, D.L. (1991) Sudden extinction of the dinosaurs: family-level patterns of ecological diversity during the latest Cretaceous, upper Great Plains, U.S.A. *Science* 254, 835–839.

Shipman, P. (1998) *Taking Wing: Archaeopteryx and the Evolution of Bird Flight*. Trafalgar Square, London.

Shipman, T.C. (2004) *Links between sediment accumulation rates and the development of alluvial architecture: Triassic Ischigualasto Formation, northwestern Argentina*. PhD thesis, University of Arizona, Tuscon.

Shubin, N.H., Olsen, P.E., Sues, H.-D. (1994) Early Jurassic small tetrapods from the McCoy Brook Formation of Nova Scotia, Canada. In: Fraser, N. C., Sues, H.-D. (eds.), *In The Shadow of the Dinosaurs*. Cambridge University Press, Cambridge, pp. 242–250.

Simms, M.J., Ruffell, A.H. (1990) Synchroneity of climatic change and extinctions in the Late Triassic. *Geology* 17, 265–268.

Simpson, E.L., Hilbert-Wolf, H.L., Wizevich, M.C. et al. (2010) Predatory digging behavior by dinosaurs. *Geology* 38, 699–702.

Skelton, P., Spicer, R.A., Kelley, S.P., Gillmour, I. (2003) *The Cretaceous World*. Open University, Cambridge University Press, Cambridge.

Smit, J. (1999) The global stratigraphy of the Cretaceous–Tertiary boundary impact ejecta. *Annual Review of Earth and Planetary Sciences* 27, 75–113.

Smith, A.B. (2001) Large-scale heterogeneity of the fossil record: implications for Phanaerozoic diversity studies. *Philosophical Transactions of the Royal Society of London, Series B* 356, 351–367.

Smith, A.B., McGowan, A.J. (2007) The shape of the marine palaeodiversity curve: how much can be predicted from the sedimentary rock record of western Europe? *Palaeontology* 50, 765–774.

Smith, A.G., Smith D.G., Funnell, B.M. (1994) *Atlas of Mesozoic and Cenozoic Coastlines*. Cambridge University Press, Cambridge.

Smith, J.B., Lamanna, M.C., Lacovara, K.J. et al. (2001) A giant sauropod dinosaur from an Upper Cretaceous mangrove deposit in Egypt. *Science* 292, 1704–1706.

Smith, J.B., Vann, D.R., Dodson, P. (2005) Dental morphology and variation in theropod dinosaurs: implications for the taxonomic identification of isolated teeth. *Anatomical Record* 285, 699–736.

Smith, M.M., Hall, B.K. (1990) Development and evolutionary origins of vertebrate skeletogenic and odontogenic tissues. *Biological Reviews* 65, 277–373.

Smith, N.D., Pol, D. (2007) Anatomy of a basal sauropodomorph dinosaur from the Early Jurassic Hanson Formation of Antarctica. *Acta Palaeontologica Polonica* 52, 657–674.

Smith-Gill, S.J. (1983) Developmental plasticity: developmental conversion versus phenotypic modulation. *American Zoologist* 23, 47–55.

Snively, E., Cox, A. (2008) Structural mechanics of pachycephalosaur crania permitted head-butting behavior. *Palaeontologia Electronica* 11(1), Article 11.1.3A.

Snively, E., Russell, A. (2002) The tyrannosaurid metatarsus: bone strain and inferred ligament function. *Senckenbergiana Lethaea* 82, 35–42.

Snively, E., Russell, A.P. (2007a) Functional variation of neck muscles and their relation to feeding style in Tyrannosauridae and other large theropod dinosaurs. *Anatomical Record* 290, 934–957.

Snively, E., Russell, A.P. (2007b) Functional morphology of the neck musculature in the Tyrannosauridae (Dinosauria: Theropoda) as determined via a hierarchial inferential approach. *Zoological Journal of the Linnean Society* 151, 759–808.

Snively, E., Theodor, J.M. (2011) Common functional correlates of head-strike behavior in the pachycephalosaur *Stegoceras validum* (Ornithischia, Dinosauria) and combative artiodactyls. *PLoS ONE* 6, e21422.

Snively, E., Russell, A.P., Powell, G.L. (2004) Evolutionary morphology of the coelurosaurian arctometatarsus: descriptive, morphometric and phylogenetic approaches. *Zoological Journal of the Linnean Society* 142, 525–553.

Snively, E., Henderson, D.M., Phillips, D.S. (2006) Fused and vaulted nasals of tyrannosaurid dinosaurs: implications for cranial strength and feeding mechanics. *Acta Palaeontologica Polonica* 51, 435–454.

Souter, T., Cornette, R., Pedraza, J., Hutchinson, J., Baylac, M. (2010) Two applications of 3D semi-landmark morphometrics implying different template designs: the theropod pelvis and the shrew skull. *Comptes Rendus Palevol* 9, 411–422.

Spoor, F., Garland, T., Krovitz, G., Ryan, T.M., Silcox, M.T., Walker, A. (2007) The primate semicircular canal system and locomotion. *Proceedings of the National Academy of Sciences USA* 104, 10808–10812.

Spotila, J.R., O'Connor, M.P., Dodson, P., Paladino, F.V. (1991) Hot and cold running dinosaurs: body size, metabolism and migration. *Modern Geology* 16, 203–227.

Stanley, S.M., Yang, X. (1994) A double mass extinction at the end of the Paleozoic Era. *Science* 266, 1340–1344.

Starck, J.M., Chinsamy, A. (2002) Bone microstructure and developmental plasticity in birds and other dinosaurs. *Journal of Morphology* 254, 232–246.

Stein, K., Csiki, Z., Curry-Rogers, K. et al. (2010) Small body size and extreme cortical bone remodeling indicate phyletic dwarfism in *Magyarosaurus dacus* (Sauropoda: Titanosauria). *Proceedings of the National Academy of Sciences USA* 107, 9258–9263.

Steinsaltz, D., Orzack, S.H. (2011) Statistical methods for paleodemography on fossil assemblages having small numbers of specimens: an investigation of dinosaur survival rates. *Paleobiology* 37, 113–125.

Stettenheim, P.R. (2000) The integumentary morphology of modern birds: an overview. *American Zoologist* 40, 461–477.

Stevens, K.A. (2002) DinoMorph: parametric modeling of skeletal structures. *Senckenbergiana Lethaea* 82, 23–34.

Stevens, K.A. (2006) Binocular vision in theropod dinosaurs. *Journal of Vertebrate Paleontology* 26, 321–330.

Stevens, K.A., Parrish, J.M. (1999) Neck posture and feeding habits of two Jurassic sauropods. *Science* 284, 798–800.

Stevens, K.A., Parrish, J.M. (2005a) Digital reconstructions of sauropod dinosaurs and implications for feeding. In: Curry-Rogers, K.A., Wilson, J.A. (eds.), *The Sauropods: Evolution and Paleobiology*. University of California Press, Berkeley, CA, pp. 178–200.

Stevens, K.A., Parrish, J.M. (2005b) Neck posture, dentition, and feeding strategies in Jurassic sauropod dinosaurs. In: Carpenter, K., Tidwell, V. (eds.), *Thunder Lizards: The Sauropodomorph Dinosaurs*. Indiana University Press, Bloomington, IN, pp. 212–232.

Stevens, K.A., Larson, P.L., Wills, E.D., Anderson, A. (2008) Rex, sit: modeling tyrannosaurid postures. In: Carpenter, K., Larson, P.L. (eds.), *Tyrannosaurus rex: The Tyrant King*. Indiana University Press, Bloomington, IN, pp. 193–203.

Stokosa, K. (2005) Enamel microstructure variation within the Theropoda. In: Carpenter, K. (ed.), *The Carnivorous Dinosaurs*. Indiana University Press, Bloomington, IN, pp. 163–178.

Sues, H.-D. (1978) Functional morphology of the dome in pachycephalosaurid dinosaurs. *Neues Jahrbuch für Geologie und Paläontologie Monatshefte* 1978, 459–472.

Sues, H.-D., Fraser, N.C. (2010) *Triassic Life on Land: The Great Transition*. Columbia University Press, New York.

Sues, H.-D., Frey, E., Martill, D.M., Scott, D.M. (2002) *Irritator challengeri*, a spinosaurid (Dinosauria: Theropoda) from the Lower Cretaceous of Brazil. *Journal of Vertebrate Paleontology* 22, 535–547.

Sullivan, C., Hone, D.W.E., Xu, X., Zhang, F. (2010) The asymmetry of the carpal joint and the evolution of wing folding in maniraptoran theropod dinosaurs. *Proceedings of the Royal Society of London, Series B* 277, 2027–2033.

Sullivan, R.M. (2003) Revision of the dinosaur *Stegoceras* Lambe (Ornithischia, Pachycephalosauridae). *Journal of Vertebrate Paleontology* 23, 181–207.

Sullivan, R.M., Williamson, T.E. (1999) A new skull of *Parasaurolophus* (Dinosauria: Hadrosauridae) from the Kirtland Formation of New Mexico and a revision of the genus. New Mexico Museum of Natural History and Science Bulletin No. 15.

Sun, G., Dilcher, D.L., Zheng, Z., Zhou, Z. (1998) In search of the first flower: a Jurassic angiosperm, *Archaefructus*, from northeast China. *Science* 282, 1692–1695.

Swinton, W.E. (1934) *The Dinosaurs: A Short History of a Great Group of Extinct Reptiles.* Murby, London.

Swofford, D.L. (2000) *PAUP: Phylogenetic Analysis Using Parsimony (and other methods)*, Version 4.10b. Released by the author.

Tanoue, K., Grandstaff, B.S., You, H.-L., Dodson, P. (2009) Jaw mechanics in Basal Ceratopsia (Ornithischia, Dinosauria). *Anatomical Record* 292, 1352–1369.

Taylor M.P. (2006) Dinosaur diversity analysed by clade, age, place and year of description. In: Barrett, P.M. (ed.), *Ninth International Symposium on Mesozoic Terrestrial Ecosystems: Short Papers.* Cambridge Publications, Cambridge, pp. 134–138.

Taylor, M.P., Wedel, M.J., Naish, D. (2009) Head and neck posture in sauropod dinosaurs inferred from extant animals. *Acta Palaeontologica Polonica* 54, 213–220.

Therrien, F., Henderson, D.M., Ruff, C.B. (2005) Bite me: biomechanical models of theropod mandibles and the implications for feeding behavior. In: Carpenter, K. (ed.), *The Carnivorous Dinosaurs.* Indiana University Press, Bloomington, IN, pp. 179–237.

Thompson, R.S., Parrish, J.C., Maidment, S.C.R., Barrett, P.M. (2011) Phylogeny of the ankylosaurian dinosaurs (Ornithischia: Thyreophora). *Journal of Systematic Palaeontology.* (July 2011), 37–41.

Thulborn, R.A. (1972) The postcranial skeleton of the Triassic ornithischian dinosaur *Fabrosaurus australis. Palaeontology* 15, 29–60.

Thulborn, R.A. (1990) *Dinosaur Tracks.* Chapman & Hall, London.

Thulborn, R.A., Wade, M. (1984) Dinosaur trackways in the Winton formation (mid-Cretaceous) of Queensland. *Memoirs of the Queensland Museum* 21, 413–517.

Tsuihiji, T. (2004) The ligament system in the neck of *Rhea americana* and its implications for the bifurcated neural spines of sauropod dinosaurs. *Journal of Vertebrate Paleontology* 24, 165–172.

Tsuihiji, T. (2005) Homologies of the transversospinalis muscles in the anterior presacral region of Sauria (Crown Diapsida). *Journal of Morphology* 263, 151–178.

Tsuihiji, T. (2007) Homologies of the longissimus, iliocostalis, and hypaxial muscles in the anterior presacral region of extant Diapsida. *Journal of Morphology* 268, 986–1020.

Tsuihiji, T. (2010) Reconstructions of the axial muscle insertions in the occipital region of dinosaurs: evaluations of past hypotheses on Marginocephalia and Tyrannosauridae using the Extant Phylogenetic Bracket approach. *Anatomical Record* 293, 1360–1386.

Tucker, M.E., Benton, M.J. (1982) Triassic environments, climates and reptile evolution. *Palaeogeography, Palaeoclimatology, Palaeoecology* 40, 361–379.

Turner, A.H., Pol, D., Clarke, J.A., Erickson, G.M., Norell, M.A. (2007a) A basal dromaeosaurid and size evolution preceding avian flight. *Science* 317, 1378–1381.

Turner, A.H., Makovicky, P.J., Norell, M.A. (2007b) Feather quill knobs in the dinosaur *Velociraptor. Science* 317, 1721.

Turner, A.H., Smith, N.D., Callery, J.A. (2009) Gauging the effects of sampling failure in biogeographic analysis. *Journal of Biogeography* 36, 612–625.

Tütken, T., Pfretzschner, H.-U., Vennemann, T.W., Sun, G., Wang, Y.D. (2004) Paleobiology and skeletochronology of Jurassic dinosaurs: implications from the histology and oxygen isotope composition of bones. *Palaeogeography, Palaeoclimatology, Palaeoecology* 206, 217–238.

Tweet, J.S., Chin, K., Braman, D.R., Murphy, N.L. (2008) Probable gut contents within a specimen of *Brachylophosaurus canadensis* (Dinosauria: Hadrosauridae) from the Upper Cretaceous Judith River Formation of Montana. *Palaios* 23, 624–635.

Tykoski, R.S., Rowe, T. (2004) Ceratosauria. In: Weishampel, D.B., Dodson, P., Osmólska, H. (eds.), *The Dinosauria*, 2nd edn. University of California Press, Berkeley, CA, pp. 47–70.

Unwin, D.M., Deeming, D.C. (2008) Pterosaur eggshell structure and its implications for pterosaur reproductive biology. *Zitteliana* B28, 199–207.

Upchurch, P. (1995) The evolutionary history of sauropod dinosaurs. *Philosophical Transactions of the Royal Society of London, Series B* 349, 365–390.

Upchurch, P. (1998) The phylogenetic relationships of sauropod dinosaurs. *Zoological Journal of the Linnean Society* 124, 43–103.

Upchurch, P., Barrett, P.M. (2000) The evolution of sauropod feeding mechanisms. In: Sues, H.-D. (ed.), *Evolution of Herbivory in Terrestrial Vertebrates*. Cambridge University Press, Cambridge, pp. 79–122.

Upchurch, P., Barrett, P.M. (2005) Phylogenetic and taxic perspectives on sauropod diversity. In: Curry-Rogers, K.A., Wilson, J.A. (eds.), *The Sauropods: Evolution and Paleobiology*. University of California Press, Berkelely, CA, pp. 104–124.

Upchurch, P., Hunn, C.A., Norman, D.B. (2002) An analysis of dinosaurian biogeography: evidence for the existence of vicariance and dispersal patterns caused by geological events. *Proceedings of the Royal Society of London, Series B* 269, 613–621.

Upchurch, P., Barrett, P.M., Dodson, P. (2004) Sauropoda. In: Weishampel, D.B., Dodson, P., Osmólska, H. (eds.), *The Dinosauria*, 2nd edn. University of California Press, Berkeley, CA, pp. 259–322.

Upchurch, P., Barrett, P.M., Galton, P.M. (2007) A phylogenetic analysis of basal sauropodomorph relationships: implications for the origin of sauropod dinosaurs. *Special Papers in Palaeontology* 77, 57–90.

Upchurch, P., Mannion, P.D., Benson, R.B.J., Butler, R.J., Carrano, M.T. (2011) Geological and anthropogenic controls on the sampling of the terrestrial fossil record: case study from the Dinosauria. In: McGowan, A.J., Smith, A.B. (eds.), *Comparing the Rock and Fossil Records: Implications for Biodiversity Studies*. Geological Society Special Publication No. 358, pp. 209–240. The Geological Society, London.

Van Valkenburgh, B., Molnar, R.E. (2002) Dinosaurian and mammalian predators compared. *Paleobiology* 28, 527–543.

Varricchio, D.J. (1993) Bone microstructure of the Upper Cretaceous theropod dinosaur *Troodon formosus*. *Journal of Vertebrate Paleontology* 13, 99–104.

Varricchio, D.J. (2001) Gut contents from a Cretaceous tyrannosaurid: implications for theropod digestive tracts. *Journal of Paleontology* 75, 401–406.

Varricchio, D.J. (2011) A distinct dinosaur life history? *Historical Biology* 23, 91–107.

Varricchio, D.J., Horner, J.R. (1993) Hadrosaurid and lambeosaurid bone beds from the Upper Cretaceous Two Medicine Formation of Montana: taphonomic and biologic implications. *Canadian Journal of Earth Sciences* 30, 997–1006.

Varricchio, D.J., Jackson, F. (2004) Cladistic analysis of eggshell characters: a phylogenetic assessment of prismatic dinosaur eggs from the Cretaceous Two Medicine Formation of Montana. *Journal of Vertebrate Paleontology* 24, 931–937.

Varricchio, D.J., Jackson, F., Borkowski, J.J., Horner, J.R. (1997) Nest and egg clutches of the dinosaur *Troodon formosus* and the evolution of avian reproductive traits. *Nature* 385, 247–250.

Varricchio, D.J., Jackson, F., Trueman, C.N. (1999) A nesting trace with eggs for the Cretaceous theropod dinosaur *Troodon formosus*. *Journal of Vertebrate Paleontology* 19, 91–100.

Varricchio, D.J., Horner, J.R., Jackson, F. (2002) Embryos and eggs for the Cretaceous theropod *Troodon formosus*. *Journal of Vertebrate Paleontology* 22, 564–576.

Varricchio, D.J., Martin, A.J., Katsura, Y. (2007) First trace and body fossil evidence of a burrowing, denning dinosaur. *Proceedings of the Royal Society of London, Series B* 274, 1361–1368.

Varricchio, D.J., Moore, J.R., Erickson, G.M., Norell, M.A., Jackson, F.D., Borkowski, J.J. (2008a) Avian paternal care had dinosaur origin. *Science* 322, 1826–1828.

Varricchio, D.J., Sereno, P.C., Zhao, X.-J., Tan, L., Wilson, J.A., Lyon, G.H. (2008b) Mud-trapped herd captures evidence of distinctive dinosaur sociality. *Acta Palaeontologica Polonica* 53, 567–578.

Vavrek, M.J., Larsson, H.C.E. (2010) Low beta diversity of Maastrichtian dinosaurs of North America. *Proceedings of the National Academy of Sciences USA* 107, 8265–8268.

Vazquez, R.J. (1992) Functional morphology of the avian wrist and the evolution of flapping flight. *Journal of Morphology* 211, 259–268.

Vickaryous, M.K., Hall, B.K. (2010) Comparative development of the crocodilian interclavicle and avian furcula, with comments on the homology of dermal elements in the pectoral apparatus. *Journal of Experimental Zoology* 314B, 196–207.

Vickaryous, M.K., Maryanska, T., Weishampel, D.B. (2004) Ankylosauria. In: Weishampel, D.B., Dodson, P., Osmólska, H. (eds.), *The Dinosauria*, 2nd edn. University of California Press, Berkeley, CA, pp. 363–392.

Vila, B., Jackson, F.D., Fortuny, J., Selle, A.G., Galobart, A. (2010) 3-D modeling of megaloolithid clutches: insights about nest construction and dinosaur behaviour. *PLoS ONE* 5, e10362.

Vinther, J., Briggs, D.E.G., Prum, R.O., Saranathan, V. (2008) The colour of fossil feathers. *Biology Letters* 4, 522–525.

Vinther, J., Briggs, D.E.G., Clarke, J., Mayr, G., Prum, R.O. (2010) Structural coloration in a fossil feather. *Biology Letters* 6, 128–131.

Voorhies, M. (1969) Taphonomy and population dynamics of an early Pliocene vertebrate fauna, Knox County, Nebraska. *University of Wyoming Contributions to Geology Special Paper* 1, 1–69.

Wagner, P.J. (1997) Patterns of morphologic diversification among the Rostroconchia. *Paleobiology* 23, 115–150.

Walsh, S.A., Barrett, P.M., Milner, A.C., Manley, G., Witmer, L.M. (2009) Inner ear anatomy is a proxy for deducing auditory capability and behaviour in reptiles and birds. *Proceedings of the Royal Society of London, Series B* 276, 1355–1360.

Wang S.C., Dodson, P. (2006) Estimating the diversity of dinosaurs. *Proceedings of the National Academy of Sciences USA* 103, 13601–13605.

Wedel, M.J. (2003a) Vertebral pneumaticity, air sacs, and the physiology of sauropod dinosaurs. *Paleobiology* 29, 243–255.

Wedel, M.J. (2003b) The evolution of vertebral pneumaticity in sauropod dinosaurs. *Journal of Vertebrate Paleontology* 23, 344–357.

Wedel, M.J. (2007) What pneumaticity tells us about "prosauropods" and vice versa. *Special Papers in Palaeontology* 77, 207–222.

Wegener, A.L. (1924) *The Origin of Continents and Oceans*. Dutton, New York.

Weinbaum, J.C., Hungerbühler, A. (2007) A revision of *Poposaurus gracilis* (Archosauria: Suchia) based on two new specimens from the Late Triassic of the southwestern U.S.A. *Paläontologische Zeitschrift* 81, 131–145.

Weishampel, D.B. (1981) Acoustic analyses of potential vocalization in lambeosaurine dinosaurs (Reptilia: Ornithischia). *Paleobiology* 7, 252–261.

Weishampel, D.B. (1984) The evolution of jaw mechanisms in ornithopod dinosaurs. *Advances in Anatomy, Embryology and Cell Biology* 87, 1–110.

Weishampel, D.B. (2004) Ornithischia. In: Weishampel, D.B., Dodson, P., Osmólska, H. (eds.), *The Dinosauria*, 2nd edn. University of California Press, Berkeley, CA, pp. 323–324.

Weishampel, D.B., Chapman, R.E. (1990) Morphometric study of *Plateosaurus* from Trossingen (Baden-Wurttemberg, Federal Republic of Germany). In: Carpenter, K., Currie, P.J. (eds.), *Dinosaur Systematics: Approaches and Perspectives*. Cambridge University Press, Cambridge, pp. 43–51.

Weishampel, D.B., Jianu, C.-M. (2000) Plant-eaters and ghost lineages: dinosaurian herbivory revisited. In: Sues, H.-D. (ed.), *Evolution of Herbivory in Terrestrial Vertebrates*. Cambridge University Press, Cambridge, pp. 123–143.

Weishampel, D.B., Norman, D.B. (1989) Vertebrate herbivory in the Mesozoic: jaws, plants and evolutionary metrics. *Geological Society of America Special Paper* 238, 87–100.

Weishampel, D.B., Grigorescu, D., Norman, D.B. (1991) The dinosaurs of Transylvania. *National Geographic Research and Exploration* 7, 196–215.

Weishampel, D.B., Barrett, P.M., Coria, R.A. et al. (2004) Dinosaur distribution. In: Weishampel, D.B., Dodson, P., Osmólska, H. (eds.), *The Dinosauria*, 2nd edn. University of California Press, Berkeley, CA, pp. 517–606.

Welles, S.P. (1954) New Jurassic dinosaur from the Kayenta Formation of Arizona. *Bulletin of the Geological Society of America* 65, 591–598.

Wheeler, P.E. (1978) Elaborate CNS cooling structures in large dinosaurs. *Nature* 275, 441–443.

White, P.D., Fastovsky, D.E., Sheehan, P.M. (1998) Taphonomy and suggested structure of the dinosaurian assemblage of the Hell Creek Formation (Maastrichtian), eastern Montana and western North Dakota. *Palaios* 13, 41–51.

Whiteside, J.H., Olsen, P.E., Eglinton, T., Brookfield, M.E., Sambrotto, R.N. (2010) Compound-specific carbon isotopes from Earth's largest flood basalt province directly link eruptions to the end-Triassic mass extinction. *Proceedings of the National Academy of Sciences USA* 107, 6721–6725.

Whiteside, J.H., Grogan, D.S., Olsen, P.E., Kent, D.V. (2011) Climatically driven biogeographic provinces of Late Triassic tropical Pangea. *Proceedings of the National Academy of Sciences USA* 108, 8972–8977.

Whitlock, J.A. (2011a) Inferences of diplodocoid (Sauropoda: Dinosauria) feeding behavior from snout shape and microwear analyses. *PLoS ONE* 6, e18304.

Whitlock, J.A. (2011b) A phylogenetic analysis of Diplodocoidea (Saurischia: Sauropoda). *Zoological Journal of the Linnean Society* 161, 872–915.

Wignall, P.B. (2001) Large igneous provinces and mass extinctions. *Earth Science Reviews* 53, 1–33.

Wilhite, R. (2003) Digitizing large fossil skeletal elements for three-dimensional applications. *Palaeontologia Electronica* 5(2), Article 5.2.8A.

Williams, V.S., Barrett, P.M., Purnell, M.A. (2009) Quantitative analysis of dental microwear in hadrosaurid dinosaurs, and the implications for hypotheses of jaw mechanics and feeding. *Proceedings of the National Academy of Sciences USA* 106, 11194–11199.

Williamson, T.E., Carr, T.D. (2002) A new genus of derived pachycephalosaurian from western North America. *Journal of Vertebrate Paleontology* 22, 779–801.

Wills, M.A., Briggs, D.E.G., Fortey, R.A. (1994) Disparity as an evolutionary index: a comparison of Cambrian and Recent arthropods. *Paleobiology* 20, 93–131.

Wilson, J.A. (1999) A nomenclature for vertebral laminae in sauropods and other saurischian dinosaurs. *Journal of Vertebrate Paleontology* 19, 639–653.

Wilson, J.A. (2002) Sauropod dinosaur phylogeny: critique and cladistic analysis. *Zoological Journal of the Linnean Society* 136, 215–275.

Wilson, J.A., Sereno, P.C. (1998) Early evolution and higher-level phylogeny of sauropod dinosaurs. *Society of Vertebrate Paleontology Memoir* 5, 1–68.

Wilson, J.A., Sereno, P.C., Srivastava, S., Bhatt, D.K., Khosla, A., Sahni, A. (2003) A new abelisaurid (Dinosauria, Theropoda) from the Lameta Formation (Cretaceous, Maastrichtian) of India. *Contributions of the Museum of Palaeontology of the University of Michigan* 31, 1–42.

Wilson, J.A., Mohabey, D.M., Peters, S.E., Head, J.J. (2010) Predation upon hatchling dinosaurs by a new snake from the Late Cretaceous of India. *PLoS Biology* 8, e1000322.

Wiman, C. (1931) *Parasaurolophus tubicen* n. sp. aus der Kreide in New Mexico. *Nova Acta Regiae Societatis Scientiarum Uppsaliensis* 4, 1–11.

Wings, O. (2007) A review of gastrolith function with implications for fossil vertebrates and a revised classification. *Acta Palaeontologica Polonica* 52, 1–16.

Wings, O., Sander, P.M. (2007) No gastric mill in sauropod dinosaurs: new evidence from analysis of gastrolith mass and function in ostriches. *Proceedings of the Royal Society of London, Series B* 274, 635–640.

Witmer, L.M. (1995a) The Extant Phylogenetic Bracket and the importance of reconstructing soft tissues in fossils. In: Thomason, J. (ed.), *Functional Morphology in Vertebrate Paleontology*. Cambridge University Press, Cambridge, pp. 19–33.

Witmer, L.M. (1995b) Homology of facial structures in extant archosaurs (birds and crocodilians) with special reference to paranasal pneumaticity and nasal conchae. *Journal of Morphology* 225, 263–327.

Witmer, L.M. (1997a) The evolution of the antorbital cavity of archosaurs: a study in soft-tissue reconstruction in the fossil record with an analysis of the function of pneumaticity. *Society of Vertebrate Paleontology Memoir* 3, 1–73.

Witmer, L.M. (1997b) Craniofacial air sinus systems. In: Currie, P.J., Padian, K. (eds.), *Encyclopedia of Dinosaurs*. Academic Press, New York, pp. 151–159.

Witmer, L.M., Ridgely, R.C. (2008) The paranasal air sinuses of predatory and armored dinosaurs (Archosauria: Theropoda and Ankylosauria) and their contribution to cephalic structure. *Anatomical Record* 291, 1362–1388.

Witmer, L.M., Ridgely, R.C. (2009) New insights into the brain, braincase, and ear region of tyrannosaurs (Dinosauria, Theropoda), with implications for sensory organization and behavior. *Anatomical Record* 292, 1266–1296.

Witmer, L.M., Ridgely, R.C., Dufeau, D.L., Semones, M.C. (2008) Using CT to peer into the past: 3D visualization of the brain and ear regions of birds, crocodiles, and nonavian dinosaurs. In: Endo, H., Frey, R. (eds.), *Anatomical Imaging: Towards a New Morphology*. Springer-Verlag, Tokyo, pp. 67–87.

Witzel, U., Preuschoft, H. (2005) Finite-element model construction for the virtual synthesis of the skulls in vertebrates: case study of *Diplodocus*. *Anatomical Record* 283A, 391–401.

Wogelius, R.A., Manning, P.L., Barden, H.E. et al. (2011) Trace metals as biomarkers for eumelanin pigment in the fossil record. *Science* 333, 1622–1626.

Wolbach, W.S., Lewis, R.S., Anders, E. (1985) Cretaceous extinctions: evidence for wildfires and search for meteoritic material. *Science* 230, 167–170.

Woodward, H.N., Lehman, T.M. (2009) Bone histology and microanatomy of *Alamosaurus sanjuanensis* (Sauropoda: Titanosauria) from the Maastrichtian of Big Bend National Park, Texas. *Journal of Vertebrate Paleontology* 29, 807–821.

Woodward, H.N., Rich, T.H., Chinsamy, A., Vickers-Rich, P. (2011) Growth dynamics of Australia's polar dinosaurs. *PLoS ONE* 6, e23339.

Wright, J.L. (1999) Ichnological evidence for the use of the forelimb in iguanodontid locomotion. *Special Papers in Palaeontology* 60, 209–219.

Wu, X.-C., Brinkman, D.B., Eberth, D.A., Braman, D.R. (2007) A new ceratopsid dinosaur (Ornithischia) from the uppermost Horseshoe Canyon Formation (upper Maastrichtian), Alberta, Canada. *Canadian Journal of Earth Sciences* 44, 1243–1265.

Xu, X., Guo, Y. (2009) The origin and early evolution of feathers: insights from recent paleontological and neontological data. *Vertebrata PalAsiatica* 47, 311–329.

Xu, X., Norell, M.A. (2004) A new troodontid dinosaur from China with avian-like sleeping posture. *Nature* 431, 838–841.

Xu, X., Zhang, F. (2005) A new maniraptoran dinosaur from China with long feathers on the metatarsus. *Naturwissenschaften* 92, 173–177.

Xu, X., Tang, Z.-L., Wang, X.-L. (1999) A therizinosauroid dinosaur with integumentary structures from China. *Nature* 399, 350–354.

Xu, X., Makovicky, P.J., Wang, X.-L., Norell, M.A., You, H.-L. (2002a) A ceratopsian dinosaur from China and the early evolution of Ceratopsia. *Nature* 416, 314–317.

Xu, X., Chieng, Y.N., Wang, X.-L., Chang, C.-H. (2002b) An unusual oviraptorosaurian dinosaur from China. *Nature* 419, 291–293.

Xu, X., Zhou, Z., Wang, X., Kuang, X., Zhang, F., Du, X. (2003) Four-winged dinosaurs from China. *Nature* 421, 335–340.

Xu, X., Norell, M.A., Kuang, X., Wang, X., Zhao, Q., Jia, C. (2004) Basal tyrannosauroids from China and evidence for protofeathers in tyrannosauroids. *Nature* 431, 680–684.

Xu, X., Forster, C.A., Clark, J.M., Mo, J. (2006) A basal ceratopsian with transitional features from the Late Jurassic of northwestern China. *Proceedings of the Royal Society of London, Series B* 273, 2135–2140.

Xu, X., Zhao, Q., Norell, M. et al. (2009a) A new feathered maniraptoran dinosaur fossil that fills a morphological gap in avian origin. *Chinese Science Bulletin* 54, 430–435.

Xu, X., Clark, J.M., Mo, J. et al. (2009b) A Jurassic ceratosaur from China helps clarify avian digital homologies. *Nature* 459, 940–944.

Xu X., Wang, K., Zhao, X.-J., Li, D. (2010) First ceratopsid dinosaur from China and its biogeographical implications. *Chinese Science Bulletin* 55, 1631–1635.

Xu, X., You, H., Du, K., Han, F. (2011a) An *Archaeopteryx*-like theropod from China and the origin of Avialae. *Nature* 475, 465–470.

Xu, X., Sullivan, C., Pittman, M. et al. (2011b) A monodactyl nonavian dinosaur and the complex evolution of the alvarezsauroid hand. *Proceedings of the National Academy of Sciences USA* 108, 2338–2342.

Yates, A.M. (2003) A new species of the primitive dinosaur *Thecodontosaurus* (Saurischia: Sauropodomorpha) and its implications for the systematics of early dinosaurs. *Journal of Systematic Palaeontology* 1, 1–42.

Yates, A.M., (2007) The first complete skull of the Triassic dinosaur *Melanorosaurus* Haughton (Sauropodomorpha: Anchisauria). *Special Papers in Palaeontology* 77, 9–55.

Yates, A.M., Kitching, J.W. (2003) The earliest known sauropod dinosaur and the first steps towards sauropod locomotion. *Proceedings of the Royal Society of London, Series B* 270, 1753–1758.

Yates, A.M., Bonnan, M.F., Neveling, J., Chinsamy, A., Blackbeard, M.G. (2010) A new transitional sauropodomorph dinosaur from the Early Jurassic of South Africa and the evolution of sauropod feeding and quadrupedalism. *Proceedings of the Royal Society of London, Series B* 277, 787–794.

You, H., Dodson, P. (2004) Basal Ceratopsia. In: Weishampel, D.B., Dodson, P., Osmólska, H. (eds.), *The Dinosauria*, 2nd edn. University of California Press, Berkeley, CA, pp. 478–493.

Young, M.T., Larvan, M.D. (2010) Macroevolutionary trends in the skull of sauropodomorph dinosaurs: the largest terrestrial animals to have ever lived. In: Elewa, A.M.T. (ed.), *Morphometrics for Nonmorphometricians*. Springer-Verlag, Heidelberg, pp. 259–269.

Zanno, L.E. (2010) Osteology of *Falcarius utahensis* (Dinosauria: Theropoda): characterizing the anatomy of basal therizinosaurs. *Zoological Journal of the Linnean Society* 158, 196–230.

Zanno, L.E., Makovicky, P.J. (2011) Herbivorous ecomorphology and specialization patterns in theropod dinosaur evolution. *Proceedings of the National Academy of Sciences USA* 108, 232–237.

Zanno, L.E., Gillette, D.D., Albright, L.B., Titus, A. L. (2009) A new North American therizinosaurid and the role of herbivory in "predatory" dinosaur evolution. *Proceedings of the Royal Society, Series B* 276, 3505–3511.

Zelditch, M.L., Swiderski, D.L., Sheets, H.D., Fink, W.L. (2004) *Geometric Morphometrics for Biologists: a Primer*. Elsevier Academic Press, New York.

Zelenitsky, D.K., Therrien, F. (2008a) Phylogenetic analysis of reproductive traits of maniraptoran theropods and its implications for egg parataxonomy. *Palaeontology* 51, 807–816.

Zelenitsky, D.K., Therrien, F. (2008b) Unique maniraptoran egg clutch from the Upper Cretaceous Two Medicine Formation of Montana reveals theropod nesting behaviour. *Palaeontology* 51, 1253–1259.

Zelenitsky, D.K., Therrien, F., Kobayashi, Y. (2009) Olfactory acuity in theropods: palaeobiological and evolutionary implications. *Proceedings of the Royal Society of London, Series B* 276, 667–673.

Zelenitsky, D.K., Therrien, F., Ridgely, R.C., McGee, A.R., Witmer, L.M. (2011) Evolution of olfaction in theropod dinosaurs. *Proceedings of the Royal Society of London, Series B* 278, 3625–3634.

Zhang, F., Zhou, Z. (2004) Leg feathers in an Early Cretaceous bird. *Nature* 431, 925.

Zhang, F., Zhou, Z., Xu, X., Wang, X. (2002) A juvenile coelurosaurian theropod from China indicates arboreal habits. *Naturwissenschaften* 89, 394–398.

Zhang, F., Zhou, Z., Dyke, G. (2006) Feathers and "feather-like" integumentary structures in Liaoning birds and dinosaurs. *Geological Journal* 41, 395–401.

Zhang, F., Zhou, Z., Xu, X., Wang, X., Sullivan, C. (2008) A bizarre Jurassic maniraptoran from China with elongate ribbon-like feathers. *Nature* 455, 1105–1108.

Zhang, F., Kearns, S.L., Orr, P.J. et al. (2010) Fossilized melanosomes and the colour of Cretaceous dinosaurs and birds. *Nature* 463, 1075–1078.

Zhao, Q., Barrett, P.M., Eberth, D.A. (2007) Social behaviour and mass mortality in the basal ceratopsian dinosaur *Psittacosaurus* (Early Cretaceous, People's Republic of China). *Palaeontology* 50, 1023–1029.

Zheng, X.-T., You, H.-L., Xu, X., Dong, Z.-M. (2009) An Early Cretaceous heterodontosaurid dinosaur with filamentous integumentary structures. *Nature* 458, 333–336.

Zhou, C.-F., Gao, K.-Q., Fox, R.C., Du, X.-K. (2007) Endocranial morphology of psittacosaurs (Dinosauria: Ceratopsia) based on CT scans of new fossils from the Lower Cretaceous, China. *Palaeoworld* 16, 285–293.

Zhou, Z., Barrett, P.M., Hilton, J. (2003) An exceptionally preserved Lower Cretaceous ecosystem. *Nature* 421, 807–814.

Index

Note: page numbers in italics refer to figures.

Dinosaur Paleobiology, First Edition. Stephen L. Brusatte.
© 2012 John Wiley & Sons, Ltd. Published 2012 by John Wiley & Sons, Ltd.

Printed and bound by CPI Group (UK) Ltd, Croydon, CR0 4YY

27/10/2024

14580162-0003